AF361937

Remote Sensing Analysis of Geologic Hazards

Remote Sensing Analysis of Geologic Hazards

Editors

Daniele Giordan
Guido Luzi
Oriol Monserrat
Niccolò Dematteis

MDPI • Basel • Beijing • Wuhan • Barcelona • Belgrade • Manchester • Tokyo • Cluj • Tianjin

Editors
Daniele Giordan
Italian National Research
Council
Italy

Guido Luzi
Technological Centre of
Telecommunications of
Catalunya
Spain

Oriol Monserrat
Technological Centre of
Telecommunications of
Catalunya
Spain

Niccolò Dematteis
Italian National Research
Council
Italy

Editorial Office
MDPI
St. Alban-Anlage 66
4052 Basel, Switzerland

This is a reprint of articles from the Special Issue published online in the open access journal *Remote Sensing* (ISSN 2072-4292) (available at: https://www.mdpi.com/journal/remotesensing/special_issues/RS_geohazards).

For citation purposes, cite each article independently as indicated on the article page online and as indicated below:

LastName, A.A.; LastName, B.B.; LastName, C.C. Article Title. *Journal Name* **Year**, *Volume Number*, Page Range.

ISBN 978-3-0365-5699-4 (Hbk)
ISBN 978-3-0365-5700-7 (PDF)

Cover image courtesy of Daniele Giordan

Contents

About the Editors . . . . **vii**

Daniele Giordan, Guido Luzi, Oriol Monserrat and Niccolò Dematteis
Remote Sensing Analysis of Geologic Hazards
Reprinted from: *Remote Sens.* **2022**, *14*, 4818, doi:10.3390/rs14194818 . . . . **1**

Erin Lindsay, Regula Frauenfelder, Denise Rüther, Lorenzo Nava, Lena Rubensdotter, James Strout and Steinar Nordal
Multi-Temporal Satellite Image Composites in Google Earth Engine for Improved Landslide Visibility: A Case Study of a Glacial Landscape
Reprinted from: *Remote Sens.* **2022**, *14*, 2301, doi:10.3390/rs14102301 . . . . **7**

Linxin Lin, Guan Chen, Wei Shi, Jiacheng Jin, Jie Wu, Fengchun Huang, Yan Chong, Yang Meng, Yajun Li and Yi Zhang
Spatiotemporal Evolution Pattern and Driving Mechanisms of Landslides in the Wenchuan Earthquake-Affected Region: A Case Study in the Bailong River Basin, China
Reprinted from: *Remote Sens.* **2022**, *14*, 2339, doi:10.3390/rs14102339 . . . . **33**

Cheila Avalon Cullen, Rafea Al Suhili and Edier Aristizabal
A Landslide Numerical Factor Derived from CHIRPS for Shallow Rainfall Triggered Landslides in Colombia
Reprinted from: *Remote Sens.* **2022**, *14*, 2239, doi:10.3390/rs14092239 . . . . **57**

Ioannis Farmakis, David Bonneau, D. Jean Hutchinson and Nicholas Vlachopoulos
Targeted Rock Slope Assessment Using Voxels and Object-Oriented Classification
Reprinted from: *Remote Sens.* **2021**, *13*, 1354, doi:10.3390/ rs13071354 . . . . **77**

Fabio Bovenga, Ilenia Argentiero, Alberto Refice, Raffaele Nutricato, Davide O. Nitti, Guido Pasquariello and Giuseppe Spilotro
Assessing the Potential of Long, Multi-Temporal SAR Interferometry Time Series for Slope Instability Monitoring: Two Case Studies in Southern Italy
Reprinted from: *Remote Sens.* **2022**, *14*, 1677, doi:10.3390/rs14071677 . . . . **99**

Shuangshuang Wu, Xinli Hu, Wenbo Zheng, Matteo Berti, Zhitian Qiao and Wei Shen
Threshold Definition for Monitoring Gapa Landslide under Large Variations in Reservoir Level Using GNSS
Reprinted from: *Remote Sens.* **2021**, *13*, 4977, doi:10.3390/rs13244977 . . . . **121**

Bochen Zhang, Songbo Wu, Xiaoli Ding, Chisheng Wang, Jiasong Zhu and Qingquan Li
Use of Multiplatform SAR Imagery in Mining Deformation Monitoring with Dense Vegetation Coverage: A Case Study in the Fengfeng Mining Area, China
Reprinted from: *Remote Sens.* **2021**, *13*, 3091, doi:10.3390/rs13163091 . . . . **145**

Lorenzo Solari, Roberto Montalti, Anna Barra, Oriol Monserrat, Silvia Bianchini and Michele Crosetto
Multi-Temporal Satellite Interferometry for Fast-Motion Detection: An Application to Salt Solution Mining
Reprinted from: *Remote Sens.* **2020**, *12*, 3919, doi:10.3390/rs12233919 . . . . **167**

Moritz Rösch and Simon Plank
Detailed Mapping of Lava and Ash Deposits at Indonesian Volcanoes by Means of VHR PlanetScope Change Detection
Reprinted from: *Remote Sens.* **2022**, *14*, 1168, doi:10.3390/rs14051168 . . . . **189**

Niccolò Dematteis and Daniele Giordan
Comparison of Digital Image Correlation Methods and the Impact of Noise in Geoscience Applications
Reprinted from: *Remote Sens.* **2021**, *13*, 327, doi:10.3390/rs13020327 . . . . **221**

About the Editors

Daniele Giordan

Daniele Giordan holds a MSc and a PhD in Earth Science at the University of Torino studying deep-seated gravitational slope deformations, monitoring systems of slope instabilities, and early warning procedures. He is senior researcher of the Research Institute for Geo-Hydrological Protection of the Italian National Research Council. Scientific coordinator of Geohazard Monitoring Group since 2015, he works in particular on developing new methodologies and systems for geo-hydrological processes monitoring. In particular, he focused his research activities on monitoring systems to improve the geological model of large slope instabilities and correct management of early warning systems. Recently, he concentrated on defining a communication strategy for proper management and dissemination of slope instabilities monitoring results. Since 2017, he is Chairman of the Commission C35 "monitoring methods and approaches in engineering geology applications" of the IAEG - International Association for Engineering Geology and the Environment. In 2019, he was elected President of the IAEG Italian National Group. He is editor of several books and special issues focused on natural hazards and engineering geology, co-author of more than 80 international papers and book chapters.

Guido Luzi

Guido Luzi graduated in Physics and holds a PhD in Electronic Systems Engineering. He is with the Geomatics Research Units of the Centre Tecnològic de Telecomunicacions de Catalunya (CTTC), Spain since January 2014. He has been involved in research activities concerning the application of spaceborne and terrestrial radar techniques since 1986, earning a wide skill in design and implementation of passive and active systems. He worked with the Department of Electronics and Telecommunications of the University of Florence, and later with the Department of Earth Sciences of the same University, dealing with various applications from the monitoring of volcanic areas to the development and experimentation of microwave sensors for the detection of vital signs, and Civil Engineering and Cultural Heritage applications. In 2010 he moved to the Institute of Geomatics, where he was involved in the design and experimentation of radar-based sensing techniques. His main research lines are focused on geophysical applications with emphasis in the observation of land surfaces through terrestrial and satellite microwave interferometry, and civil structures monitoring. He holds a long-time experience in Development and Research and technology transfer matured working in research institutions with a consolidated link to industrial entities, aiming at the development and experimentation of new ideas and related prototypes. He has authored or co-authored of more than seventy papers published in peer reviewed international journals, and more than two hundred of international Proceeding contributions. He acts as referee for different journals, among which: IEEE TGRS, GRSL, JSTAR, IJRS, Remote Sensing. He took part in several International projects funded by ESA and EC, participating in outstanding Remote Sensing campaigns as AGRISCATT87, AGRISCATT88, MACEurope, European research projects as EPOCH, MEDALUS II, ENVIRONMENT, GALAHAD, APhoRISM, HEIMDALL and GIMS. He is IEEE Senior member.

Oriol Monserrat

Oriol Monserrat holds a PhD in aerospace science and technology from the Polytechnic University of Catalonia (2012) and a degree in mathematics from the University of Barcelona (2004). In 2003 he started working as a researcher in the Active Remote Sensing Unit of the Geomatics Institute. Since January 2014 works as a Head of the Remote Sensing Department of the Division of Geomatics at the Technological Centre of Telecommunications of Catalunya (CTTC-www.cttc.es). His research activity is related to the analysis of satellite, airborne and terrestrial remote sensing data and the development of scientific and technical applications using mainly active sensors, such as the Synthetic Aperture Radar (SAR) radar, Real aperture Radar (RAR) and laser scanners. In addition, he has experience with passive sensors. From the point of view of the application, Dr. Monserrat is specialized in the measurement and monitoring of deformations using SAR interferometry techniques (InSAR). His most relevant research includes: (i) the development of algorithms and analysis methods for SAR interferometry, differential SAR interferometry (DInSAR) and terrestrial SAR; (ii) processing and analysis of remote sensing data; (iii) terrestrial remote sensing techniques: terrestrial laser scanner, SAR and RAR; iv) the measurement of deformations with Remote Sensing Data. This last aspect is directly related to applications related with natural hazards risk management, and in particular, with landslide and Subsidence hazards.

Niccolò Dematteis

Niccolò Dematteis is research assistant with the Research Institute for Geo-hydrological Protection (Italy) since 2016. He holds a master degree in physics and obtained a PhD in Earth Science in 2020. His scientific interests focus on glacial hazard monitoring and assessment. To this end, he worked on developing original techniques to measure glacier morpho-dynamics, using terrestrial and satellite optical images and data fusion approaches. He is member of the Glacier and Permafrost Hazards in Mountains (GAPHAZ) group. He acts as a reviewer for various journals and is member of the editorial boards of Journal of Applied Remote Sensing and Earth System Science Data. He is author of fifteen scientific papers published on peer-reviewed journals and more than thirty conference proceedings and book chapters.

Editorial

Remote Sensing Analysis of Geologic Hazards

Daniele Giordan [1], Guido Luzi [2], Oriol Monserrat [2] and Niccolò Dematteis [1],*

[1]   Research Institute for Geo-Hydrological Protection, Italian National Research Council, 10135 Turin, Italy
[2]   Geomatics Research Unit, Centre Tecnològic de Telecomunicacios de Catalunya, 08860 Castelldefels, Spain
*   Correspondence: niccolo.dematteis@irpi.cnr.it

Citation: Giordan, D.; Luzi, G.; Monserrat, O.; Dematteis, N. Remote Sensing Analysis of Geologic Hazards. *Remote Sens.* **2022**, *14*, 4818. https://doi.org/10.3390/rs14194818

Received: 16 September 2022
Accepted: 22 September 2022
Published: 27 September 2022

1. Introduction

In recent decades, classical survey techniques (i.e., field measurements and aerial remote sensing) have evolved, and with the advent of new technologies—e.g., terrestrial radar interferometry [1,2], digital time-lapse cameras [3], terrestrial and aerial laser scanners [4,5] and platforms, e.g., UAV [6,7]—remote sensing systems have become popular and widely used in geosciences. Contactless devices are not invasive and allow measuring without accessing the investigated area. This is an excellent advantage as earth surface processes often occur in remote areas and can be potentially dangerous or difficult to access [8]. Satellite and aerial remote sensing offer the possibility of surveying large areas, using hyperspectral optical [8,9], synthetic aperture radar (SAR) [10,11] and thermal infrared [12,13] images and altimetric lasers [14]. The progressive rise in available public and private satellite constellations has permitted individuals to reach very high-resolution images at weekly to daily revisit time. On the other hand, ground-based surveys usually have higher acquisition frequency and spatial resolution compared to satellite systems, and they are able to observe the evolution of fast processes and their possible paroxysmal phase, e.g., volcanic eruptions [15,16], glacier instabilities [17], landslides [18,19], and floods [20,21]. For their characteristics, proximal sensing applications are often used in monitoring activities at a short revisit time, as they can provide real-time or near-real-time information [22]. Therefore, they can be of great support in early warning procedures and risk assessment and management [23,24]. Combined with aerospace sensors, contactless terrestrial devices are particularly suitable for data-fusion techniques, multi-scale approaches and supporting numerical model analysis [25–28].

Satellite and terrestrial remote sensing are of paramount importance in specific tasks of geologic hazard analysis. This Special Issue has collected ten papers concerned with recent and upcoming advances in remote sensing applications in geologic hazard analysis. In particular, this Special Issue includes studies about satellite and terrestrial contactless devices for detecting, monitoring and analyzing geologic processes, as well as new data-processing and warning techniques (Figure 1).

Figure 1. Word cloud of the abstracts of the contributions of this Special Issue.

2. Contributions of the Special Issue

Satellite and terrestrial remote sensing allow for the analysis of different geological processes and are of paramount importance in specific tasks of hazard analysis. In particular, this Special Issue collects contributions that are concerned with (i) detecting and mapping potential hazards, (ii) acquiring data for analysis of susceptibility and triggering factors, (iii) monitoring the evolution of the process, and (iv) developing strategies of warning and early warning. The considered geological processes are landslides, mining-induced subsidence and land deformations, volcanoes, glacier flows and dune migration. The adopted techniques are varied and include, but are not limited to, optical sensors, SAR, GNSS and multi-sensor approaches (Figure 2). Usually, multi-sensor-based studies use satellite optical images to detect the areas of interest and multitemporal SAR interferometry (MTInSAR) to measure ground deformation in such areas.

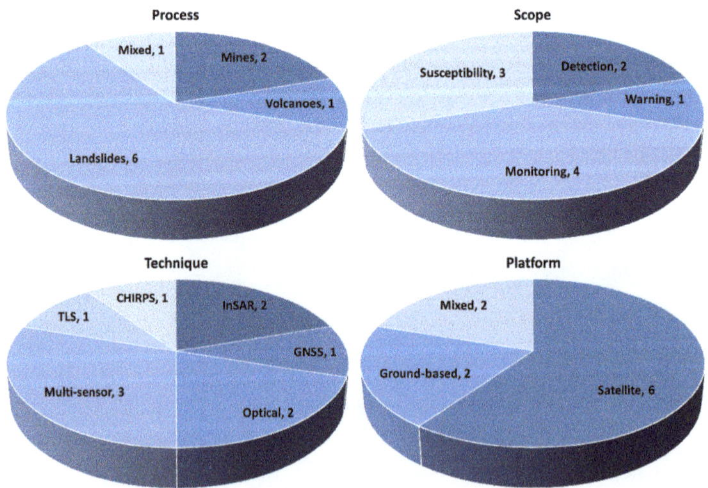

Figure 2. Distributions of the topics dealt with in this Special Issues, according to different schemes of classification: (i) the investigated process, (ii) the scope of the study, (iii) the adopted technique (TLS is terrestrial laser scanner, while CHIRPS is the Climate Hazards group Infrared Precipitation with Stations dataset) and (iv) the platform adopted.

In the following, we will describe the contributions present in this Special Issue. They are organized according to: first, the investigated geologic process, and second, the scope of the study (i.e., detection, monitoring, process analysis and warning).

3. Landslides

Landslides are the most-frequently investigated process in this Special Issue. Six papers examine this phenomenon. However, the studies focus on different themes; some are dedicated to landslide detection [29] and others to triggering factors and susceptibility analysis [30–32], monitoring [33] and early warning procedures [34].

3.1. Detection and Mapping

Lindsay et al. [29] focused their work on landslide mapping. They used optical (Sentinel-2) and SAR (Sentinel-1) satellite images from Google Earth Engine to map the landslides triggered by a rainstorm in western Norway. To detect the landslides, they manually analyzed the differential Normalized Difference Vegetation Index and the SAR amplitude difference of VV and VH polarizations (from the Sentinel-2 and Sentinel-1, respectively). Their results showed that, by using a stack of pre-event images to improve the signal-to-noise ratio, the number of detected landslides increased from 14 (registered by the Norwegian Landslides Inventory) to 120. Moreover, they found that optical images performed better than SAR amplitude images with the aim of landslide detection.

3.2. Process Analysis and Susceptibility

Three works are dedicated to the analysis of triggering factors and susceptibility to landslides and rock slopes. The surveys were conducted from ground, satellite or combined approaches.

Lin et al. [30] studied the spatiotemporal evolution pattern and driving mechanisms of landslides in the Bailong River Basin (China), where a strong earthquake occurred in 2008. They analyzed the period 2007–2020, mapping the occurred landslides using optical images from various sources and measuring the landslide deformation using MTInSAR. They identified three stages of landslide triggering: the earthquake (2008), the coupled earthquake–rainfall (2008–2017), and the rainfall (2017–present) driving stages. In particular, they observed that the landslides in the limestone area were more responsive to the earthquake, while the loess–phyllite-dominated sectors were mainly controlled by rainfall.

The study of Cullen et al. [31] was located in Colombia between 2016 and 2019. They used an inventory of 346 rainfall-induced landslide events and the Climate Hazards group Infrared Precipitation with Stations (CHIRPS) dataset, which combines satellite-based rainfall data and ground-based gauge measurements to produce global rainfall data at a $0.05° \times 0.05°$ resolution. They formulated new dynamic variables based on the rainfall occurrence and amount during dry and wet periods, and determined an original landslide triggering factor (LTF). Comparing their LTF with canonical event–duration threshold, they found that LTF performed better in 81% of cases.

Farmakis et al. [32] analyzed point clouds (PCs) of rock slopes acquired by a terrestrial laser scanner (TLS) in British Columbia (Canada). They partitioned the PCs into voxels based on local dimensionality, orientation, and topology and built an automatic decision tree that utilized geometrical, topological, and contextual information and enabled the classification of a multi-hazard railway rock slope into classes involved in landslide risk management. Their results demonstrated precision similar to more complex machine learning algorithms and manual knowledge-based analysis.

3.3. Monitoring Activities

Bovenga et al. [33] applied MTInSAR to Sentinel-1 and COSMO-SkyMed to measure the velocity time series of two landslides that occurred in 2013 and 2019 in southern Italy, which caused damages to buildings and roads. Their results evidenced the presence of non-linear displacements in correspondence of some key infrastructures. They concluded that the analysis of accelerations and decelerations of persistent scatter objects corresponding to structures affected by recent stabilization measures helps to shed new light in relation to known events that occurred in the area of interest.

3.4. Warning Procedures

Wu et al. [34] developed a threshold-based early warning procedure for the Gapa Landslide (Southwest China), which was reactivated by the impoundment of a large water reservoir. They observed that the landslide deformation was strongly related to the fluctuations in reservoir water levels; thus, a crucial water level was also defined to reduce false warnings from the velocity threshold alone. The current monitoring system is composed of six permanent GNSS receivers and one water level station. A warning procedure can be activated in case of a velocity exceeding 4 mm day^{-1} and a water level higher than 1820 a.s.l. (i.e., reservoir depth > 70 m).

4. Mines

Zhang et al. [35] treated the measurement of ground deformation caused by mining activities in the Fengfeng area (Eastern China) in 2015 and 2016. They adopted MTInSAR applied to multiple satellite SAR images, i.e., TerraSAR-X, Sentinel-1, Radarsat-2, and PALSAR-2, thus increasing the data adopted to build the time series. They focused their study on vegetated areas, where MTInSAR observations are usually critical. They measured subsidence of almost 800 mm year^{-1}, and registered a root mean squared deviation of 83 mm year^{-1} compared to terrestrial observations.

Solari et al. [36] applied MTInSAR in a mining area in southern Italy to measure the deformation time series caused by subsidence in the period 2016–2018. They found areas with deformation rates up to 250 mm year^{-1}. Moreover, they manually mapped more than 100 sinkholes that occurred in the area between 1956 and 2018 using nine orthoimages acquired across this period, and they analyzed the deformation in correspondence of the sinkholes. In their work, quite homogeneous subsidence rates (10–20 mm year^{-1}) were measured, except for the more recent sinkholes, where the velocities were more heterogeneous and higher (up to 80 mm year^{-1}).

5. Volcanoes

The work of Rosch and Plank [37] concerned the mapping of lava and ash deposits using PlanetScope optical images. This study introduced an object-oriented classification for mapping lava flows in vegetated and unvegetated areas during several eruptive phases of three Indonesian volcanoes. A change detection investigation was combined with the analysis of variations in texture and brightness, with hydrological runoff modelling and with analysis of thermal anomalies derived from Sentinel-2 or Landsat-8. The results showed good agreement with the reports of the Global Volcanism Program and showed the benefits of PlanetScope images for volcano daily monitoring and eruption risk assessment.

6. Glaciers and Sand Dunes

Dematteis and Giordan [38] conducted a methodological study that compared the performances of fifteen digital image correlation functions and proposed a new similarity index (DOT). They analyzed the outcomes considering four template sizes and thirteen types and levels of noise applied on a shaded relief of a digital elevation. They conducted the same comparison on optical images of a glacier (acquired by a terrestrial camera) and the dunes of the Bodélé Depression (Chad) (acquired by Sentinel-2). Overall, they observed that using orientation images provided the best performances in the presence of shadows or snow patches, adopting either frequency-based or DOT correlations.

7. Conclusions

This Special Issue consists of ten papers that used remote sensing (either from ground or aerospace) to analyze geologic hazards. The applied techniques are varied and include

GNSS, TLS, SAR, optical images and rainfall datasets. On several occasions, multi-sensor approaches were adopted. The scopes are different as well, and they can be summarized in four classes: (i) detection, (ii) susceptibility and triggering factors analysis, (iii) monitoring, and (iv) early warning. Considering all the contributions, this Special Issue demonstrates the high benefit of using remote sensing to analyze geologic hazards and represents a valuable advance of the research in this field.

Author Contributions: Writing—original draft preparation, N.D.; writing—review and editing, D.G., G.L., O.M. and N.D. All authors have read and agreed to the published version of the manuscript.

Funding: This research received no external funding.

Acknowledgments: The Guest Editors of this Special Issue would like to thank all authors who have contributed to this volume for sharing their scientific results and for their excellent collaboration. Special gratitude will go to the community of distinguished reviewers for their constructive inputs. The Remote Sensing editorial team is acknowledged for its support during all phases related to the successful completion of this issue.

Conflicts of Interest: The authors declare no conflict of interest.

References

1. Monserrat, O.; Crosetto, M.; Luzi, G. A review of ground-based SAR interferometry for deformation measurement. *ISPRS J. Photogramm. Remote Sens.* **2014**, *93*, 40–48. [CrossRef]
2. Caduff, R.; Schlunegger, F.; Kos, A.; Wiesmann, A. A review of terrestrial radar interferometry for measuring surface change in the geosciences. *Earth Surf. Process. Landf.* **2015**, *40*, 208–228. [CrossRef]
3. Travelletti, J.; Delacourt, C.; Allemand, P.; Malet, J.P.; Schmittbuhl, J.; Toussaint, R.; Bastard, M. Correlation of multi-temporal ground-based optical images for landslide monitoring: Application, potential and limitations. *ISPRS J. Photogramm. Remote Sens.* **2012**, *70*, 39–55. [CrossRef]
4. Baldo, M.; Bicocchi, C.; Chiocchini, U.; Giordan, D.; Lollino, G. LIDAR monitoring of mass wasting processes: The Radicofani landslide, Province of Siena, Central Italy. *Geomorphology* **2009**, *105*, 193–201. [CrossRef]
5. Giordan, D.; Allasia, P.; Manconi, A.; Baldo, M.; Santangelo, M.; Cardinali, M.; Corazza, A.; Albanese, V.; Lollino, G.; Guzzetti, F. Morphological and kinematic evolution of a large earthflow: The Montaguto landslide, southern Italy. *Geomorphology* **2013**, *187*, 61–79. [CrossRef]
6. Colomina, I.; Molina, P. Unmanned aerial systems for photogrammetry and remote sensing: A review. *ISPRS J. Photogramm. Remote Sens.* **2014**, *92*, 79–97. [CrossRef]
7. Giordan, D.; Adams, M.S.; Aicardi, I.; Alicandro, M.; Allasia, P.; Baldo, M.; De Berardinis, P.; Dominici, D.; Godone, D.; Hobbs, P.; et al. The use of unmanned aerial vehicles (UAVs) for engineering geology applications. *Bull. Eng. Geol. Environ.* **2020**, *79*, 3437–3481. [CrossRef]
8. Leprince, S.; Berthier, E.; Ayoub, F.; Delacourt, C.; Avouac, J.P. Monitoring earth surface dynamics with optical imagery. *Eos* **2008**, *89*, 1–2. [CrossRef]
9. Van der Meer, F.; Freek, D. Multi-and hyperspectral geologic remote sensing: A review. *Int. J. Appl. Earth Obs. Geoinf.* **2012**, *14*, 112–128. [CrossRef]
10. Cigna, F.; Osmanoğlu, B.; Cabral-Cano, E.; Dixon, T.H.; Ávila-Olivera, J.A.; Garduño-Monroy, V.H.; DeMets, C.; Wdowinski, S. Monitoring land subsidence and its induced geological hazard with Synthetic Aperture Radar Interferometry: A case study in Morelia, Mexico. *Remote Sens. Environ.* **2012**, *117*, 146–161. [CrossRef]
11. Shen, X.; Wang, D.; Mao, K.; Anagnostou, E.; Hong, Y. Inundation extent mapping by synthetic aperture radar: A review. *Remote Sens.* **2019**, *11*, 879. [CrossRef]
12. Gerhards, M.; Schlerf, M.; Mallick, K.; Udelhoven, T. Challenges and future perspectives of multi-/Hyperspectral thermal infrared remote sensing for crop water-stress detection: A review. *Remote Sens.* **2019**, *11*, 1240. [CrossRef]
13. Weng, Q. Thermal infrared remote sensing for urban climate and environmental studies: Methods, applications, and trends. *ISPRS J. Photogramm. Remote Sens.* **2009**, *64*, 335–344. [CrossRef]
14. Neumann, T.A.; Martino, A.J.; Markus, T.; Bae, S.; Bock, M.R.; Brenner, A.C.; Brunt, K.M.; Cavanaugh, J.; Fernandes, S.T.; Hancock, D.W.; et al. The Ice, Cloud, and Land Elevation Satellite—2 mission: A global geolocated photon product derived from the advanced topographic laser altimeter system. *Remote Sens. Environ.* **2019**, *233*, 111325. [CrossRef]
15. Casagli, N.; Tibaldi, A.; Merri, A.; Del Ventisette, C.; Apuani, T.; Guerri, L.; Fortuny-Guasch, J.; Tarchi, D. Deformation of Stromboli Volcano (Italy) during the 2007 eruption revealed by radar interferometry, numerical modelling and structural geological field data. *J. Volcanol. Geotherm. Res.* **2009**, *182*, 182–200. [CrossRef]
16. Honda, K.; Nagai, M. Real-time volcano activity mapping using ground-based digital imagery. *ISPRS J. Photogramm. Remote Sens.* **2002**, *57*, 159–168. [CrossRef]

17.   Dematteis, N.; Giordan, D.; Troilo, F.; Wrzesniak, A.; Godone, D. Ten-Year Monitoring of the Grandes Jorasses Glaciers Kinematics. Limits, Potentialities, and Possible Applications of Different Monitoring Systems. *Remote Sens.* **2021**, *13*, 3005. [CrossRef]

18.   Uhlemann, S.; Smith, A.; Chambers, J.; Dixon, N.; Dijkstra, T.; Haslam, E.; Meldrum, P.; Merritt, A.; Gunn, D.; Mackay, J. Assessment of ground-based monitoring techniques applied to landslide investigations. *Geomorphology* **2016**, *253*, 438–451. [CrossRef]

19.   Pecoraro, G.; Calvello, M.; Piciullo, L. Monitoring strategies for local landslide early warning systems. *Landslides* **2019**, *16*, 213–231. [CrossRef]

20.   Lo, S.W.; Wu, J.H.; Lin, F.P.; Hsu, C.H. Visual sensing for urban flood monitoring. *Sensors* **2015**, *15*, 20006–20029. [CrossRef]

21.   Perks, M.T.; Russell, A.J.; Large, A.R.G. Technical note: Advances in flash flood monitoring using unmanned aerial vehicles (UAVs). *Hydrol. Earth Syst. Sci.* **2016**, *20*, 4005–4015. [CrossRef]

22.   Casagli, N.; Catani, F.; Del Ventisette, C.; Luzi, G. Monitoring, prediction, and early warning using ground-based radar interferometry. *Landslides* **2010**, *7*, 291–301. [CrossRef]

23.   Intrieri, E.; Gigli, G.; Mugnai, F.; Fanti, R.; Casagli, N. Design and implementation of a landslide early warning system. *Eng. Geol.* **2012**, *147–148*, 124–136. [CrossRef]

24.   Manconi, A.; Giordan, D. Landslide early warning based on failure forecast models: The example of the Mt. de la Saxe rockslide, northern Italy. *Nat. Hazards Earth Syst. Sci.* **2015**, *15*, 1639–1644. [CrossRef]

25.   Malet, J.P.; Ferhat, G.; Ulrich, P.; Boetzlé, P. The French National Landslide Observatory OMIV–Monitoring surface displacement using permanent GNSS, photogrammetric cameras and terrestrial LiDAR for understanding the landslide mechanisms. In Proceedings of the 3rd Joint International Symposium on Deformation Monitoring (JISDM), Vienne, Austria, 7 May 2016; pp. 1–7.

26.   Casagli, N.; Frodella, W.; Morelli, S.; Tofani, V.; Ciampalini, A.; Intrieri, E.; Raspini, F.; Rossi, G.; Tanteri, L.; Lu, P. Spaceborne, UAV and ground-based remote sensing techniques for landslide mapping, monitoring and early warning. *Geoenviron. Disasters* **2017**, *4*, 9. [CrossRef]

27.   Lollino, P.; Giordan, D.; Allasia, P. The Montaguto earthflow: A back-analysis of the process of landslide propagation. *Eng. Geol.* **2014**, *170*, 66–79. [CrossRef]

28.   Musa, Z.N.; Popescu, I.; Mynett, A. A review of applications of satellite SAR, optical, altimetry and DEM data for surface water modelling, mapping and parameter estimation. *Hydrol. Earth Syst. Sci.* **2015**, *19*, 3755–3769. [CrossRef]

29.   Lindsay, E.; Frauenfelder, R.; Rüther, D.; Nava, L.; Rubensdotter, L.; Strout, J.; Nordal, S. Multi-Temporal Satellite Image Composites in Google Earth Engine for Improved Landslide Visibility: A Case Study of a Glacial Landscape. *Remote Sens.* **2022**, *14*, 2301. [CrossRef]

30.   Lin, L.; Chen, G.; Shi, W.; Jin, J.; Wu, J.; Huang, F.; Chong, Y.; Meng, Y.; Li, Y.; Zhang, Y. Spatiotemporal Evolution Pattern and Driving Mechanisms of Landslides in the Wenchuan Earthquake-Affected Region: A Case Study in the Bailong River Basin, China. *Remote Sens.* **2022**, *14*, 2339. [CrossRef]

31.   Cullen, C.A.; Al Suhili, R.; Aristizabal, E. A Landslide Numerical Factor Derived from CHIRPS for Shallow Rainfall Triggered Landslides in Colombia. *Remote Sens.* **2022**, *14*, 2239. [CrossRef]

32.   Farmakis, I.; Bonneau, D.; Hutchinson, D.J.; Vlachopoulos, N. Targeted rock slope assessment using voxels and object-oriented classification. *Remote Sens.* **2021**, *13*, 1354. [CrossRef]

33.   Bovenga, F.; Argentiero, I.; Refice, A.; Nutricato, R.; Nitti, D.O.; Pasquariello, G.; Spilotro, G. Assessing the Potential of Long, Multi-Temporal SAR Interferometry Time Series for Slope Instability Monitoring: Two Case Studies in Southern Italy. *Remote Sens.* **2022**, *14*, 1677. [CrossRef]

34.   Wu, S.; Hu, X.; Zheng, W.; Berti, M.; Qiao, Z.; Shen, W. Threshold definition for monitoring Gapa Landslide under large variations in reservoir level using GNSS. *Remote Sens.* **2021**, *13*, 4977. [CrossRef]

35.   Zhang, B.; Wu, S.; Ding, X.; Wang, C.; Zhu, J.; Li, Q. Use of multiplatform sar imagery in mining deformation monitoring with dense vegetation coverage: A case study in the fengfeng mining area, china. *Remote Sens.* **2021**, *13*, 3091. [CrossRef]

36.   Solari, L.; Montalti, R.; Barra, A.; Monserrat, O.; Bianchini, S.; Crosetto, M. Multi-temporal satellite interferometry for fast-motion detection: An application to salt solution mining. *Remote Sens.* **2020**, *12*, 3919. [CrossRef]

37.   Rösch, M.; Plank, S. Detailed Mapping of Lava and Ash Deposits at Indonesian Volcanoes by Means of VHR PlanetScope Change Detection. *Remote Sens.* **2022**, *14*, 1168. [CrossRef]

38.   Dematteis, N.; Giordan, D. Comparison of digital image correlation methods and the impact of noise in geoscience applications. *Remote Sens.* **2021**, *13*, 327. [CrossRef]

remote sensing

Article

Multi-Temporal Satellite Image Composites in Google Earth Engine for Improved Landslide Visibility: A Case Study of a Glacial Landscape

Erin Lindsay [1,*], Regula Frauenfelder [2], Denise Rüther [3], Lorenzo Nava [4], Lena Rubensdotter [5,6], James Strout [2] and Steinar Nordal [1]

[1] Department of Civil and Environmental Engineering, Norwegian University of Science and Technology (NTNU), 7491 Trondheim, Norway; steinar.nordal@ntnu.no
[2] Norwegian Geotechnical Institute (NGI), 0806 Oslo, Norway; regula.frauenfelder@ngi.no (R.F.); james.michael.strout@ngi.no (J.S.)
[3] Department Environmental Sciences, Western Norway University of Applied Sciences (HVL), 5063 Bergen, Norway; denise.christina.ruther@hvl.no
[4] Machine Intelligence and Slope Stability Laboratory, Department of Geosciences, University of Padova, 35129 Padua, Italy; lorenzo.nava@phd.unipd.it
[5] Geohazard and Earth Observation, Norwegian Geological Survey (NGU), 7040 Trondheim, Norway; lena.rubensdotter@ngu.no
[6] Department of Arctic Geology, University Centre of Svalbard (UNIS), 9171 Longyearbyen, Norway
* Correspondence: erin.lindsay@ntnu.no; Tel.: +47-91564607

Citation: Lindsay, E.; Frauenfelder, R.; Rüther, D.; Nava, L.; Rubensdotter, L.; Strout, J.; Nordal, S. Multi-Temporal Satellite Image Composites in Google Earth Engine for Improved Landslide Visibility: A Case Study of a Glacial Landscape. *Remote Sens.* **2022**, *14*, 2301. https://doi.org/10.3390/rs14102301

Academic Editors: Oriol Monserrat, Guido Luzi, Daniele Giordan and Niccolò Dematteis

Received: 25 March 2022
Accepted: 2 May 2022
Published: 10 May 2022

Publisher's Note: MDPI stays neutral with regard to jurisdictional claims in published maps and institutional affiliations.

Abstract: Regional early warning systems for landslides rely on historic data to forecast future events and to verify and improve alarms. However, databases of landslide events are often spatially biased towards roads or other infrastructure, with few reported in remote areas. In this study, we demonstrate how Google Earth Engine can be used to create multi-temporal change detection image composites with freely available Sentinel-1 and -2 satellite images, in order to improve landslide visibility and facilitate landslide detection. First, multispectral Sentinel-2 images were used to map landslides triggered by a summer rainstorm in Jølster (Norway), based on changes in the normalised difference vegetation index (NDVI) between pre- and post-event images. Pre- and post-event multi-temporal images were then created by reducing across all available images within one month before and after the landslide events, from which final change detection image composites were produced. We used the mean of backscatter intensity in co- (VV) and cross-polarisations (VH) for Sentinel-1 synthetic aperture radar (SAR) data and maximum NDVI for Sentinel-2. The NDVI-based mapping increased the number of registered events from 14 to 120, while spatial bias was decreased, from 100% of events located within 500 m of a road to 30% close to roads in the new inventory. Of the 120 landslides, 43% were also detectable in the multi-temporal SAR image composite in VV polarisation, while only the east-facing landslides were clearly visible in VH. Noise, from clouds and agriculture in Sentinel-2, and speckle in Sentinel-1, was reduced using the multi-temporal composite approaches, improving landslide visibility without compromising spatial resolution. Our results indicate that manual or automated landslide detection could be significantly improved with multi-temporal image composites using freely available earth observation images and Google Earth Engine, with valuable potential for improving spatial bias in landslide inventories. Using the multi-temporal satellite image composites, we observed significant improvements in landslide visibility in Jølster, compared with conventional bi-temporal change detection methods, and applied this for the first time using VV-polarised SAR data. The GEE scripts allow this procedure to be quickly repeated in new areas, which can be helpful for reducing spatial bias in landslide databases.

Keywords: multi-temporal image composite; change detection; Jølster; landslide database; Sentinel-2; Sentinel-1; Google Earth Engine; NDVI; glacial landscape

1. Introduction

The frequency and intensity of severe precipitation events are expected to continue increasing as a result of climate change [1,2]. In turn, precipitation-triggered disasters, including landslides and floods, are also becoming more frequent [3]. Landslide early warning systems (LEWSs) aim to reduce loss of life and damage, by giving people advanced warning so that emergency responders may prepare resources and warn the public to avoid high-risk areas [4]. Existing LEWSs rely on knowledge of past events in order to improve the understanding of landslide triggering processes, develop warning thresholds and prediction models, and finally verify warnings issued and evaluate performance [5]. Landslide inventory data are also needed to produce landslide susceptibility and hazard maps used for spatial planning, and emergency responders need detailed information to assess damages and save lives [6].

However, for operators of LEWSs, collecting landslide data and verifying warnings is a difficult and tedious task [7]. Generally, the registration of events and preparation of landslide inventories are not performed systematically [8]. Presently, single-landslide event data are collected mainly from ground observations, including reports from road and rail authorities, and to a lesser degree from the public, or mined from social media and news reports [9,10]. For triggering events (e.g., extreme precipitation or earthquake) with multiple resulting landslides, mapping using aerial or satellite optical images may be undertaken on a case-by-case basis, depending on motivation and resources [11–13]. Such mapping is most commonly limited to areas that are already known to have been affected by landslides (e.g., [14–16]), or to verify the exact location of landslides that may have been reported without precise coordinates, for example, in media reports [17]. Systematic mapping is generally not yet carried out over large areas or to identify previously unknown landslides in remote areas.

The resulting landslide databases, however, tend to show clusters of observations around linear infrastructure and populated areas, thus not giving a realistic representation of the true spatial distribution of landslides, as can be seen in Figure 1. Additionally, landslide inventories tend to be rather incomplete. For example, in Europe, a comparison between landslide density based on landslide records from the Geological Surveys of Europe and the European landslide susceptibility map (ELSUS 1000 v1) reveals that only three countries (Poland, Italy, and Slovakia) have inventories considered to be over 50% complete [18]. These issues with spatial bias and missing data limit how the landslide data can be used for predictive models [19].

Using earth observation images (EO images) to detect unknown single landslide events is akin to finding a needle in a haystack. Hence, such surveys tend to focus on mapping landslides within areas that are known to have multiple landslides, due to a specific triggering event [20]. However, the ongoing and rapid developments in the fields of computer vision and cloud computing, as well as in the quality, range, and availability of EO images, mean that continuous, automated landslide monitoring may soon be feasible. Today, deep-learning models enable machines to accomplish image recognition tasks that were not possible a few years ago. Such a system is being developed for monitoring changes in forests and land cover (e.g., Global Forest [21], and Continuous Change Detection and Classification (CCDC) of land cover [22]). These operate at a global scale, using 30 m resolution Landsat images and the processing power of Google Earth Engine (GEE). Globally trained (also known as generalised) models for detecting landslides from EO images have begun to emerge recently [23,24], building on the results of locally trained models (e.g., [25,26]). These are not yet capable of outperforming humans in the tasks of landslide detection or mapping, and preliminary trials have found that they did not perform well in a glacial landscape [27]. With the exception of two landslides occurring in Fiordland, New Zealand [24], the globally trained models for landslide detection mentioned above have not, for the most part, included events from fjord landscapes. Based on the progress being made in other image recognition tasks, with additional training data and development, there appears to be good potential for computers to outperform humans in these tasks in the near future.

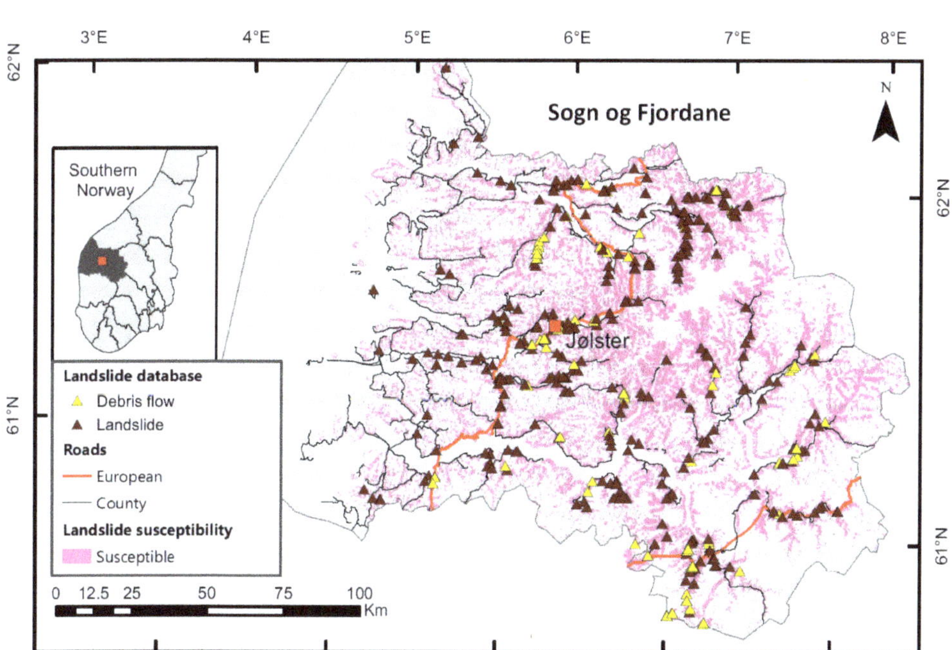

Figure 1. Historic landslide events in the former county of Sogn og Fjordane (now part of Vestland county), as registered in the Norwegian national landslide database (NLDB) on www. skredregistrering.no (accessed: 18 November 2019). Spatial bias is evident, as the events appear clustered along major roads, instead of being evenly distributed across susceptible areas (source: Norwegian Geological Survey, NGU). The location of the study area, Jølster, is shown with the red square.

While research continues on fine-tuning machine-learning model architectures and classifiers, other options for improving landslide detectability using EO data are immediately available. Google Earth Engine (GEE) is a cloud-based platform, designed to simplify geospatial analyses, by improving the process of accessing and processing data. With a free researcher account, GEE provides instant access to 37 years of satellite images, with a plethora of publicly available geospatial datasets, including optical image collections (e.g., MODIS, Landsat, and Sentinel-2), synthetic aperture radar (SAR) data from Sentinel-1, land cover classifications, and precipitation data, to name a few. Analysis of large areas while detecting changes over time can be performed very efficiently, using Google's cloud infrastructure [28].

Several researchers have begun to use GEE for improving landslide detection. HazMapper is a GEE app designed to facilitate the mapping of natural hazard events [29]. For landslides, images showing the relative difference in the normalised difference vegetation index (rdNDVI) calculated from cloud-free composites using Sentinel-2 or Landsat images can be used to map landslides for an event of interest (e.g., earthquake, or extreme precipitation event) [29]. SAR data can also be used for landslide detection [30]. SAR data have an advantage over optical images for rapid landslide detection, as the sensor is cloud-penetrating and measures the earth's surface independently of lighting conditions. However, it is rather underutilised due to the level of expertise needed to pre-process SAR data [31] and generally lower visibility of landslides. GEE opens doors for more widespread exploration of SAR data, as the Sentinel-1 GRD image collection has already accomplished much of the pre-processing [32]. A method for landslide detection using SAR data on GEE has been proposed by [31] using VH-polarised Sentinel-1 data to detect changes in radar backscatter that could be associated with landslides.

One aspect that both these methods have in common is the use of multi-temporal image stacks from pre- and post-events that are used to create difference images, from which landslides can be identified (Figure 2). Outside of GEE, such analysis would generally require downloading, storing, and locally processing vast quantities of data that could easily exceed the storage capacity of a personal computer (e.g., [33]). Hence, most change detection methods for landslide detection have used a single set of pre- and post-event images to make a difference image [34]. The advantage of multi-temporal images for change detection is that they reduce noise, compared with bi-temporal images, thus allowing for clearer visualisation of changes on the earth's surface. For optical images, the greenest-pixel method accomplishes this by selecting the maximum value NDVI pixel within the stack to produce an aggregate image. This effectively removes cloudy pixels and smoothens over temporarily reduced NDVI signals from agricultural activity. In comparison, for SAR images, speckle noise is reduced by aggregating across multi-temporal stacks. This is achieved without loss of spatial resolution, as occurs in the more common practice of speckle filtering across spatially neighbouring pixels from a single date image [35].

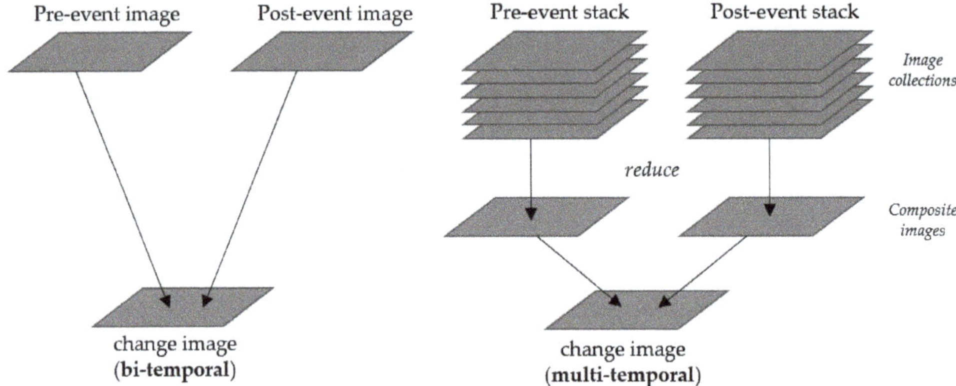

Figure 2. Approaches for creating bi-temporal and multi-temporal change detection images.

The main objective of this study was to investigate how multi-temporal image-composite approaches can be used to improve landslide visibility in change detection images, thereby improving spatial bias and completeness of landslide databases. As part of this study, a landslide inventory for Jølster was created from Sentinel-2 images, then verified with fieldwork, drone images, and helicopter observations. Then, we explored the difference in landslide visibility between bi-temporal and multi-temporal change detection approaches, with Sentinel-1 images with VV- and VH-polarised data, and with NDVI in Sentinel-2 images. We also aimed to improve knowledge of how these approaches perform in a glacial landscape, considering the role of local vegetation. We used the Jølster case study as an example of a multiple landslide event occurring in a glacially formed landscape, at a northern latitude.

The main questions involved are as follows:

1. To what extent can spatial bias in the NLDB be improved using EO images and change detection approaches to detect landslides?
2. How can multi-temporal, image-composite change detection approaches using GEE improve landslide visibility, compared with bi-temporal change detection approaches?
3. In which data types (S2-NDVI, S1-VV, S1-VH) are the landslides most visible?

This study focused on the performance of manual approaches. Results of tests of the globally trained landslide detection models for the Jølster case study will be presented in a separate paper [27]. The term landslide is applied generally throughout, to refer to rapid landslides that occur in soil, i.e., debris flows and debris avalanches. However, the techniques discussed could potentially also be used for deeper landslides occurring due to bedrock failure, so long as there is a removal of surface vegetation. Details on triggers, landslide classification, and failure mechanisms in Jølster are available in [36].

2. Case Study: Jølster Rainstorm Event on 30 July 2019, Western Norway

On 30 July 2019, the town of Vassenden, west of Lake Jølster, in western Norway, was hit by a heavy summer rainstorm. The maximum recorded rainfall of the event was 113 mm in 24 h, exceeding the 200-year event magnitude at the two nearest precipitation weather stations, Botnen and Haukedalen [37]. This resulted in numerous landslides and floods, causing severe impacts on infrastructure, damage to private property, and one fatality (Figures 3 and 4). The majority of landslides occurred on open slopes, initiating as debris avalanches [36]. Some of these developed downslope into channelised or partially channelised debris flows, including the large landslide in Vassenden, shown in Figure 3D,E.

Over 150 people were evacuated from the area, and Highway E39—the main transport route between the capital city Oslo and the city of Bergen—was closed in six places due to flooding and debris. Works to remove debris and secure the area following this event cost 17 million kroner (ca. 1.65 million euro) [38].

The study area shown in Figure 5 extends 22 km W–E from Førde to Årnes and 18 km N–S. The landscape consists of steep glacial valleys, lakes, and mountains over 1500 m.a.s.l. The bedrock geology in the area is predominantly relatively competent Precambrian basement orthogneiss, affected locally by the Caledonian orogeny (400 Ma). Geomorphologically, the study area is strongly altered by the glacial history of the region. The Scandes mountain range is composed of very varying bedrock types but in all areas has been subjected to repeated hard physical erosion by ice sheets, stripping almost all sediments (including weathering residuals) from the bedrock. At the last glacial maximum, ice sheets reached over 1800 m a.s.l. [39]. The area was likely deglaciated around 12-11 ka [40], with much of the moraine in the region formed during this period, although there may have been a remnant ice sheet in the valley for some time after.

Figure 3. Photo collage of the largest landslides in the study: (**A**) the fatal Årnes debris avalanche (length 850 m); (**B**) evidence of historic debris with avalanche deposits (buried humus layers) in the western-most eroded debris flow channel at Slåtten (NGU); (**C**) overview of the Slåtten debris flows (lengths between 850 and 1100 m); (**D,E**) the compound debris avalanche/flow at Vassenden (length 1500 m), in July 2019, and following remedial work in June 2020 (Source: Hallstein Dvergsdal/Firda Tidend, 2019, and Oddleif Løset/NRK, 2020, in [38], used with permission).

Figure 4. Selection of photos of 'shallow' landslides triggered by the 30 July 2019 extreme precipitation event: (**A**) Novabakken overview on the northern side of Tindefjellet mountain; (**B**) superficial landslide deposits from landslide in thin soil; (**C–E**) shallow landslides in the Svidalen valley with erosion of grass-bound topsoil and thin soil profiles, down to bedrock in places and superficial deposits; (**F,G**) small landslide scarps on Tindefjellet mountain, west of the main Vassenden landslide. These occurred at higher elevations, in areas with thin soil cover, and sparse vegetation. In many cases, the erosion is limited to grass-bound topsoil that has eroded right from bedrock, and only a thin layer of material is deposited over vegetation, partially covering but not burying it. The landslide naming system stems from the landslide inventory based on detailed field mapping [36].

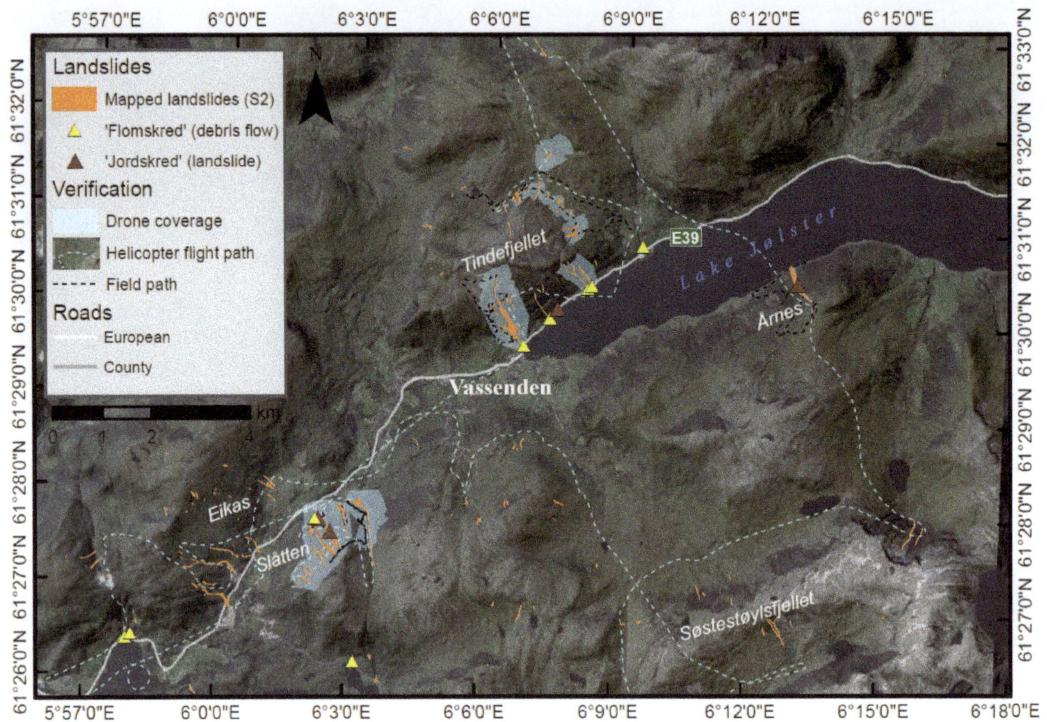

Figure 5. Study area showing landslides mapped with Sentinel-2 dNDVI images (orange polygons), registered landslides (triangles; source: NLDB, 18 December 2019), and ground-truth locations, including drone survey areas (shaded light blue) and helicopter and field visit GPS tracks (green and black dashed lines, respectively). Background: Sentinel-2 image.

The landscape consists of U-shaped valleys with flat, angled, polished bedrock surfaces that are favourable for the formation of sliding planes. Other signatures of the strong linear, glacial erosion in competent rock are visible in the area, including roche moutonnée, crag, and tail features. These types of bedrock features are exposed in the upper half of the Jølster landscape. A few remnant small-scale brittle faults are visible in the landscape [41]. Most of the study area has a surface cover of glacial moraine sediments [42]. The moraine is often highly consolidated on slopes overlying the bedrock. At high altitudes in the mountains, small pockets of moraine can be found within depressions in the exposed bare bedrock. A loose veneer of colluvium extends downslope on the valley sides, increasing to several metres thick at the valley bottom.

Buried hummus layers show evidence of previous landslides, seen in the stratigraphy exposed by a debris flow at Slåtten (Figure 3B). There are at least 40 historical landslide events registered as points in the Norwegian national landslide database (NLDB, at www. skredregistering.no, accessed on 18 January 2022, hence referred to as *registered events*) in the study area, and several instances of rock falls. Registered events are dated from 1625 to 2019. The majority of the registered events in the NLDB are derived from observations along the road network.

The modern climate of the study area is temperate, with relatively mild winters and wet summers due to the proximity to the coast. Glaciers are still active in the region, with Jostedalsbreen National Park, located 30 km east of the study area. Vegetation in the study area ranges from sparse moss and shrubs or light birch forest at high elevations, to spruce forest and agricultural fields lower in the valleys. Roads and built areas are mainly located in the flatter main valleys.

3. Data Used

The main input data were satellite images from Sentinel-1 and -2 missions of the European Space Agency (ESA), acquired either by downloading from Copernicus Open Access Hub (COAH, https://scihub.copernicus.eu/dhus/#/home, accessed 10 August 2019) or used directly within the GEE code editor (https://code.earthengine.google.com/, accessed 10 August 2019). Sentinel-1 SAR data are dual-polarised (VV and VH), with C-band (5.405 GHz), and available in Level 1 single-look complex (SLC) or ground-range-detected (GRD) scenes. We used data acquired in interferometric wide (IW) acquisition mode. Terrain corrections were applied using a 10 m resolution digital terrain model (DTM), downloaded from Høydedata, as shown in Table 1. For Sentinel-2, we used the Level 2A product, bottom of atmosphere (BOA) reflectance images, pre-processed from Level 1C products using sen2cor. In addition, historic landslide data from the NLDB and road data were used in the analysis of spatial bias, with source details shown in Table 1.

Table 1. Additional data used for pre-processing and geospatial analyses.

Data	ID/Description/Source Link	Source
DTM, 10 m	6800_1 to 6800_4, accessed 10 August 2019, raster: https://hoydedata.no/LaserInnsyn/	Høydedata
Registered landslides	'Skredhendelser', accessed 18 November 2019, point data with attributes: https://nedlasting.nve.no/gis/	NVE, RegObs
Roads	'Vbase', accessed 10 August 2019, polyline data: https://kartkatalog.geonorge.no/	NMA

DTM—digital terrain map. NVE—Norwegian Directorate of Water and Energy. NMA—Norwegian Mapping Authority.

4. Methods

This section describes the approaches used to produce change detection images for the Jølster case study, from which new landslides were manually mapped or detected. These are summarised in Table 2. Broadly, these are split according to the following features:

- Technique: bi-temporal or multi-temporal (Figure 2);
- Input data: Sentinel-1 SAR backscatter intensity values using either cross (VH) or co-polarised (VV) data, or Sentinel-2 optical data using NDVI values.

Table 2. The four change detection image composites used for manual landslide detection.

Image	Value	Pre-Image Dates	Post-Image Dates
S2-BT	NDVI	28.07.2019	02.02.2019
S2-MT	Max NDVI	01.07.2019–29.07.2019 (*n* = 12)	31.07.2019–30.08.2019 (*n* = 12)
S1-BT	VV & VH	25.07.2019	31.07.2019
S1-MT	Mean VV & mean VH	01.07.2019–29.07.2019 (*n* = 25)	31.07.2019–30.08.2019 (*n* = 27)

Sentinel-2 (S2), Sentinel-1 (S2), bi-temporal (BT), multi-temporal stack (MT), normalised difference vegetation index (NDVI), vertical–vertical polarisation (VV), vertical–horizontal polarisation (VH).

Bi-temporal approaches refer to change detection images produced by subtracting a single pre-event image from a single post-event image. By contrast, multi-temporal methods refer to change detection images produced in Google Earth Engine, using one month of images from before and after the events, composited into a single set of pre-and post-event images, prior to the difference being calculated.

In order to avoid introducing noise from the seasonal effects occurring in September, we used a relatively short time period to filter images. One month of images was deemed appropriate for our case study, due to the following reasons:

a.   There was a relatively high frequency of image acquisition at a northern latitude location;

b.   The optical images were not all cloud-covered in this period;

c.   With the coming of fall, the conditions changed significantly between August and September in the study area, including more shadows on north-facing slopes and a reduction in green vegetation at high altitudes.

For repeating this study in other locations, a longer time period could be used, for instance, when there is persistent cloud cover or lower frequency of image acquisition, so long as the vegetation and lighting conditions do not vary significantly.

4.1. Methods for Producing the Change Detection Images

4.1.1. Sentinel-2 Bi-Temporal (S2-BT) Image

Level 2A tiles from the closest dates before and after 30 July 2019, with less than 50% cloud cover, were downloaded from COAH. The acquisition window between the pre- and post-event tiles used for the dNDVI map was five days (28 July and 2 August), with cloud covers of 0.64% and 27.19%, respectively. The difference images were produced using SNAP 7.0 software as follows: first, the pre- and post-event images were cropped to the area of interest using a spatial subset. These image pixels were then aligned using the collocation raster tool. Finally, the band math tool was used to calculate the dNDVI using the following equation:

$$dNDVI = \frac{B8_{RM} - B4_R}{B8_R + B4_R} - \frac{B8_S - B4_S}{B8_S + B4_S} \tag{1}$$

where B8 is the near-infrared band, B4 is the red band, and the subsets R and S correspond to reference (post-event) and secondary (pre-event). The result is a raster with values from -2 to 2, where negative values show a loss of vegetation. The results are displayed as a single-band, black-and-white image. For display purposes, the colour scale was stretched across a range including 90% of the values $[-0.6, 0.1]$.

4.1.2. Sentinel-2 Multi-Temporal (S2-MT) Image

In GEE, first, separate pre- and post-event image collections were produced by filtering the entire Sentinel-2 level 2A collection, by location and date. Next NDVI bands were added to all images in the filtered collection. Then, 'greenest-pixel' pre- and post-event composite images were created by taking the maximum NDVI value for each pixel, within the image collection, using the quality mosaic function. Finally, a difference image (dNDVI) was produced by subtracting the pre-event composite from the post-event composite.

4.1.3. Sentinel-1 Bi-Temporal (S1-BT)

Pre- and post-event scenes were downloaded from COAH. The selected scenes were acquired in interferometric wide (IW) acquisition mode, with Level 1 single-look complex (SLC) format, dual polarisation (VV and VH), and from ascending paths. The pre-event scene was from 25 July 2019, i.e., five days before the event, while the post-event scenes were from 31 July, 6 August, and 12 August, respectively, i.e., 1, 7, and 13 days after the onset of the event (cf. Table 1 for scene IDs). Three composite images were produced with the different post-event images using the steps outlined below. The downloaded SLC products were in slant-range geometry and geocoded prior to further analysis. Pre-processing steps included orbit-state vector refinement and atmospheric noise removal. Then, the scenes were multi-looked to suppress speckle noise and geocoded using a 10 m resolution digital terrain model (DTM). The final images were radar backscatter images, with a spatial ground resolution of approximately 10 m × 10 m, stored in GeoTIFF format. In order to facilitate the visual interpretation of the imagery, we applied multi-temporal change detection using enriched colour composites, which display changes in backscatter as red–green–blue (RGB) composites, following an approach developed for snow-avalanche detection [43,44]. These RGB composites include the pre-event scene in red and blue bands, and the post-event scene (with landslide activity) in the green band: (R, G, B) = (pre, post, pre). Finally, the composite was stretched from its full dynamic range $[-35, 10]$ dB to $[-25, 5]$ dB in order to enhance the dominant intensity ranges. The best of the three composite images was selected for further analyses, and the remaining two were discarded.

4.1.4. Sentinel-1 Multi-Temporal (S1-MT)

Using GEE, Sentinel-1 GRD scenes were used. These are available as pre-processed images, in which steps such as calibration and ortho-correction have already been performed using the Sentinel-1 Toolbox [32]. Again, two separate image collections, one pre-event and one post-event, were created by filtering the entire Sentinel-1 collection by location, date, polarisation (VV, and VH), satellite acquisition geometry (the mean of ascending and descending images were calculated separately, then combined into a single image using a median) and acquisition mode (IW). Next, a topographic correction function [45] was applied to all images within the collections, using a 10 m resolution digital terrain model (DTM) that was uploaded as a GEE asset (details Table 1). Then, the two image collections were reduced to two single mean images. Finally, RGB composites were created for each of the VV, and VH polarisations, with the pre-event image in the red and green channels, and the post-event image in the blue channel. For visualisation, the S1-MT image was displayed with a stretch range of $[-21, 0.5]$, with a gamma value of 0.65.

4.2. Analyses of the Different Approaches

4.2.1. Preliminary Landslide Mapping S2-BT and Spatial Bias Analysis

Preliminary landslide mapping was conducted using the conventional S2-BT approach. This was completed within a week of the landslide occurrences and took approximately three hours. Polygons delimiting assumed landslides, typically represented by dark elongated clusters of pixels, were drawn manually in SNAP. In analysing spatial bias, a comparison was made of the minimum distance between landslides and roads, between the S2-BT landslide inventory, and the landslides reported in the national database, RegObs. The Vbase roads dataset was used. The minimum distance was estimated using the Near (analysis toolbox) tool in ArcMap, using the reported landslide points from RegObs, and points generated from the centre of the mapped landslide polygons using the Feature-to-Point (data management toolbox) tool.

4.2.2. Preliminary Field Mapping and Verification

Verification of the preliminary map was conducted through field visits (in August and October 2019) and a helicopter flight (October 2019). The preliminary map was then updated to include several small landslide scarps observed during the helicopter flight, which were visible in the image but had not been included in the preliminary mapping due to uncertainty. Finally, a comparison was made with a detailed map of the landslides produced by co-author D. Rüther. The detailed field map by Rüther was initially produced by visual comparison of pre- and post-event Sentinel-2 images, then updated based on extensive field investigations (in August and September 2019, and May, June, and August 2020), along with drone orthophotos. A selection of photos from this fieldwork is shown in Figure 4, and a detailed report of field observations is also available [36]. This inventory shows the landslides divided into release, transport, and deposition zones and has a much higher resolution than the S2-BT inventory outlines.

4.2.3. Comparison between the Four Manual Mapping Approaches

The verified landslide inventory produced from the S2-BT image was used as a baseline for comparing the three other approaches. The other images were first visually inspected for visible landslides. Then, the S2-BT landside inventory was used to systematically search for traces of the prior known landslides in each of the other images. Following the classification system used by [46], the prior known landslides were classified into one of three sets—namely, in *Set 1*, landslides were not visible in the SAR-RGB image; in *Set 2*, landslides were hardly visible, recognisable only given prior knowledge of the location; in *Set 3*, landslides were clearly visible.

4.2.4. Landslide Visibility in VV and VH Polarisations and Effect of Local Incidence Angle

To analyse the difference in landslide visibility in VV and VH polarisations, first, a qualitative approach for comparison was used, by examining a selection of individual landslides in subplots in detail. Next, time-series data showing the backscatter intensity values of two landslides, with low and high local-incident angles were analysed.

5. Results

5.1. Analysis of Spatial Bias with Preliminary Landslide Mapping Using Sentinel-2 dNDVI

Figure 5 shows the spatial distribution of landslides mapped from the S2-BT image and registered landslides from the NLDB, available at www.skredregistrering.no (downloaded from NVE, 18 December 2019), that occurred on 30 July 2019 in Jølster, Norway. During verification, minor differences were found between the two inventories due to the coarser, 10 m resolution of the S2-BT image (e.g., clustering small landslides into one large), compared with the higher-resolution inventory of Rüther; however, overall, they were relatively consistent.

It also shows the locations that were investigated in more detail, using helicopter or drone, or by foot. Based on the consistency of the observations, along with the short acquisition window (five days) between pre- and post-event images, the authors have high confidence in the results of the Sentinel-2 mapping, within the limits of the 10 m spatial resolution. A total of 120 landslides ranging from <0.01 km^2 to nearly 1 km^2 in area (Figure 6) were mapped across the area, over an eightfold increase from the 14 registered landslides. The majority of these (>80%) were small landslides (<0.01 km^2). The results of the spatial bias analysis are presented in Figure 6. It is seen that nearly all of the registered landslides were located within 500 m of a road (100% of events from Jølster and 94.9% of all landslides for the whole county), while only one-third of the mapped landslides were within 500 m of a road. While detailed analyses of trigger and failure mechanisms are beyond the scope of this paper, we note again that the majority of the landslides occurred on open slopes and shallowly eroded till overlying polished bedrock. The spatial distribution of landslides appeared to be strongly controlled by the location of the highest rainfall intensities of the highly localised thunderstorm [27], with minor local topographic variations (submetre scale) determining exact initiation zones [36].

5.2. Comparison of Landslide Visibility between the Approaches

Figure 7 shows the dNDVI images, S2-BT, and S2-MT, covering the study area. The landslides are very clear in both images; however, the cloud-free greenest-pixel composite, S2-MT, had less noise from clouds, high water levels in rivers, and agricultural fields. The co-polarised (VV) SAR-RGB composite images, S1-BT and S1-MT, are shown in Figure 8. Again, a significant improvement in landslide visibility is observed in the multi-temporal image, S1-MT, as well as significantly less background noise and reduced terrain artefacts (mainly foreshortening), compared with S1-BT. However, other signals that appear bright green must still be differentiated from landslides. These include changes in agricultural areas and snow cover, as well as some topographical features and general speckling. Different types of vegetation can also be seen in this image. Commercially planted spruce appears as bright areas at lower elevations in the valleys, while birch and alpine scrub vegetation are darker, typically at higher elevations. Farmland and grass appear as darker green or purple patches in the lower valleys. Landslides must be distinguished from other bright green areas, for example, due to new snow cover on the mountain tops (Figure 8, S2-MT, SE corner), or reaping of grass fields.

To quantitatively compare the detectability of landslides between these images, the landslide inventory produced using the S2-BT image was used as a baseline for comparison. The landslide visibility in S2-MT was not significantly different, compared with S2-BT, so this was not included in the following results. Each landslide was searched for systematically within the Sentinel-1 images and classified in an adaptation of the system used by [46], according to their detectability. As shown in Table 3, the detectability of landslides was improved significantly using the S1-MT image, compared with the S1-BT image; however, the landslides considered detectable, respectively, represent only 7.5% and 43% of the 120 landslides mapped with the S2-BT image.

Table 3. Results of landslide detection using the S1-BT and S1-MT images, with respect to the 120 landslides mapped with the S2-BT image.

Set:	1—Undetectable	2—Detectable with Prior Knowledge	3—Detectable without Prior Knowledge
S1-BT	111	5	4
S1-MT	68	41	11

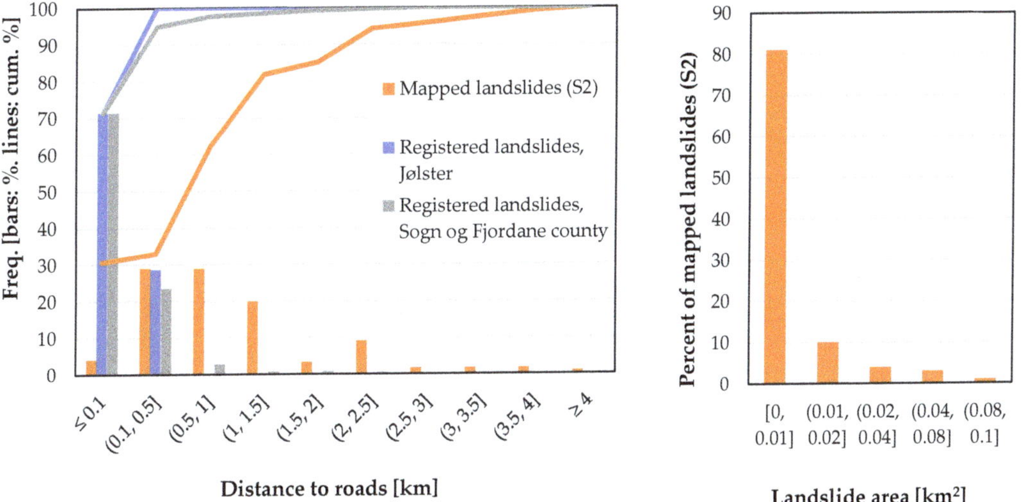

Figure 6. Left: Histogram showing size distribution of the landslides mapped with Sentinel-2 images. Right: Histogram and cumulative percentage, showing the shortest distance between a landslide to a road, for landslides mapped with Sentinel-2 images (n = 120), and landslides registered in the RegObs database for Jølster (n = 14), and Sogn og Fjordane county (n = 392, see Figure 1).

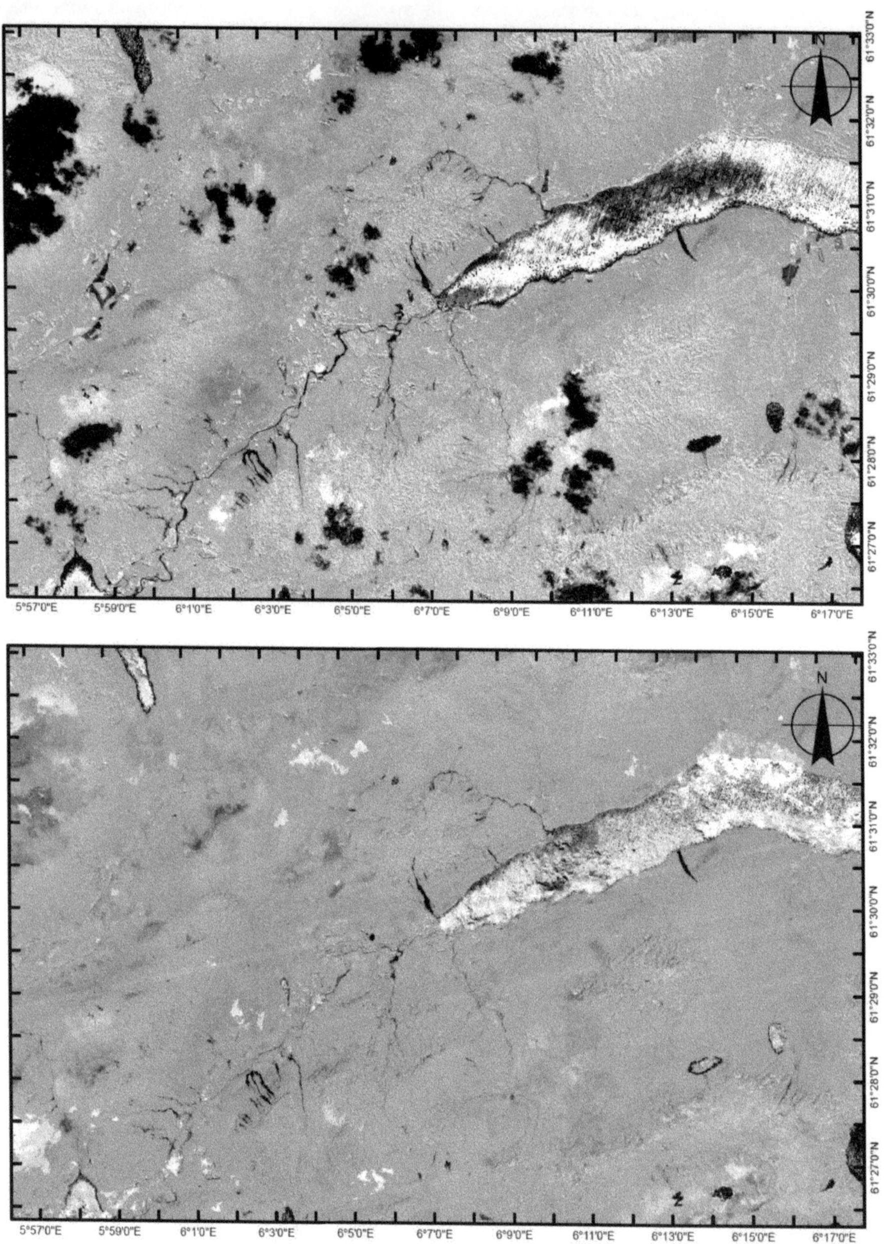

Figure 7. Sentinel-2 dNDVI images of study area: (**Top**) S2-BT, a single set of pre- and post-event images; (**Bottom**) S2-MT, cloud-free greenest-pixel composite, created using Google Earth Engine, with all images from within one month before and after the landslide events. Black corresponds to areas with vegetation loss, i.e., negative NDVI.

Figure 8. Sentinel-1 SAR RGB composite images of study area: (**Top**) S1-BT, a single set of pre- and post-event images; (**Bottom**) S1-MT, median composite, created using Google Earth Engine, with all images from within one month before and after the landslide events. Fresh landslide debris is characterised by an increase in backscatter in green. The numbered squares show the locations of subplots used in Figure 9—(A) Vassenden, (B) Slåtten, (C) Årnes, and (D) Tindefjellet.

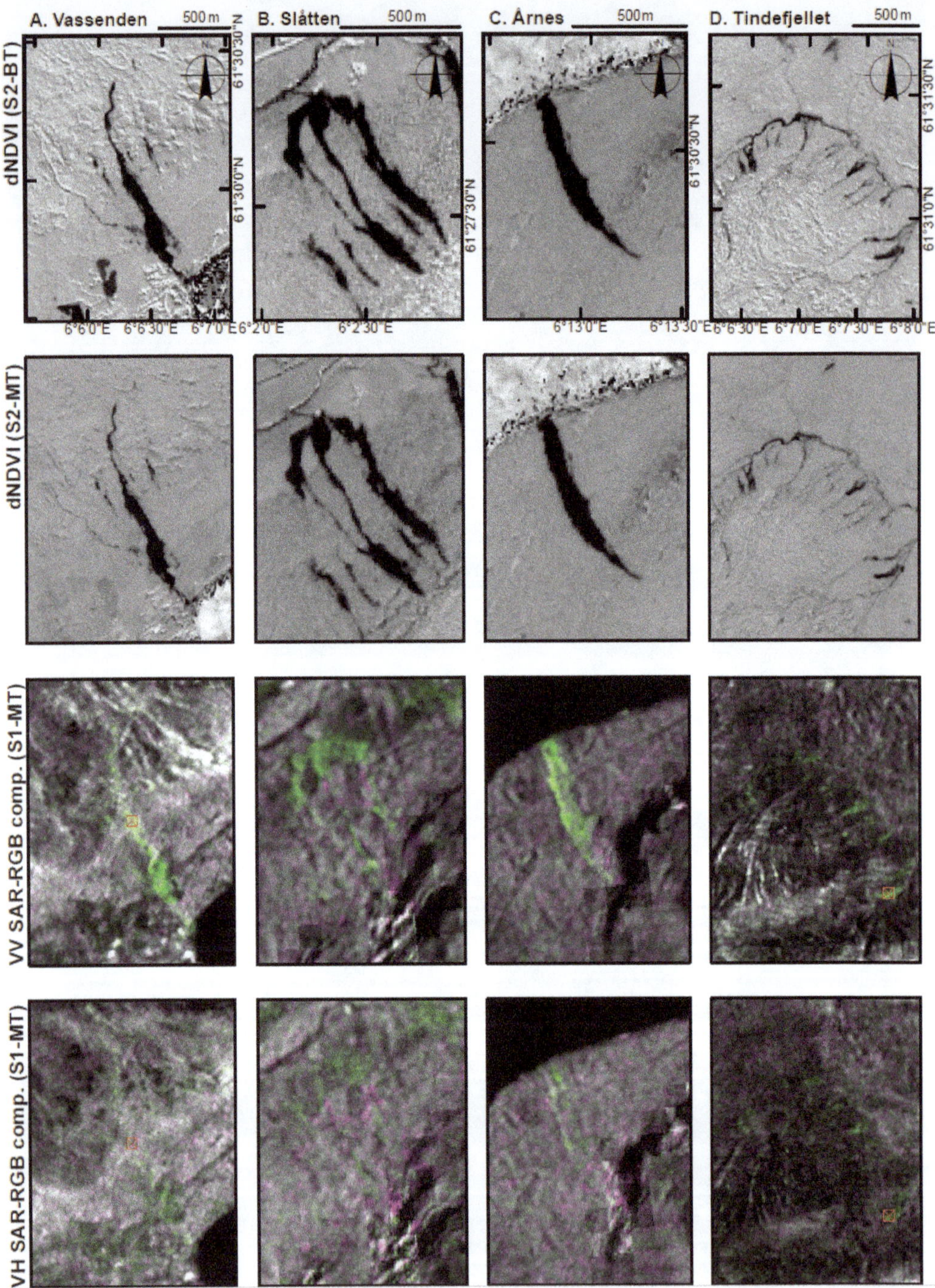

Figure 9. Comparison of landslide visibility in S2-BT, S2-MT S1-MT-VV, and S1-MT-VH images. Locations of subplots include (**A**) Vassenden, (**B**) Slåtten, (**C**) Årnes, and (**D**) Tindefjellet. The red markers show the locations of time-series data in Figure 10.

Figure 9 shows a subset of the landslides that were detectable in the S1-BT image, including, from top to bottom, S2-BT, S2-MT, and S1-MT in VV polarisation, and S1-MT in VH polarisation. A short description of the landslide visibility in the S2 images is given, followed by more detailed observations for the S1 images. In the S2 images, it is seen that the landslides are clearly distinguishable from the background, and even the small, shallow landslides on Tindefjellet (column 4) stand out rather clearly.

Overall, the S2-MT image was considered the best for detecting landslides. The difference in landslide visibility, compared with BT, is not significant; however, the reduction in noise from clouds and agriculture shows improvement over the BT image. There is some noise from agriculture visible in the S2-BT subplot of Vassenden, which is significantly reduced in the S2-MT image. In true colour Sentinel-2 level 2A images (not shown), distinguishing new landslides from older landslides is difficult. Landslide visibility is also limited by shadows, particularly on the north-facing slopes, although this is improved significantly in the Level 2A image, compared with the Level 1C image. Furthermore, the smaller landslides in Tindefjellet are not as easily distinguishable from the background vegetation as they are in the dNDVI image.

5.3. Landslide Visibility in VV and VH Polarisations

In the lower two rows of Figure 9, S1-MT images are shown, using VV and VH polarisations. The landslides shown in the subplots are described in more detail as follows:

A. The Vassenden landslide appears clearly in VV polarisation from top to tail; however, it is barely visible in the VH polarisation. In VH polarisation, the landslide is not easily distinguishable from surrounding vegetated areas, especially grass. The upper section is slightly visible in the VH image; however, this would not be picked out without prior knowledge.

B. Apart from an area with image distortion due to terrain correction at the initiation zone of the landslides, the large Slåtten landslides are mostly visible in the S1-MT VV image. However, the boundaries between the separate landslides are less clear than those in the S2 images. The smaller landslides to the west of the large ones are barely visible. Again, in the VH image, the deposit areas of the landslides are barely distinguishable from the surrounding grass. However, some of the eroded channels are slightly visible within the landslide area in the western and centre of the large landslides.

C. At Årnes, the landslide appears clearly in the VV polarisation aside from the distortion in the initiation zone, while in the VH polarisation, again only some of the channelised areas are visible.

D. At Tindefjellet, the small landslides are visible in both VV and VH polarisations, although appear brighter in VV.

The difference in visibility is shown quantitatively in Figure 10 and Table 4. Figure 10 shows a general increase in backscatter intensity as the land cover changes from vegetated to non-vegetated. Overall, the backscatter intensity at Tindefjellet is higher than at Vassenden. It is seen that, for the landslide sampled on Tindefjellet, there is a clear difference between the mean backscatter intensity in both VV (5.8 dB) and VH (6.4 dB), with the greatest change within VH. For the landslide sampled at Vassenden, the differences in means are lower than those at Tindefjellet, for both VV (2.2 dB) and VH (1.2 dB); however, the largest difference is in VV.

Table 4. Mean, standard deviation, and difference in means, from the data shown in Figure 10.

Landslide	Tindefjellet				Vassenden			
Polarisation	VH		VV		VH		VV	
Surface Type	Veg.	Landslide	Veg.	Landslide	Veg.	Landslide	Veg.	Landslide
Mean	−20.7	−14.0	−11.0	−5.0	−14.3	−12.9	−7.0	−5.0
std. dev	2.2	1.6	3.4	2.4	2.3	2.8	1.3	3.5
Diff. Mean	6.7		6.0		1.4		2.1	

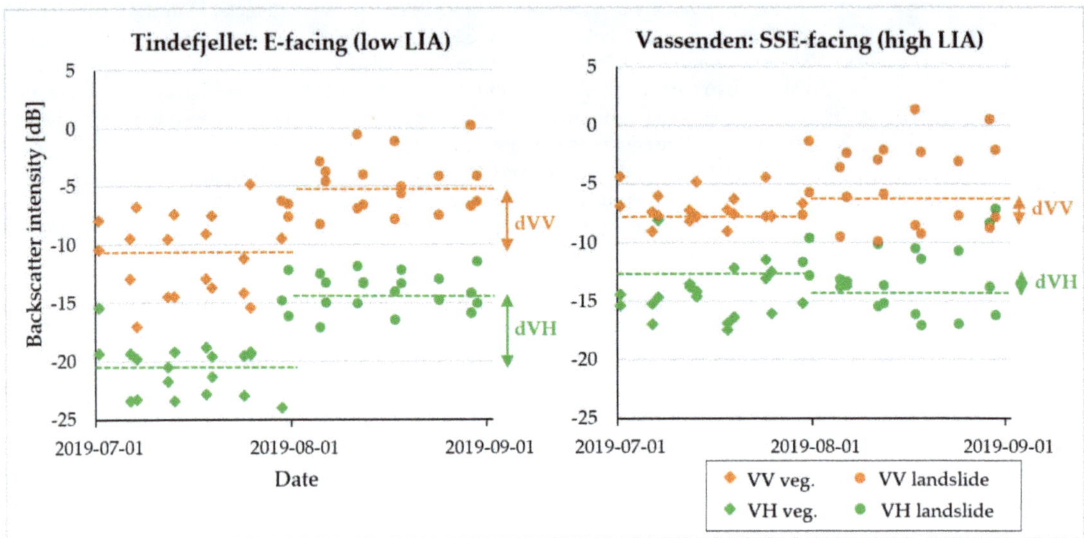

Figure 10. Sentinel-1 data time-series plots from pixels within landslides that changed from alpine scrub and birch to bare soil on 30 July 2019. VV refers to the co-polarised backscatter and VH to cross-polarised backscatter. Tindefjellet (6.129393, 61.514124); Vassenden (6.105213, 61.500352). Dashed lines show the mean for the month, also shown in Table 4.

6. Discussion

6.1. Improving Landslide Visibility with Multi-Temporal Composites and Decreasing Spatial Bias

Detection of landslides in remote areas is a necessary step in creating non-spatially biased landslide inventories, from which predictive models used in landslide early warning systems are tuned and verified, and which inform landslide risk and hazard evaluations. The results from this case study showed great potential for using freely available, medium resolution (10 m) Sentinel-2 images for landslide mapping. S2-dNDVI images could be produced from both Level 1C and Level 2A products, and landslides could be easily mapped at a 10 m resolution, regardless of slope orientation, vegetation type, and erosion depth. Using this approach, the number of landslides detected was increased by over 8 times, and size information could be quickly obtained. Furthermore, the spatial bias was significantly reduced in this case study, with only one-third of landslides located within 500 m of a road, compared with 95% for the region. The additional information needed for the verification of landslide warnings (type and consequences) was not determined in this study. However, such information could possibly be added to the remotely mapped landslides, by comparison with reported landslides or building maps to determine consequences, or by investigating the proximity to channels and landslide shape for determining types.

The multi-temporal image composites produced with GEE significantly improved the clarity of S1 and S2 images, compared with bi-temporal images, by reducing noise from speckle and clouds, respectively. Furthermore, the code is fast to run, images can easily be reproduced for different dates and locations, no software is required, and there is no need to download and store large quantities of data. For the Jølster case study, the detectability of landslides in the S2-MT images was greater than that in the S1-MT images. However, the S2-MT approach is very seasonally dependent. Changes in snow cover, darkness, lack of green vegetation, loss of vegetation due to non-landslide processes, and long periods of cloudy weather are the main limitations of this method.

On the other hand, the visibility of landslides in the S1-MT (VV) image for Jølster exceeded our expectations. The large landslides in the study area were clearly visible and, in some cases, could be mapped with reasonable accuracy. Furthermore, even some of the small shallow landslides that were not in the NLDB were visible, although mostly on east-facing slopes. This method has considerably more potential for rapid detection of landslides that occur outside of the summer season than the S2-MT approach, due to the lack of green vegetation needed for effective dNDVI results. Some quick tests beyond the study area (e.g., the event shown at https://www.regobs.no/Registration/278548, accessed 18 January 2022) showed that this method could be used to detect landslides that occurred in winter, i.e., when there is no cloud-free data available. However, due to noise from changes in snow cover, this approach seems more useful for verifying reported landslides where only approximate locations are given (e.g., from news reports), rather than for mapping landslides over large areas.

6.2. How the Approaches Differ to Similar Methods

Our approach for creating the S2-MT images is very similar to that used in the HazMapper GEE app [29], with one difference: the HazMapper app uses a normalised percentage value of dNDVI, while we used the conventional dNDVI. Additionally, with the HazMapper GEE app, the results are visualised using a red-to-blue colour scale, while we used black and white. Both work well for detecting landslides. However, an advantage of HazMapper over the approach in this paper is that HazMapper tools are available in an app format, which is useful for those who are not comfortable using scripts.

With regard to Sentinel-1, our approach for creating the S1-BT image was inspired by the method in [46], with some differences in the pre-processing and visualisation. However, our results were in stark contrast with those of the study in ref [46], according to which 83% of landslides (n. test cases = 27/32) were considered detectable. By comparison, in this study fewer than 10% of the S2-BT mapped landslides were considered detectable in the S1-BT composite (9/120). This discrepancy is largely explainable due to the size of landslides included in [46] where the majority of successful cases were over 0.05 km^2 in area, and only four examples of landslides smaller than this were included. The Jølster landslide inventories, by contrast, have a majority of landslides with less than 0.01 km^2 in size. The largest landslides from the Jølster study (over 0.03 km^2) were also all at least partially visible in the S1-RGB composite. The smaller landslides did not have any direct consequences for the local population, so taking this into account, these findings do not disaffirm the conclusions of [46], i.e., that Sentinel-1 data have the potential as sources to identify rapid landslides in emergency situations when there is cloud cover.

With further regard to [46], a major limitation of their method was the speckle filtering; they discussed that better results could have been obtained with parameter tuning on a case-by-case basis. In our S1-MT image, no specific speckle filtering was applied; however, we found that averaging pixel values over time effectively eliminates the need for this step. Indeed, aside from the time saved by using the GEE data catalogue, this is the greatest advantage of the S1-MT approach, compared with using single images—the speckle is significantly reduced by averaging across time instead of space, without compromising spatial resolution and losing detail on object boundaries.

A caveat, however, remains, in that this method is most appropriate when pixel reflectivity can be assumed to be temporally homogenous. Where seasonal changes significantly affect the scattering properties of the ground cover over time, e.g., from loss of leaves or snow cover, then better results may be obtained by using pre-event images from the same month in the previous year, instead of the previous month. This was also observable in [31], in which the accuracy of predictions generally increased, by averaging across an increasing number of images (with the rate of increase showing exponential decay with an increasing number of images), except for when fall and winter images are included.

The generation of the S1-MT image was relatively similar to the method presented by [31]. The differences between the methods included slightly different terrain correction calculations (a tutorial covering SAR basics in GEE by Dr. Eric Bullock is available: https://www.youtube.com/watch?v=JZbLokRI8as&t=1213s, accessed 18 January 2022) and visualisation of results. We found the RGB-composite provided more contextual terrain information for visual identification of landslides than the single-band difference results. However, as input for automatic change detection, the single-band format is possibly more appropriate.

Finally, we note that the SAR approaches appeared to work particularly well in the glacial landscape in Jølster. Most of the landslides occurred on open slopes and, therefore, are relatively visible in relation to their surroundings. The smoothly eroded valley sides also likely resulted in better terrain corrected SAR images than areas with younger, more tectonically complex, faulted bedrock morphology. Additionally, the globally available DEMs in Google Engine are currently only available at 30 m resolution, so areas with more frequent angular ridges and valleys may have reduced performance, compared with our results. The few faults in our study area do not appear to have any effect on landslide distribution; rather, their location appears to be strongly related to the distribution of heavy precipitation, with initiation influenced by very localised bedrock variations [36]. In tectonically active landscapes, narrower, angular valleys, and fault-controlled sliding planes, it could be more difficult to see landslide occurrences. This may be improved in the near future with higher resolution DTMs and SAR data.

6.3. Polarimetric Scattering Properties of Vegetation and Landslides

Perhaps the most interesting difference from [31], however, was that we tested landslide visibility in both VV and VH polarisations and found much better results using VV (Figure 9), while they tested only VH. Their reason for using VH was that cross-polarisations are more sensitive to biomass forest structure and would, therefore, be useful for identifying landslides in vegetated areas. They encouraged other studies to explore the use of other polarisations. Similarly, from a forestry perspective, ref [47] wrote *"cross-polarised [VH] observations are essential for identifying changes from volume to surfaces"*. On the other hand, ref [46] also tested landslide detection with both VV and VH, stating, *"we noticed a better definition of the changes in the VH polarisation when landslides occurred in previously deeply vegetated areas, while VV offers sometimes more definition and contrast in presence of glaciers or snow"*.

In order to better understand the discrepancy between our results and the expectations that VH would be more useful for landslide detection, we present a conceptual model in Figure 11 and discuss some key elements affecting SAR backscatter intensity response. Backscattering intensity in VV and VH polarisations depends on the reflections of surface (bare soil) and volumetric (vegetation) scattering mechanisms. For landslides with planar scarps, the angle of the landslide surface relative to the plane of the radar determines the relative density of the signal received, and the degree to which the signal is reflected toward or away from the sensor. For channelised landslides, with high LIA, changes in vegetation on the side of the channel that faces towards the sensor are more likely to be detected. VH backscatter intensity is less affected by the LIA due to the 3D scattering behaviour of vegetation. The difference between VV and VH is more pronounced at smaller local incidence angles (up to 30°), while it becomes difficult to distinguish between forest and landslides at higher incidence angles (over 60°).

The surface scattering mechanism and local angle of incidence are most relevant in determining the intensity of backscatter in co- and cross-polarisation. Dielectric properties, including surface water content, are also relevant [48]. However, we assumed that this did not vary significantly between the locations being compared.

Landslides and vegetation both showed components of surface and volumetric scattering mechanisms (Table 4). The landslides behaved as expected [47], as predominantly a rough surface with some volumetric scattering, reflecting a strong co-polarised (VV) signal and a low to moderately low cross-polarised (VH) signal back to the receiver. However, for vegetation, instead of a predominantly volumetric scattering response, we observed that the vegetated areas also gave a stronger signal in VV than in VH polarisation. This is considered to be due to the sparseness of vegetation, allowing more reflection from the ground surface.

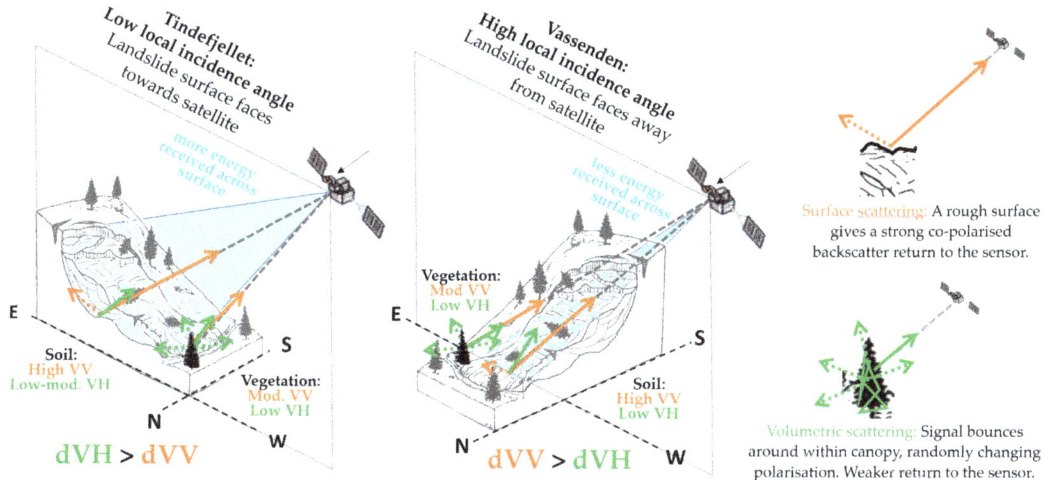

Figure 11. Conceptual model showing how local incidence angle (LIA) affects the visibility of landslides in VV and VH polarisations.

The intensity of backscatter was reduced at a higher angle of incidence in the Vassenden example, compared with the Tindefjellet landslide. This is due to the landslide receiving a lower density of energy across the surface. The difference in backscatter intensity was most pronounced in vegetated areas, both in VV and VH polarisations, while bare soil produced similar strength signals in both polarisations. This is presumably because the very rough surface of fresh landslides provides an abundance of reflective surfaces, while the reflective strength of backscatter from vegetation seems to be more affected by the initial density of energy received.

These findings, however, are opposite to the conclusions of a study on the polarimetric scattering properties of landslides in forested areas [49], where it was found " … *in landslide areas, polarimetric indices ps [VV], pv [VH] change drastically with the local incidence angle, whereas in forested areas, those indices are stable regardless of the local incidence angle change*". Nevertheless, the difference could be due to the use of L-band PALSAR-2 data, with longer wavelengths ($\lambda \sim 25$ cm) that penetrate deeper into the canopy than the Sentinel-1 C band ($\lambda = 5.6$ cm). Furthermore, the forested areas referred to by [49] in Japan appear to be much denser and woodier than the vegetation areas in our comparison. Thus, more investigation is needed on the influence of vegetation type and wavelength on the visibility of landslides in cross- and co-polarised SAR data.

6.4. Recommendations and Future Research Directions

The change detection images using S1 and S2 can be easily reproduced for different date ranges and locations using GEE. For manual landslide detection, the S2-MT approach is recommended when green vegetation is present, and a combined S1-MT and S2-MT approach is recommended in other cases, or when a satisfactory cloud-free composite cannot be obtained. Further research is needed to understand how well these methods can be applied in different vegetation and seasonal conditions. However, it is likely that the SAR-RGB composites could be used to detect landslides when the S2-dNDVI image does not yield good results, i.e., when clouds, snow, or darkness obscure the vegetation, or due to a seasonal lack of green vegetation. Alternatively, for mapping purposes, S2-MT images can be created using images from summer periods in consecutive years, in order to detect landslides that occurred outside of the summer months. However, the longer time interval introduces additional negative dNDVI signals that can make detecting landslides more difficult.

Ongoing developments in data availability and pre-processing of images will provide many more options to explore. These include a harmonised optical collection combining Sentinel-2 and Landsat images [50], and the NISAR satellites due to be launched in 2023 with L- and S-band SAR capabilities [51].

In working towards developing a system for continuous detection of landslides over large areas, the most appropriate platform is GEE, as multiple datasets (e.g., optical, SAR, soil moisture, precipitation, slope, land cover type) can be combined and analyses performed quickly over large areas. Furthermore, there is a possibility to run the full process of image pre-processing, classifier training, and running the classification model within this platform. We are currently creating a database of verified landslides visible in satellite images across Norway, and we hope that the landslide inventory produced in this study will be included in training datasets in the future development of globally trained landslide detection models.

7. Conclusions

This study sought to investigate satellite-image-based change detection approaches for landslide detection, in order to improve the spatial bias and completeness of landslide inventories. A preliminary landslide inventory was produced using a conventional NDVI-based change detection approach, subsequently verified by helicopter, drone, and field observations. We demonstrated how spatial bias and inventory completeness could be improved using this method, compared with current practice. We then compared how landslide visibility in change detection images could be further improved with multi-temporal image composites instead of bi-temporal (i.e., pre- and post-event pair of images) methods, using both Sentinel-1 and Sentinel-2 images within Google Earth Engine. Finally, we examined in detail how the effect of polarity on landslide visibility in SAR images. The main findings are summarised as follows:

1. The landslide event inventory for Jølster produced in this study consists of 120 landslides mapped from the Sentinel-2 bi-temporal change detection image with a 10 m resolution. This represents a significant improvement in inventory completeness, with the initial 14 landslides reported in the NLDB for this event.

2. Spatial bias towards roads was also significantly reduced, with the percentage of landslides located within 500 m of a road reduced from 100% in the NLDB from ground-based landslide reporting to ca. 30% in the remotely mapped landslide inventory.

3. Landslide visibility was improved for both Sentinel-1 and Sentinel-2, using multi-temporal image composites instead of bi-temporal composites. For Sentinel-1, this was due to noise reduction from speckle and the removal of clouds for Sentinel-2.

4. Landslides appeared most clearly in the S2 dNDVI images. For this case study, as a relatively cloud-free image was available very soon after the event, there was not a great advantage improvement in landslide visibility observed in using the MT approach, compared with the BT approach for Sentinel-2 data.

5. On the other hand, significant improvements in the clarity of the Sentinel-1 image were achieved by applying this method, with the number of detectable landslides increasing from 9 in the S1-BT image to 52 in the S1-MT image. The MT image composites were also significantly faster to produce than the BT images, without the need for downloading large quantities of data. In the S1-MT and S1-BT images, respectively, 52 and 9 out of the 120 mapped landslides were considered detectable. We note that, although the rates of landslide detection using S1-MT images were lower than that in other studies using BT methods (e.g., 83% in [46]), the average size of the landslides in our set was significantly smaller than those investigated in the aforementioned study, and several of these were visible in the S1-MT image but not the S1-BT image.

6. Contrary to other studies, landslides in our investigation area appeared much more clearly using VV polarisation, compared with VH polarisation. We presented a conceptual model to help explain these results.

Our results indicate that (a) manual or automated landslide detection could significantly be improved by analysing multi-temporal image composites, using freely available earth observation images and Google Earth Engine, and (b) a systematic application of these methods could significantly improve spatial bias and landslide inventory completeness.

Author Contributions: Conceptualisation, E.L.; methodology, E.L. and R.F.; software, E.L. and R.F.; validation, E.L., D.R. and R.F.; formal analysis, E.L.; investigation, E.L., D.R., R.F. and L.R.; resources, E.L., D.R., R.F. and L.R.; data curation, E.L. and D.R.; writing—original draft preparation, E.L.; writing—review and editing, E.L., R.F. and L.N.; visualisation, E.L. and D.R.; supervision, R.F., L.R., J.S. and S.N.; project administration, E.L.; funding acquisition, E.L., D.R. and R.F. All authors have read and agreed to the published version of the manuscript.

Funding: This research was funded by the Research Council of Norway and several partners through the Centre for Research-based Innovation Klima 2050 (Grant No 237859), and by the NGI core research funding provided by the Research Council Norway (GBV 2020–2022). Denise Rüther's contribution was funded by NORADAPT (Norwegian Research Centre on Sustainable Climate Change Adaption), and Lena Rubensdotter's by NGU (Norges Geologiske Undersøkelse).

Data Availability Statement: The GEE scripts are available on the following GitHub page: https://github.com/erin-ntnu/Change-detection-images-GEE (accessed 1 May 2022).

Acknowledgments: We would like to thank journalists Hallstein Dvergsdal from Firda Tidend, and Oddleif Løset from NRK, for providing images of the Vassenden landslide, Dagrun Arsten, for assisting with the interpretation of agricultural activity, and co-supervisors James Strout and Graziella Devoli for much appreciated support and feedback. This paper contains modified Copernicus Sentinel data [2020] processed by Sentinel Hub.

Conflicts of Interest: The authors declare no conflict of interest.

References

1. Field, C.B.; Barros, V.; Stocker, T.F.; Dahe, Q. *Managing the Risks of Extreme Events and Disasters to Advance Climate Change Adaptation: Special Report of the Intergovernmental Panel on Climate Change*; Cambridge University Press: Cambridge, UK, 2012.
2. Myhre, G.; Alterskjær, K.; Stjern, C.W.; Hodnebrog, Ø.; Marelle, L.; Samset, B.H.; Sillmann, J.; Schaller, N.; Fischer, E.; Schulz, M.; et al. Frequency of extreme precipitation increases extensively with event rareness under global warming. *Sci. Rep.* **2019**, *9*, 1–10. [CrossRef] [PubMed]
3. Hanssen-Bauer, I.; Drange, H.; Førland, E.J.; Roald, L.A.; Børsheim, K.Y.; Hisdal, H.; Lawrence, D.; Nesje, A.; Sandven, S.; Sorteberg, A.; et al. Climate in Norway 2100: Background information to NOU Climate Adaptation. In *Klima i Norge 2100: Bakgrunnsmateriale til NOU Klimatilplassing*; Norsk Klimasenter: Stavanger, Norway, 2009.
4. UNISDR (United Nations International Strategy for Disaster Reduction). *Terminology on Disaster Risk Reduction*; Geneva, Switzerland, 2009. Available online: http://www.unisdr.org (accessed on 1 May 2022).
5. Krøgli, I.K.; Devoli, G.; Colleuille, H.; Boje, S.; Sund, M.; Engen, I.K. The Norwegian forecasting and warning service for rainfall- and snowmelt-induced landslides. *Nat. Hazards Earth Syst. Sci.* **2018**, *18*, 1427–1450. [CrossRef]
6. Guzzetti, F.; Mondini, A.C.; Cardinali, M.; Fiorucci, F.; Santangelo, M.; Chang, K.-T. Landslide inventory maps: New tools for an old problem. *Earth Sci. Rev.* **2012**, *112*, 42–66. [CrossRef]
7. Devoli, G.; Colleuille, H.; Sund, M.; Wasrud, J. Seven Years of Landslide Forecasting in Norway—Strengths and Limitations. In *Understanding and Reducing Landslide Disaster Risk: Monitoring and Early Warning*; Casagli, N., Tofani, V., Sassa, K., Bobrowsky, P.T., Takara, K., Eds.; Springer International Publishing: Cham, Germany, 2021; Volume 3, pp. 257–264. ISBN 978-3-030-60311-3.
8. Devoli, G.; (NVE, Oslo, Norway). Personal Communication, 2022.
9. Taylor, F.E.; Malamud, B.D.; Freeborough, K.; Demeritt, D. Enriching Great Britain's National Landslide Database by searching newspaper archives. *Geomorphology* **2015**, *249*, 52–68. [CrossRef]
10. Kirschbaum, D.B.; Adler, R.; Hong, Y.; Hill, S.; Lerner-Lam, A. A global landslide catalog for hazard applications: Method, results, and limitations. *Nat. Hazards* **2010**, *52*, 561–575. [CrossRef]
11. Damm, B.; Klose, M. The landslide database for Germany: Closing the gap at national level. *Geomorphology* **2015**, *249*, 82–93. [CrossRef]
12. Foster, C.; Pennington, C.V.L.; Culshaw, M.G.; Lawrie, K. The national landslide database of Great Britain: Development, evolution and applications. *Environ. Earth Sci.* **2012**, *66*, 941–953. [CrossRef]
13. Zhao, C.; Lu, Z. Remote Sensing of Landslides—A Review. *Remote Sens.* **2018**, *10*, 279. [CrossRef]
14. Massey, C.I.; Townsend, D.T.; Lukovic, B.; Morgenstern, R.; Jones, K.; Rosser, B.; de Vilder, S. Landslides triggered by the MW7.8 14 November 2016 Kaikōura earthquake: An update. *Landslides* **2020**, *17*, 2401–2408. [CrossRef]

15. Bessette-Kirton, E.K.; Cerovski-Darriau, C.; Schulz, W.H.; Coe, J.A.; Kean, J.W.; Godt, J.W.; Thomas, M.A.; Hughes, K.S. Landslides triggered by Hurricane Maria: Assessment of an extreme event in Puerto Rico. *Geol. Soc. Am. Today* **2019**, *29*, 4–10.

16. Meena, S.R.; Tavakkoli Piralilou, S. Comparison of earthquake-triggered landslide inventories: A case study of the 2015 Gorkha earthquake, Nepal. *Geosciences* **2019**, *9*, 437. [CrossRef]

17. Devoli, G.; Jarsve, K.T.; Mongstad, H.H.J.; Sandboe, K.S.; Are, O. *Control of Registered Mass-Movement Events and Determination of Quality Level*; Report 31/2022; NVE: Oslo, Norway, 2020; 73p, Available online: https://publikasjoner.nve.no/rapport/2020/rapport2020_31.pdf (accessed on 1 May 2022).

18. Herrera, G.; Mateos, R.M.; Garcia-Davalillo, J.C.; Grandjean, G.; Poyiadji, E.; Maftei, R.; Filipciuc, T.-C.; Auflič, M.J.; Jež, J.; Podolszki, L.; et al. Landslide databases in the Geological Surveys of Europe. *Landslides* **2018**, *15*, 359–379. [CrossRef]

19. Steger, S.; Brenning, A.; Bell, R.; Glade, T. The influence of systematically incomplete shallow landslide inventories on statistical susceptibility models and suggestions for improvements. *Landslides* **2017**, *14*, 1767–1781. [CrossRef]

20. Van Den Eeckhaut, M.; Hervás, J. State of the art of national landslide databases in Europe and their potential for assessing landslide susceptibility, hazard and risk. *Geomorphology* **2012**, *139–140*, 545–558. [CrossRef]

21. Strickland, E. Big data comes to the forest. *IEEE Spectr.* **2014**, *51*, 11–12. [CrossRef]

22. Arévalo, P.; Bullock, E.L.; Woodcock, C.E.; Olofsson, P. A Suite of Tools for Continuous Land Change Monitoring in Google Earth Engine. *Front. Clim.* **2020**, *2*, 26. [CrossRef]

23. Prakash, N.; Manconi, A.; Loew, S. A new strategy to map landslides with a generalized convolutional neural network. *Sci. Rep.* **2021**, *11*, 1–15. [CrossRef]

24. Tehrani, F.S.; Santinelli, G.; Herrera, M.H. Multi-Regional landslide detection using combined unsupervised and supervised machine learning. *Geomat. Nat. Hazards Risk* **2021**, *12*, 1015–1038. [CrossRef]

25. Ghorbanzadeh, O.; Blaschke, T.; Gholamnia, K.; Meena, S.R.; Tiede, D.; Aryal, J. Evaluation of Different Machine Learning Methods and Deep-Learning Convolutional Neural Networks for Landslide Detection. *Remote Sens.* **2019**, *11*, 196. [CrossRef]

26. Đurić, D.; Mladenović, A.; Pešić-Georgiadis, M.; Marjanović, M.; Abolmasov, B. Using multiresolution and multi-temporal satellite data for post-disaster landslide inventory in the Republic of Serbia. *Landslides* **2017**, *14*, 1467–1482. [CrossRef]

27. Lindsay, E.; Frauenfelder, R.; Nava, L.; Furuseth, J.I.; Nordal, S. Applying ML-models for landslide detection on a northern, glacially-formed landscape: Jølster case study. *Remote Sens.* **2022**. *in preparation*.

28. Gorelick, N.; Hancher, M.; Dixon, M.; Ilyushchenko, S.; Thau, D.; Moore, R. Google Earth Engine: Planetary-scale geospatial analysis for everyone. *Remote Sens. Environ.* **2017**, *202*, 18–27. [CrossRef]

29. Scheip, C.M.; Wegmann, K.W. HazMapper: A global open-source natural hazard mapping application in Google Earth Engine. *Nat. Hazards Earth Syst. Sci.* **2021**, *21*, 1495–1511. [CrossRef]

30. Mondini, A.C.; Guzzetti, F.; Chang, K.-T.; Monserrat, O.; Martha, T.R.; Manconi, A. Landslide failures detection and mapping using Synthetic Aperture Radar: Past, present and future. *Earth Sci. Rev.* **2021**, *216*, 103574. [CrossRef]

31. Handwerger, A.L.; Huang, M.-H.; Jones, S.Y.; Amatya, P.; Kerner, H.R.; Kirschbaum, D.B. Generating landslide density heatmaps for rapid detection using open-access satellite radar data in Google Earth Engine. *Nat. Hazards Earth Syst. Sci.* **2022**, *22*, 753–773. [CrossRef]

32. Google Sentinel-1 Algorithms-Earth Engine-Google Developers. Available online: https://developers.google.com/earth-engine/guides/sentinel1 (accessed on 1 May 2022).

33. Ochtyra, A.; Marcinkowska-Ochtyra, A.; Raczko, E. Threshold- and trend-based vegetation change monitoring algorithm based on the inter-annual multi-temporal normalized difference moisture index series: A case study of the Tatra Mountains. *Remote Sens. Environ.* **2020**, *249*, 112026. [CrossRef]

34. Mondini, A.C.; Guzzetti, F.; Reichenbach, P.; Rossi, M.; Cardinali, M.; Ardizzone, F. Semi-automatic recognition and mapping of rainfall induced shallow landslides using optical satellite images. *Remote Sens. Environ.* **2011**, *115*, 1743–1757. [CrossRef]

35. Di Martino, G.; di Simone, A.; Iodice, A.; Riccio, D.; Ruello, G. Assessing Performance of Multi-temporal SAR Image Despeckling Filters via a Benchmarking Tool. In Proceedings of the IGARSS 2020-2020 IEEE International Geoscience and Remote Sensing Symposium, Waikoloa, HI, USA, 26 September–2 October 2020; pp. 1536–1539.

36. Ruther, D.; Hefre, H.; Rubensdotter, L. Extreme precipitation-induced landslide event on June 30, 2019 in Jølster, western Norway. *Nor. J. Geol.* **2022**, *submitted*.

37. Meteorologisk Institutt. *Intense Byger Med Store Konsekvenser i Sogn og Fjordane 30. Juli 2019*; Meteorologisk Institutt: Bergen, Norway, 2019.

38. Løset, O. For Ett År Siden Gikk Det 30 Skred i Jølster, Førde Og Gloppen. Slik Ser Det Ut Nå. Available online: https://www.bt.no/nyheter/lokalt/i/K3OLk5/slik-ser-det-ut-ett-aar-etter-jordskredene (accessed on 1 May 2022).

39. Olsen, L.; Sveian, H.; Bergstrøm, B.; Ottesen, D.; Rise, L. Quaternary Glaciations and Their Variations in Norway and on the Norwegian Continental Shelf. In *Quaternary Geology of Norway*; Olsen, L., Fredin, O., Olesen, O., Eds.; Norges Geologiske Undersøkelse: Trondheim, Norway, 2013; pp. 27–78. ISBN 0801-5961.

40. Hughes, A.L.C.; Gyllencreutz, R.; Lohne, Ø.S.; Mangerud, J.; Svendsen, J.I. The last Eurasian ice sheets–a chronological database and time-slice reconstruction, DATED-1. *Boreas* **2016**, *45*, 1–45. [CrossRef]

41. Norges Geologiske Undersøkelse. Bedrock. Available online: https://geo.ngu.no/kart/berggrunn_mobil/ (accessed on 1 May 2022).

42. Norges Geologiske Undersøkelse. Løsmasser-Nasjonal Løsmassedatabase. Available online: http://geo.ngu.no/kart/losmasse_mobil/ (accessed on 1 May 2022).
43. Wiesmann, A.; Wegmuller, U.; Honikel, M.; Strozzi, T.; Werner, C.L. Potential and Methodology of Satellite Based SAR for Hazard Mapping. In Proceedings of the IGARSS 2001. Scanning the Present and Resolving the Future, IEEE 2001 International Geoscience and Remote Sensing Symposium (Cat. No.01CH37217), Sydney, NSW, Australia, 9–13 July 2001; Volume 7, pp. 3262–3264.
44. Eckerstorfer, M.; Malnes, E. Manual detection of snow avalanche debris using high-resolution Radarsat-2 SAR images. *Cold Reg. Sci. Technol.* **2015**, *120*, 205–218. [CrossRef]
45. Vollrath, A.; Mullissa, A.; Reiche, J. Angular-Based Radiometric Slope Correction for Sentinel-1 on Google Earth Engine. *Remote Sens.* **2020**, *12*, 1867. [CrossRef]
46. Mondini, A.; Santangelo, M.; Rocchetti, M.; Rossetto, E.; Manconi, A.; Monserrat, O. Sentinel-1 SAR Amplitude Imagery for Rapid Landslide Detection. *Remote Sens.* **2019**, *11*, 760. [CrossRef]
47. Kellndorfer, J.; Flores-Anderson, A.I.; Herndon, K.E.; Thapa, R.B. Using SAR Data for Mapping Deforestation and Forest Degradation. In *SAR Handbook: Comprehensive Methodologies for Forest Monitoring and Biomass Estimation*; National Space Science and Technology Center: Hunstville, AL, USA, 2019; pp. 65–79.
48. Mouginis-Mark, P. Effects of Surface Cover. Available online: http://satftp.soest.hawaii.edu/space/hawaii/vfts/kilauea/radar_ex/page4.html (accessed on 1 May 2022).
49. Shibayama, T.; Yamaguchi, Y.; Yamada, H. Polarimetric Scattering Properties of Landslides in Forested Areas and the Dependence on the Local Incidence Angle. *Remote Sens.* **2015**, *7*, 15424–15442. [CrossRef]
50. Claverie, M.; Ju, J.; Masek, J.G.; Dungan, J.L.; Vermote, E.F.; Roger, J.-C.; Skakun, S.V.; Justice, C. The Harmonized Landsat and Sentinel-2 surface reflectance data set. *Remote Sens. Environ.* **2018**, *219*, 145–161. [CrossRef]
51. National Aeronautics and Space Administration. Quick Facts. Available online: https://nisar.jpl.nasa.gov/mission/quick-facts/ (accessed on 1 May 2022).

Article

Spatiotemporal Evolution Pattern and Driving Mechanisms of Landslides in the Wenchuan Earthquake-Affected Region: A Case Study in the Bailong River Basin, China

Linxin Lin [1,2], Guan Chen [1,2,3,*], Wei Shi [2,4], Jiacheng Jin [1,2], Jie Wu [1,2], Fengchun Huang [1,2], Yan Chong [2,4], Yang Meng [5], Yajun Li [1,2,3] and Yi Zhang [1,2,3]

[1] School of Earth Sciences, Lanzhou University, Lanzhou 730000, China; linlx20@lzu.edu.cn (L.L.); jinjch21@lzu.edu.cn (J.J.); wuj21@lzu.edu.cn (J.W.); huangfch21@lzu.edu.cn (F.H.); yajunli@lzu.edu.cn (Y.L.); zhangyigeo@lzu.edu.cn (Y.Z.)
[2] Technology & Innovation Centre for Environmental Geology and Geohazards Prevention, Lanzhou University, Lanzhou 730000, China; shiw19@lzu.edu.cn (W.S.); zhongy2018@lzu.edu.cn (Y.C.)
[3] Gansu Geohazards Field Observation and Research Station, Lanzhou University, Lanzhou 730000, China
[4] College of Earth and Environmental Sciences, Lanzhou University, Lanzhou 730000, China
[5] Lanzhou Country Garden School, Lanzhou 730000, China; mengyang@brightscholar.com
* Correspondence: gchen@lzu.edu.cn

Citation: Lin, L.; Chen, G.; Shi, W.; Jin, J.; Wu, J.; Huang, F.; Chong, Y.; Meng, Y.; Li, Y.; Zhang, Y. Spatiotemporal Evolution Pattern and Driving Mechanisms of Landslides in the Wenchuan Earthquake-Affected Region: A Case Study in the Bailong River Basin, China. *Remote Sens.* **2022**, *14*, 2339. https://doi.org/10.3390/rs14102339

Academic Editors: Oriol Monserrat, Daniele Giordan, Guido Luzi and Niccolò Dematteis

Received: 30 March 2022
Accepted: 10 May 2022
Published: 12 May 2022

Publisher's Note: MDPI stays neutral with regard to jurisdictional claims in published maps and institutional affiliations.

Abstract: Understanding the spatiotemporal evolution and driving mechanisms of landslides following a mega-earthquake at the catchment scale can lead to improved landslide hazard assessment and reduced related risk. However, little effort has been made to undertake such research in the Wenchuan earthquake-affected region, outside Sichuan Province, China. In this study, we used the Goulinping valley in the Bailong River basin in southern Gansu Province, China, as an example. By examining the multitemporal inventory, we revealed various characteristics of the spatiotemporal evolution of landslides over the past 13 years (2007–2020). We evaluated the activity of landslides using multisource remote-sensing technology, analyzed the driving mechanisms of landslides, and further quantified the contribution of landslide evolution to debris flow in the catchment. Our results indicate that the number of landslides increased by nearly six times from 2007 to 2020, and the total volume of landslides approximately doubled. The evolution of landslides in the catchment can be divided into three stages: the earthquake driving stage (2008), the coupled driving stage of earthquake and rainfall (2008–2017), and the rainfall driving stage (2017–present). Landslides in the upstream limestone area were responsive to earthquakes, while the middle–lower loess–phyllite-dominated reaches were mainly controlled by rainfall. Thus, the current landslides in the upstream region remain stable, and those in the mid-downstream are vigorous. Small landslides and mid-downstream slope erosion can rapidly provide abundant debris flow and reduce its threshold, leading to an increase in the frequency and scale of debris flow. This study lays the foundation for studying landslide mechanisms in the Bailong River basin or similar regions. It also aids in engineering management and landslide risk mitigation under seismic activity and climate change conditions.

Keywords: landslide; evolution characteristics; state of activity; earthquake; rainfall; the Bailong River basin

1. Introduction

Landslides cause catastrophic and significant economic and human losses worldwide [1]. In recent years, the incidence of landslides has increased owing to climate change and seismic activity [2–4]. Fragile rocks, fault structures, frequent earthquakes, and heavy and concentrated rainfall during monsoon periods make the eastern margin of the Tibetan Plateau a notable hotspot for landslide risk [5,6]. The Wenchuan earthquake in the region posed a great threat to people's lives and property safety, triggered more than 15,000 landslides, and caused the deaths of more than 20,000 people in Sichuan Province [7–9]. Except for Sichuan Province, the Bailong River basin is one of the areas most seriously affected by the Wenchuan earthquake [10]. More than 800 landslides were triggered by seismic shaking along the Zhouqu–Wudu section [11,12]. Meanwhile, the earthquake reactivated the ancient landslides, such as the Sanjiadi and Hongtupo landslides [13,14]. According to the State Council Information Office of the People's Republic of China (https://www.scio.gov.cn (accessed on 15 March 2022)), the Wenchuan earthquake caused 10,523 casualties (including 365 deaths) in the Gansu Province and the three hardest-hit areas were in the Bailong River basin. However, the landslide evolution and driving mechanisms of the catchment or region scale in this region are unclear, thus warranting the undertaking of relevant research.

Clarifying the spatiotemporal evolution pattern of landslides is vital for landslide-prone areas to better predict and mitigate disasters [15,16]. Most studies of rainfall- or earthquakes-induced landslides were focused on identifying the spatial pattern and potential risk of landslides, whereas the dynamic evolution processes of landslides at catchment or regional scale was ignored [17–19]. However, the evolution characteristics of landslides at the catchment scale can be well-reflected by spatiotemporal landslide inventories [20–23]. The spatial and temporal completeness of multitemporal inventories is of utmost importance in the study of landslide evolution at the catchment scale. Recently, many researchers have devoted themselves to researching the spatiotemporal evolution of landslides covering different types and mechanisms at the catchment scale following the Wenchuan earthquake to evaluate the long-term impacts of the earthquake on geological hazards [16,24–28]. In areas affected by such large earthquakes, the threat of landslides persists in the years following the earthquake, as intensified landslides and landscape erosion are induced by coseismic mass wasting [23,24]. However, these studies are mainly concentrated in Sichuan Province, while the effects on the Bailong River basin, which was also severely affected by the Wenchuan earthquake, remain poorly understood.

The classification of landslide activity can reflect the driving mechanisms of landslides and guide land-use planning, thus effectively reducing social and economic costs [29]. The evaluation of landslide activity is mainly based on optical images or interferometric synthetic aperture radars (InSAR) [24,25,30]. Optical remote sensing has great potential for monitoring seasonal changes in landslides [31]. In the last decade, unmanned aerial vehicles (UAV) have been widely used in landslide studies, offering users a more convenient way to obtain high-spatial-resolution optical images [32]. Even so, it is difficult to quantitatively analyze optical remote-sensing data and determine the process of landslide activity [22]. InSAR is a powerful technique with a comprehensive area coverage and high sensitivity to surface displacement [33,34]. It has good monitoring capability for quantitative analysis, but has limitations in terms of monitoring fast deformation areas, and some regions are decorrelated [22,31]. Using a single remote-sensing technology is often inefficient in accurately classifying the activity of all landslides in the region; thus, it is necessary to combine multiple remote-sensing techniques to enhance the accuracy of evaluating landslide activity.

Understanding the relationship between landslides and their driving mechanisms forms the basis for predicting future landslides and assessing landslide hazards [35–37]. Previous studies have shown that rainfall and earthquakes are the two principal mechanisms that induce landslides [37–39]. Several attempts have been made to establish relationships between rainfall levels and landslides [40–44]. In earthquake-affected areas, long-term slope destabilization was predicted because of monsoonal climatic conditions and the seismodynamic setting [45–48]. The spatial distribution and topographic context of rainfall-triggered and coseismic landslides in the same study area may differ [9,49]. Therefore, the coupling of the postearthquake effect and rainfall has been widely studied [45–47]. During the wet season, vast amounts of deposits loosened by landslides may lose stability due to rainfall infiltration and can easily evolve into deadly debris flows and trigger impressive chains of geohazards [23,49]. Hence, clarifying landslide spatial likelihood and driving mechanisms is critical for mitigating risk through proper countermeasures and reconstruction strategies.

In this study, we explored the spatiotemporal evolution pattern and driving mechanisms of landslides in a valley of the Bailong River basin, China, in the Wenchuan earthquake-affected region. To this end, we generated a multitemporal landslide inventory map and used multisource remote-sensing technology to classify landslide activity. Furthermore, we quantified the potential contribution of landslides to debris flows in this catchment. The novelty of this study is that combined multiple remote-sensing approaches to investigate the evolution pattern of landslides and the potential contribution of landslides to debris flow at the catchment scale in the Wenchuan earthquake-affected area outside Sichuan Province. It provides a valuable basis for constraining the control of landscape evolution and improving landslide risk management.

2. Study Area

The Bailong River basin is located in the transition zone between the Tibetan Plateau, Loess Plateau, and Sichuan basin in China, featuring high-relief and steeply-incised valleys [50,51]. Across the region, faults are extensively developed, and neotectonic movements are frequent, resulting in the widespread formation of fragmented rocks coupled with frequent downpours, making the region one of the four most disastrous areas in China [52]. There are 171 tributary valleys in the middle reaches of the Bailong River basin, and the intermediate-sized tributary valleys account for a primary proportion in this section. Most of them have hypsometric integrals values between 0.41 and 0.60, and relief between 1645 and 2405 m [53].

The Goulinping (GLP) valley is located in the middle section of the Bailong River basin. It is a representative small–medium tributary valley in this area in terms of relief, drainage, and hypsometric integrals [53]. It has high mountains and narrow valleys, with a watershed area of 20.15 km^2 and a relief of 1947 m. The main channel is the first-level branch of the Bailong River, with a length of approximately 6.1 km and average ratio of approximately 17%, as shown in Figure 1. It is sourced from a high-relief carboniferous limestone area and subsequently passes through a low-relief Silurian phyllite area covered by Quaternary loess, before finally entering the Bailong River. The tectonic structure in the GLP was developed, and the Jiudun fault, which is the secondary fault of the Diebu–Bailongjiang fault zone, passes through the watershed. The rock layers have many folds, faults, and joints, and their lithology is weak and broken, providing prerequisites for geohazards, as is shown in Figure 2.

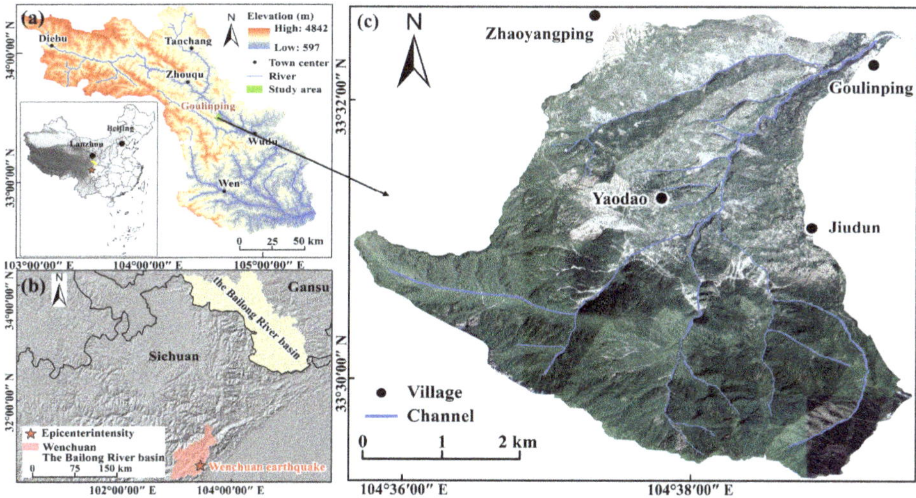

Figure 1. Location of the study area. (**a**) Location of the Bailong River basin; (**b**) location of the Wenchuan earthquake epicenter; (**c**) location of the GLP (the optical image was produced in 2019).

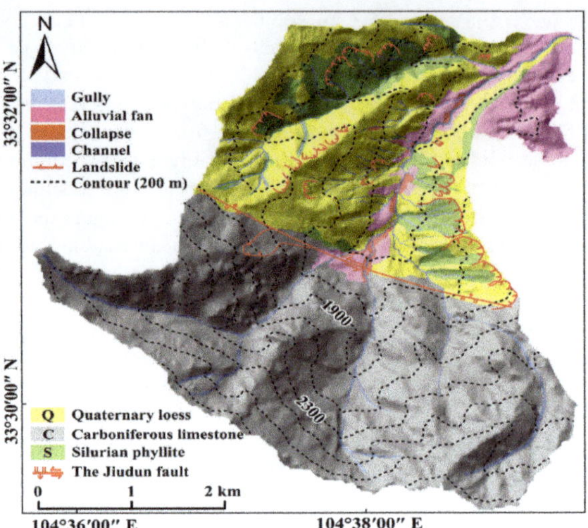

Figure 2. Geomorphic map of the GLP.

Climatically, the monsoon strongly influences the study area, which is warm and humid in summer and cold and dry in winter [52]. The annual precipitation of the area is 487.2 mm, 75 to 85% of which falls between May and September [53]. The average annual precipitation varies across the valley, with the largest cumulative precipitation occurring in the upper–middle reaches and decreasing downstream (Figure 3). It should be emphasized that there have been several extreme rainfall events in recent years, the most serious being in August 2020, with an annual precipitation of nearly 800 mm. In this region, the average precipitation of August in the past 30 years was 59.4 mm, while it quadrupled to 258.9 in August 2020. Climatic conditions have triggered geohazards in the study area.

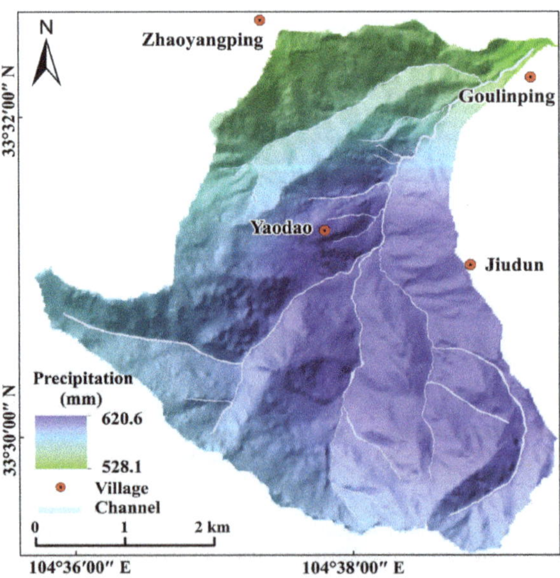

Figure 3. Spatial distribution of rainfall in the GLP.

GLP has an active crust and is affected by surrounding mega-earthquakes, such as the 1654 Lixian M8.0 earthquake, the 1879 Southern Wudu (Wenxian) M8.0 earthquake, the 2008 Wenchuan M8.0 earthquake, and the 2017 Jiuzhaigou M7.0 earthquake. The study area is located in the VI intensity of the 1654 Lixian earthquake and the X intensity of the 1879 Southern Wudu (Wenxian) earthquake (Figure 4a,b). These two earthquakes triggered massive landslides in history, of which the 1879 Southern Wudu (Wenxian) earthquake had a more serious impact [54,55]. The mountain ecosystems have a strong resiliency following major geologic disturbances [56]. The 1879 Southern Wudu (Wenxian) earthquake occurred more than 100 years ago. As the landscape has evolved, the scars can be covered with vegetation, which makes them difficult to identify on remote-sensing images [57]. Due to the long history and sparse data of the 1879 Southern Wudu (Wenxian) earthquake, it is hard to make explicit its coseismic landslides, whereas it can be speculated that parts of the landslides in the study area before 2008 were affected by this earthquake. According to the China Earthquake Administration survey, the intensity of the Wenchuan earthquake in the study area is VII, while that of the Jiuzhaigou earthquake is less than VI (Figure 4c,d). The coseismic displacement of the Wenchuan earthquake measured using GPS is 53 mm [58]. Therefore, the Wenchuan earthquake is considered a mega-earthquake that has had a great impact on the study area in recent years. The research focuses on the landslide evolution after the Wenchuan earthquake.

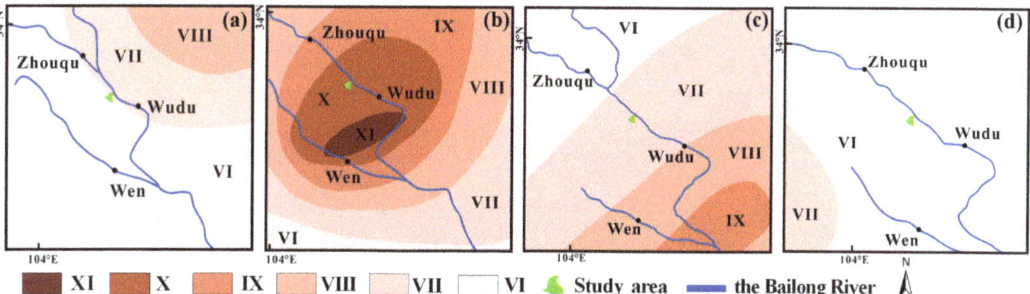

Figure 4. The intensity of different earthquakes in the study area. (**a**) The intensity of the 1654 Lixian earthquake (Adapted with permission from Ref. [54]. 2017, Yuan et al.); (**b**) the intensity of the 1879 Southern Wudu (Wenxian) earthquake (Adapted with permission from Ref. [55]. 2014, Yuan et al.); (**c**) the intensity of the 2008 Wenchuan earthquake (the data comes from the China Earthquake Administration); (**d**) the intensity of the 2017 Jiuzhaigou earthquake (the data comes from the China Earthquake Administration).

3. Materials and Methods

3.1. Data

In this study, we used optical images from 2007 to 2020 (Table 1) for the landslide inventory, extraction of geomorphic factors based on digital elevation model (DEM) data with a resolution of 12.5 m from the advanced land-observing satellite (ALOS), small baseline subset (SBAS) InSAR calculated using SAR data from ENVISAT ASAR 2003–2010 and Sentinel-1A 2014–2020, precise orbit determination (POD) data, and a resolution of 30 m SRTM data; detailed information of the SAR data is presented in Table 2, and the coverages are shown in Figure 5. We used the moisture balance drought index (MBDI) to indicate regional soil moisture [59]. Sunshine, air temperature, and precipitation data were obtained from the Wudu meteorological station, while the rest of the precipitation data were obtained from rain gauges installed by our team in the study area, the detailed information of precipitation data is showed in Table 3.

Table 1. Optical image data sources and information.

Image No.	Source	Acquisition Data	Spatial Resolution (m)	Type
1	IKONOS-2	24 November 2007	1.0 m	Panchromatic
2	SPOT	16 May 2008	2.5 m	Multi spectral
3	ZY03	11 October 2013	2.1 m	Pan-sharpened
4	UAV	22 April 2014	0.5 m	Multi spectral
5	Google Earth	1 November 2017	0.5 m	Multi spectral
6	Google Earth	31 July 2019	0.5 m	Multi spectral
7	UAV	30 October 2019	0.2 m	Multi spectral
8	UAV	20 August 2020	0.1 m	Multi spectral

Table 2. ENVISAT ASAR data and Sentinel-1A data information.

Parameters	ENVISAT ASAR	Sentinel-1A
Band	C	C
Wavelength (cm)	5.6	5.6
Incidence angle θ (°)	22.8	39.2
Heading angle γ (°)	−165	−167
Track	018	62
Polarization	VV	VV
Number of images used	32	135
Orbit direction	Descending	Descending
Acquisition time	13 August 2003 to 15 September 2010	9 October 2014 to 1 October 2020

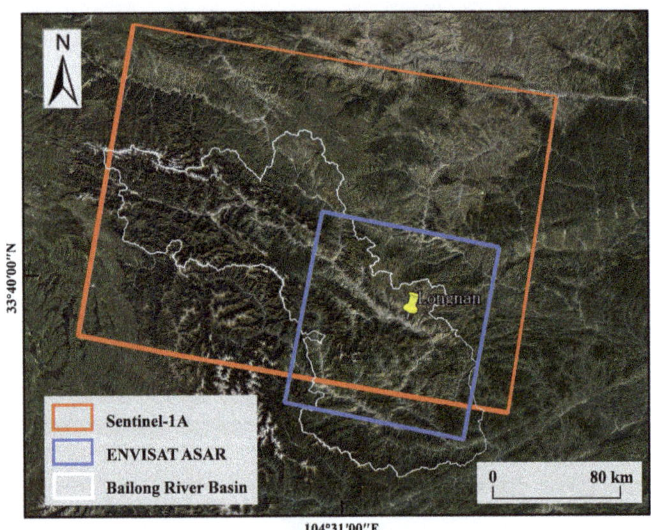

Figure 5. ENVISAT ASAR data and Sentinel-1A data coverage.

Table 3. Annual precipitation data of the study area.

Year	Annual Precipitation (mm)	Year	Annual Precipitation (mm)	Year	Annual Precipitation (mm)
1996	339.9	2005	410.1	2014	494.2
1997	270.5	2006	337.9	2015	448.6
1998	479.8	2007	457.3	2016	465.2
1999	397.2	2008	489.4	2017	570.1
2000	450	2009	508	2018	492.4
2001	377.1	2010	338.3	2019	574.3
2002	414.7	2011	546.7	2020	759.8
2003	519.8	2012	422	Mean annual precipitation (mm)	
2004	411.6	2013	618.7	463.7	

3.2. Landslide Inventory

In this study, landslide catalogues were interpreted using multiple optical images from 2007 to 2020 and verified by field investigation in 2020. Owing to the limited image data, different data sources and time gaps exist. Landslides are interpreted according to image features, including shape, size, color, tone, spot, and texture [25,60–65]. Owing to the differences in the resolution of optical images from different sources, we considered only landslide areas larger than 100 m^2, to avoid the impact on the subsequent statistical analysis. During the field investigation, the landslide cataloguing results were verified and supplemented. In addition, we collected the basic characteristic parameters of the landslides. According to the landslide classification criteria of Varnes (1978) and Hungr et al. (2014), landslides are classified into fall, slide, and flow, as is shown in Figure 6 [66,67].

Figure 6. Classification of landslides. **A**: The location of landslides; (**a**) fall; (**b**) slide; (**c**) flow.

3.3. SBAS–InSAR Technology

The SBAS–InSAR technology uses a multimaster image to obtain the interferogram of a short time and spatial baseline and extract the distributed point that maintains coherence in a period, which is suitable for the deformation monitoring of natural surfaces [68,69]. In this study, SBAS–InSAR technology based on SARscape software was used to process the ENVISAT ASAR datasets from 2003 to 2010 and Sentinel-1A datasets from 2014 to 2020. The SAR data were divided into three groups, and the connection graphs are shown in Figure 7; this was generated according to the maximum temporal baseline of 1000 days (2003–2008), 100 days (2008–2010), and 160 days (2014–2020) and the maximum normal baseline of 45%. The Goldstein filtering method was used to improve the signal-to-noise ratio of the differential interferogram, and the image pairs with low coherence, poor unwrapping effect, and serious influence of atmosphere and terrain were deleted. To eliminate the influence of the residual terrain phase on the results, 20 ground control points (GCPs), which were coherently higher than 0.85 and confirmed in the stable region by field investigation, were selected in the study area. The unwrapping phase obtained after removing the terrain phase was used to derive the remaining height and initial displacement using a linear model, and the singular value decomposition method was applied to search the least-squares solution for each coherent pixel and estimate the nonlinear deformation. After subtracting the estimated atmospheric artefacts and orbital ramps, the ultimate time-series deformation was inverted [70].

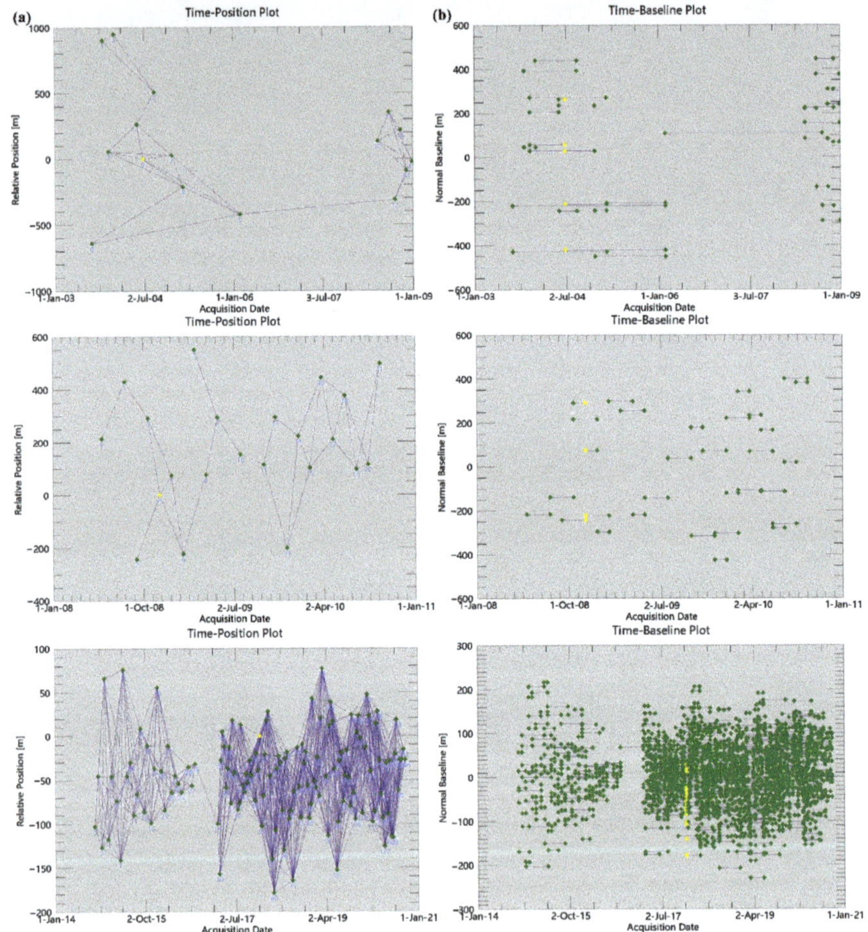

Figure 7. Spatial and temporal baseline of the SAR image interferogram. (**a**) Time–position of SAR image interferometric pairs; (**b**) time–baseline of SAR image interferometric pairs. (The yellow points denote the master image, the green points denote the slave images, and the blue lines represent interferometric pairs.)

4. Results

4.1. Spatiotemporal Evolution of Landslides

Figures 8 and 9a,b show the spatiotemporal inventory of landslides, changes in the number and area of landslides, and changes in the number and area of different types of landslides. The landslide area ranged from 102 m² to 192,208 m². Seventy-one landslides were developed in 2007, covering 8.78% of the study area and concentrated in the loess and weathered phyllite in the middle–lower reaches; their type was mainly slides. Compared to 2007, the number and area of landslides doubled in 2008. Most of the added landslides were developed in the upstream limestone area, mainly in falls and flows. The falls were developed on the steep slope with good free-face condition and the flows were developed on the concave slope covered by debris deposits; the number and area of landslides in subsequent years exhibited a moderate rise, and the spatial distribution of new landslides gradually shifted to the middle–lower reaches. Compared with previous years, the number and area increased significantly in 2020, with a total of 408 landslides developing and covering 17.7% of the study area. The new landslides are concentrated in the middle and lower reaches, and their spatial distribution is consistent with that in 2007; the type of these was mainly slides.

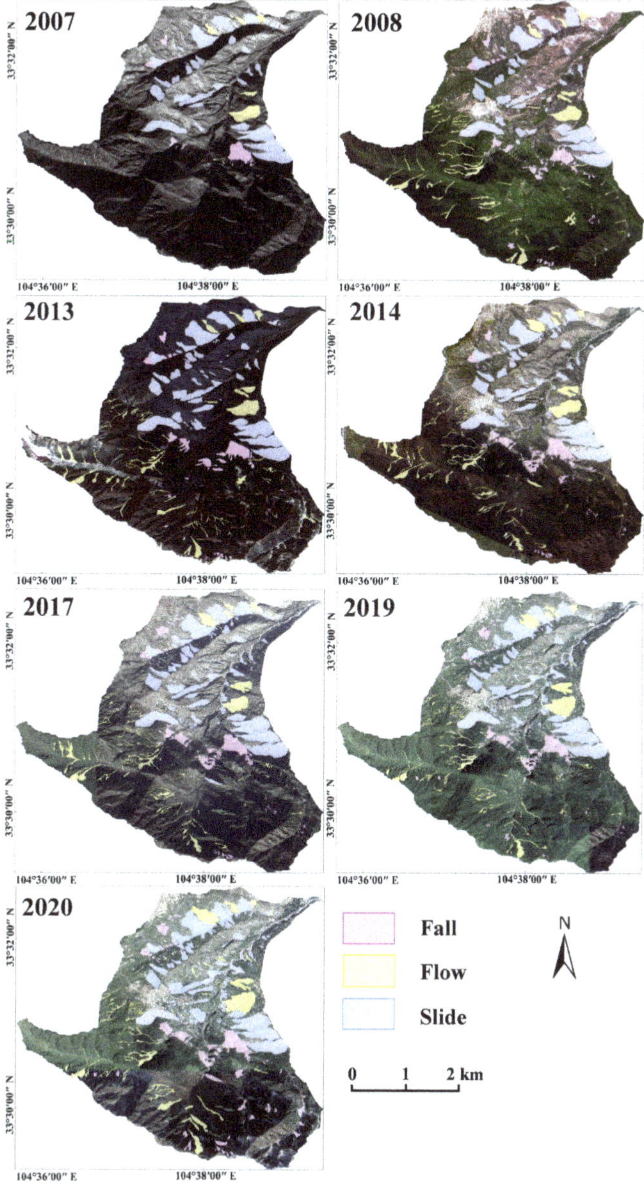

Figure 8. Spatiotemporal inventory of landslides.

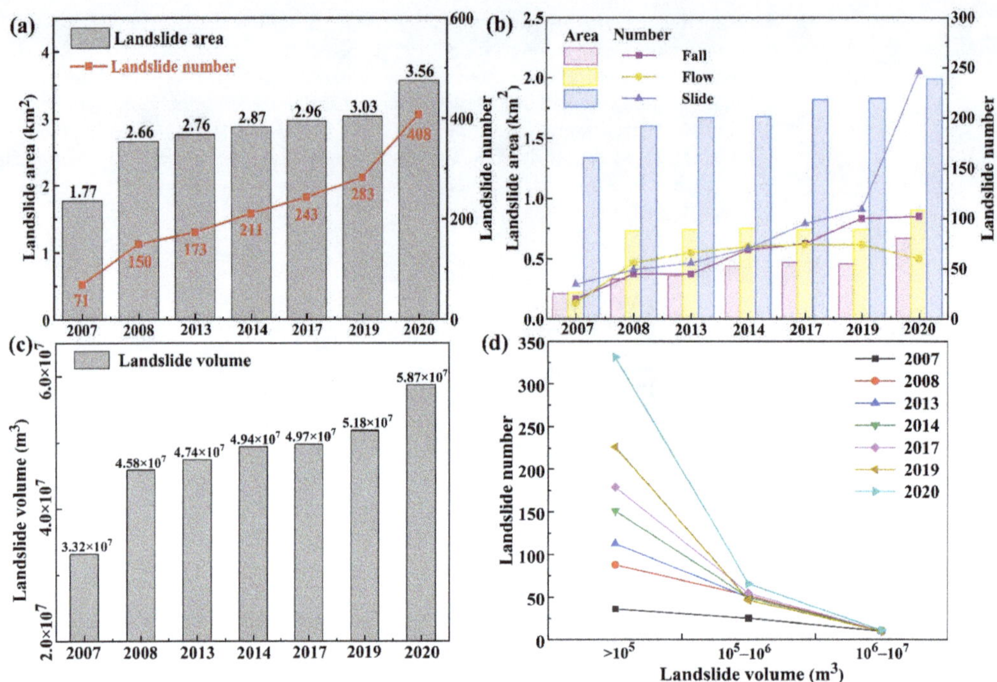

Figure 9. Landslide change trends. (**a**) Changes in the number and area of landslides; (**b**) changes in the number and area of different types of landslides; (**c**) changes in the volume of landslides; (**d**) changes in the number of landslides with different volumes.

We estimated the landslide volume based on the field survey, literature data, and area–volume geometric scaling relationships, defined as:

$$V = aA^b \tag{1}$$

where V is the landslide volume, A is the landslide area, a is a fit parameter, and b is the power-law exponent [71]. The method of calculating the volume of landslides was used in many studies [72–74]. We used 60 landslide data in an area range of 1.0×10^1–6.33×10^5 m^2 and volume range of 1.7×10^1–2.26×10^7 m^3 for fitting and obtained the fitting formula with $V = 0.5505A^{1.3163}$ ($R^2 = 0.9538$) to estimate the remaining landslide quantity in the study area. The accuracy of the total landslide volume depends on the quality of the individual volume measurements and the completeness of the landslide inventory. The landslide depth is an important parameter for landslide volume [75]. Most of the landslides are shallow in the study area; thus, we measured the depth of the landslide scarp with a laser rangefinder to represent its depth [76,77]. The volume of landslides measured in the field was obtained as the product of landslide area and the average landslide depth, meaning that the volume of landslides was probably overestimated [71]. Therefore, the area–volume empirical formula has certain uncertainties. Owing to the limitations of remote-sensing image resolution and terrain, some boundaries of small landslides which were located in unreachable mountain areas cannot be interpreted and verified precisely. Meanwhile, the connected small landslides were possibly delineated with a larger boundary. Therefore, the areas of small landslides may be enlarged and the volume was probably overestimated. However, the uncertainties do not affect the variation trend of landslide volume. Figure 9c,d show that the total volume of landslides increased from 3.32×10^7 m^3 (2007) to 5.87×10^7 m^3 (2020). Moreover, the volume of most landslides was less than

10^5 m^3, and this number increased annually. The number of landslides increased at a constant rate from 2007 to 2019, and the growth rate in 2020 was relatively large compared to that in the previous period. A rapid growth was evident in the number of landslides with a volume between 10^5 and 10^6 m^3 in 2008 and 2020, while the number of landslides in other years varied little. However, the number of landslides with a volume between 10^6 and 10^7 m^3 remained unchanged for many years.

4.2. Effect of Terrain and Geomorphic Factors on Landslides

Terrain and geomorphic conditions control the distribution, scale, and types of landslides. The spatial distribution of landslides in various terrains and geomorphic factors in the study area is shown in Figure 10. Landslides were mainly distributed at 1300–1900 m, and their peak value shifted from to 1500–1700 m to 1300–1500 m from 2019. Simultaneously, the number of landslides higher than 1900 m increased after 2008 (Figure 10a). The landslides were mostly distributed at a slope gradient of 30–40°. However, compared to other years, the number of landslides at a slope gradient of 40–50° increased in 2008. Additionally, the number of landslides at a slope gradient of 20–30° increased considerably in 2020 (Figure 10b). Landslides occurred in all slope aspects but were mainly distributed in the SE direction in all years (Figure 10c). In terms of geomorphic distribution, the landslides were concentrated on slopes consisting of loess, phyllite, and limestone (Figure 10d). In 2007, the most landslides occurred in loess slopes. After 2008, the number of landslides that developed on limestone slopes increased markedly and then declined in 2020. In 2020, the number of landslides that developed in loess slopes reached a record high, and the geomorphic distribution trend of landslides recovered from a rate consistent with that of 2007.

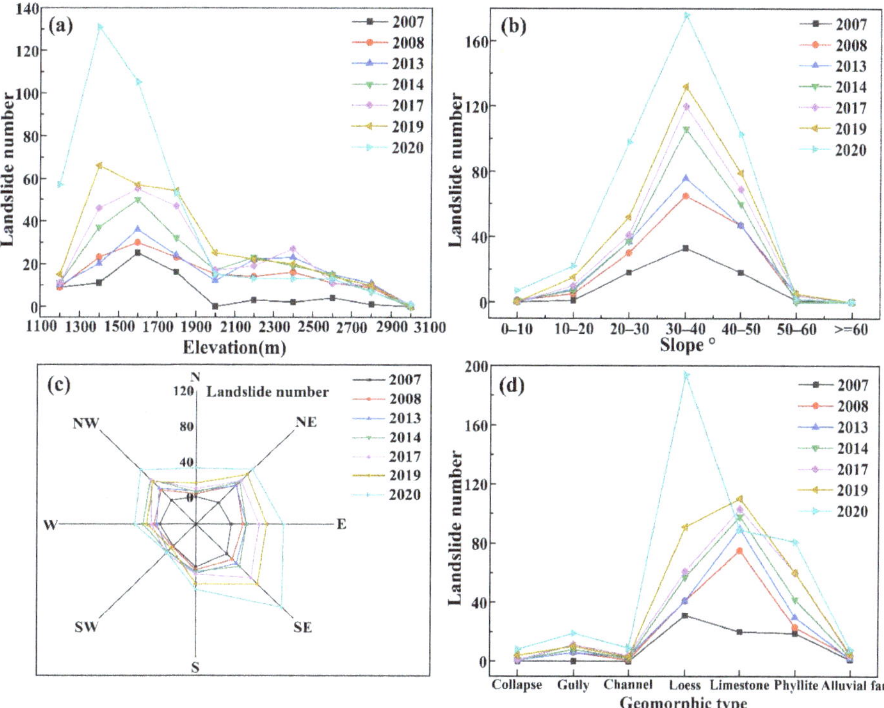

Figure 10. Various characteristics of landslide distribution. (**a**) Elevation distribution of landslides; (**b**) slope distribution of landslides; (**c**) aspect distribution of landslides; (**d**) geomorphic distribution of landslides.

4.3. Landslide Activity

4.3.1. InSAR Results

By processing ENVISAT ASAR and Sentinel-1A data with the SBAS–InSAR technique, we obtained the surface deformation rates in the line-of-sight (LOS) of the study area from 2003 to 2020. Owing to the different data used, we transformed the results along the LOS into the slope direction based on the methods of Zhao et al. (2012) [78]. The formula used was the following:

$$V_{Slope} = \frac{V_{Los}}{(-\sin\theta\cos\alpha,\ \sin\theta\sin\alpha,\ \cos\theta)(-\sin\beta\cos\varphi,\ -\cos\beta\cos\varphi,\ \sin\varphi)} \tag{2}$$

where V_{Slope} is the deformation in the slope direction; V_{Los} is the deformation in the LOS direction; θ is the incidence angle; α is the satellite flight azimuth; β is the slope angle above the horizontal surface; ϕ is the slope azimuth. According to the formula, when the numerator approaches zero, the V_{Slope} tends to infinity. Therefore, we used 0.3 to replace the value of 0 to 0.3 and used -0.3 to replace the value of -0.3 to 0. When the V_{Slope} was greater than 0, the material moved upwards along the slope direction. This situation generally does not exist, except for the rock layer reverse. Therefore, we removed V_{Slope} values greater than zero (Figure 11). The results show that from 2003 to 2006 (Figure 11a), in the phyllite slope in the north of Jiudun (Figure 11d), the maximum deformation rate was -60 mm/a, and the deformation region was small. The deformation was more significant from 2008 to 2010 (Figure 11b) than during the previous stage. The deformation region was enlarged within the red circle and was more severe, reaching -105 mm/a. After 2014 (Figure 11c), the slopes near Jiudun village and some parts of the branch ditch gradually formed major deformations. Compared with the previous stage, the deformation of the slope in the northern part of Jiudun was exacerbated, and the maximum deformation rate increased to -120 mm/a. Based on the field investigation, flow landslide with serious erosion is marked with a red circle in Figure 10d. Additionally, we found the seepage of rock mass in the lower part of the landslide (Figure 11e), which indicates that pore or fissure water were developed in the landslide, the strength of rock decreased, and the deformation of the slope was aggravated.

4.3.2. Landslide Activity Based on Optical Remote-Sensing and InSAR Technology

In this study, we classified landslides as active, dormant, reactivated, and new according to the definitions of landslide activity by UNESCO-WP/WLI (1993) [79]. Using SBAS–InSAR technology, we chose peak velocities to represent landslide motion and evaluated landslide activity by comparing the historical (V_H) and present (V_P) representative velocities of each landslide [30,80]. Thus, we gave the maximum possible sliding or reactivation range. Owing to the lack of SAR data, the SBAS–InSAR results from 2003 to 2006 represented the landslide activity before the earthquake. Based on the SBAS–InSAR results, the stable threshold was set to -16 mm/a through statistical analysis of the results and field investigations [81,82]. The complex relief and rapid deformation in some parts of the study area will lead to the incoherence of InSAR results [73]. A reliable and robust interpretation of landslide motion rates includes those phenomena with a density of coherent targets (CTs) of up to 20–30 km^2 and at least three within the landslide [30,83]. Thus, when the number of CTs was insufficient, a comparison of the two optical images before and after was used to supplement the activity of some landslides [24]. The classification criteria are presented in Table 4.

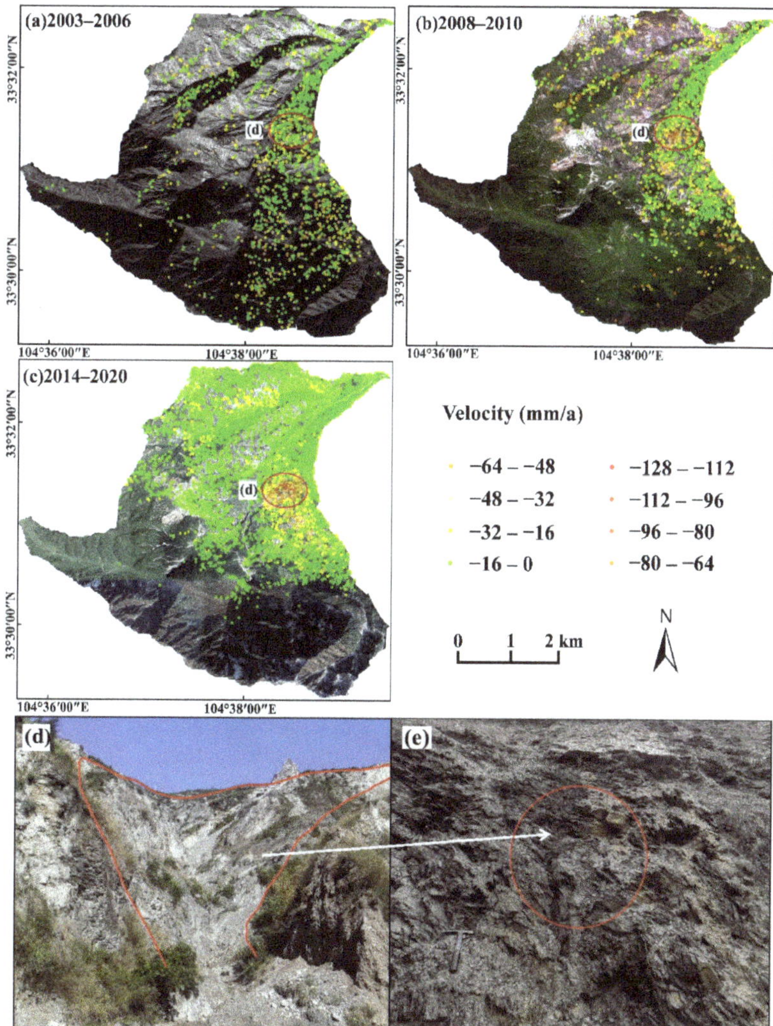

Figure 11. Deformation rate in the slope direction and severely deformed slope. (**a**) Deformation from 2003 to 2006; (**b**) deformation from 2008 to 2010; (**c**) deformation from 2014 to 2020; (**d**) severely deformed landslide; (**e**) position of water seepage.

Table 4. Classification criteria of landslide activity in this study.

Landslide State	Sufficient Number of CTs	Insufficient Number of CTs
Active	$V_P < -16$ mm/a and $V_H < -16$ mm/a	<1/3 vegetated
Reactivated	$V_P > -16$ mm/a and $V_H < -16$ mm/a	Vegetated reduction
Dormant	$V_P > -16$ mm/a	Vegetated
New	Landslide that does not occur on a pre-existing landslide.	

The distribution of landslide activity is shown in Figure 12, and the quantitative statistics are shown in Figure 13. The results showed that 25 active landslides and 46 dormant landslides were initially detected (2007). The active landslides were concentrated near Jiudun village in the middle–upper reaches, and sporadic active landslides were also observed on both sides of the main and secondary channels. In 2008, many dormant landslides were reactivated, and abundant new landslides were found. Regarding landslide distribution, new landslides were located in high mountain limestone. From 2017 to 2019, the number of dormant landslides exhibited a steady upward trend. The restored stable landslides were mainly coseismic landslides of the Wenchuan earthquake, while the concentrated areas of active, new, and reactivated landslides gradually recovered to the same level as that before the Wenchuan earthquake. However, many dormant landslides were reactivated in 2020. The number of active and new landslides increased, and the new landslides were mainly distributed in the middle–lower section. Based on the spatiotemporal distribution of active landslides, the landslides triggered by the Wenchuan earthquake were mainly located on the upstream and steep slopes, while the landslides induced by seasonal rainfall were mainly concentrated in the middle–lower reaches. The activity of coseismic landslides generated by the 2008 Wenchuan earthquake gradually decreased. Numerous coseismic landslides recovered to stability from 2017 to 2019, indicating that the postseismic effect of the Wenchuan earthquake on landslides lasted nearly 10 years. During these 10 years, the gradual weakening of the postseismic effect gradually led to the distribution of active landslides to the middle and lower reaches of the basin.

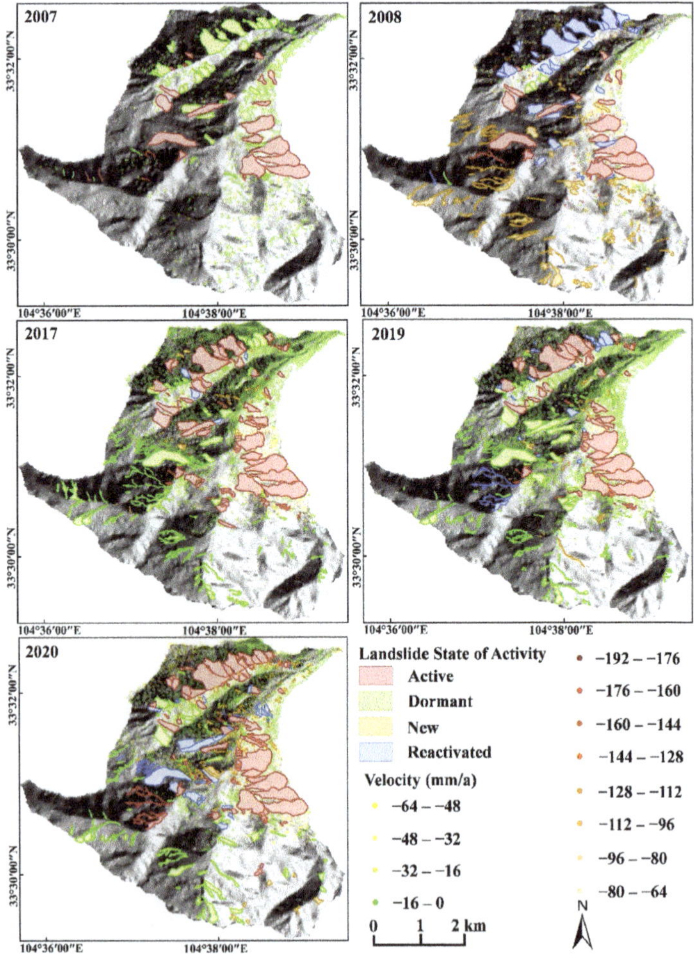

Figure 12. Distribution of landslide activity in different years.

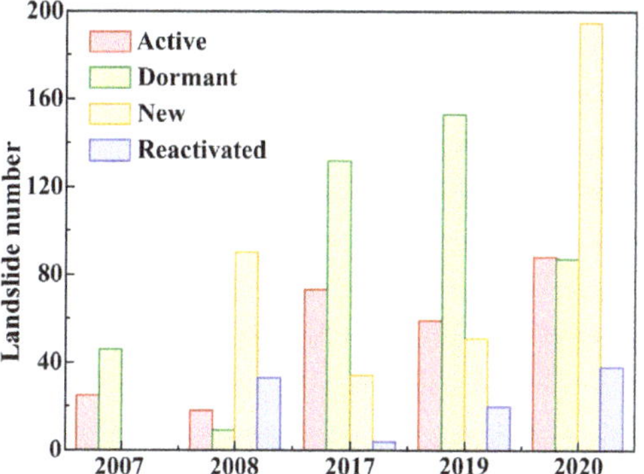

Figure 13. Number of landslide activity types in different years.

4.3.3. Characteristics of Landslide Time Series

To quantify the variation in the landslide displacement rate over time and better reveal the process and stage of landslide deformation, we selected some representative landslides to study the characteristics of the landslide time series. Based on different landslide types and the distribution of the SBAS–InSAR results, we selected 11 landslides from 2008 to 2010 and 25 landslides from 2014 to 2020 as representative landslide sets in the study area. We selected the 75th percentile CTs values within the landslide body to observe temporal variation [84]. Landslides are caused by surface and subsurface saturation. Therefore, the landslide representative set velocity was normalized to analyze the relationship between the change in landslide velocity and MBDI, as shown in Figure 14. The time-series rate change of the landslides shows that the deformation presented continuous and seasonal changes. On a long-term scale, the deformation was related to the MBDI, and the landslide movement was most significant from May to October each year. On a short-term scale, there was hysteresis in the movement of landslides caused by water gradually infiltrating the slope through various pathways. The velocity of some landslides peaked after the Wenchuan earthquake, and it gradually decreased until the MBDI increased. In addition, some landslides did not reach their peak movement rate immediately after the earthquake, but their velocities increased significantly. Under the combined effect of water, landslide velocity peaked in August and September 2008 and then gradually decreased, showing seasonal cycle changes affected by MBDI. Owing to the heterogeneity of landslides, their response to soil moisture varied; however, it was clear that extreme rainfall events in August 2020 resulted in an increase in most landslide movement rates in the study area.

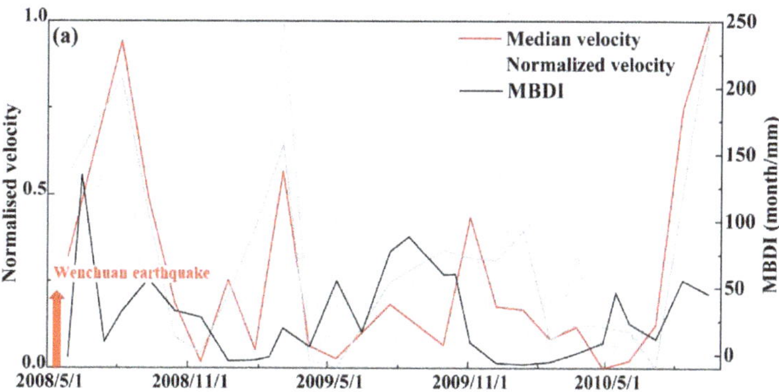

Figure 14. *Cont.*

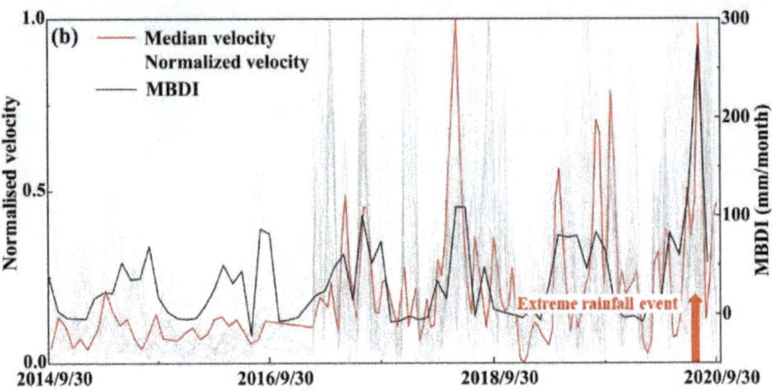

Figure 14. Normalized velocity time series and the moisture balance drought index (MBDI). (a) Eleven representative landslides from 2008 to 2010; (b) twenty-five representative landslides from 2014 to 2020.

5. Discussion

5.1. Landslide Driving Mechanism and Evolution Pattern

After the earthquake, the landslide density of the areas at elevations above 1900 m exhibited an incremental trend (Figure 10a). A comparison of the data prior to and after the earthquake showed that few pre-earthquake landslides were distributed in locations with a slope in the 40–50°, while the opposite was true for coseismic and postseismic landslides (Figure 10b). Moreover, there were numerous new landslides at the ridge or areas with local variations in topography and morphology (Figure 8). Various studies have justified that the role of topographic amplification of ground motion accelerates the initiation of slope failures and significantly modifies across different parts of hillslopes [85–88]. The probability of slope failure becomes larger along ridgelines and changes in slope [86–88]. Furthermore, the contribution of amplification effects also appears at steep slopes and higher altitudes [85]. Thus, compared with the middle–lower reaches of the river valley, the Wenchuan earthquake has a greater impact on the upstream. After the earthquake, cracks formed on the slopes. The formation of cracks made slopes vulnerable to failure by subjecting them to rain seepage. Under the coupling effect of rainfall, the shear capacity of the slopes gradually decreased and eventually formed landslides. As these cracks led to the development of landslides, the postearthquake effects wore off. Therefore, the number of landslides in the upper limestone area increased between 2008 and 2014. From 2017 to 2019, the activity of landslides upstream decreased, and many coseismic landslides recovered and stabilized. It can be inferred that the study area was significantly affected by the postearthquake effects from 2008 to 2014, while the postearthquake effects gradually weakened after 2014, and the coseismic landslides were stable until approximately 2017.

As is shown in Figure 15, the number of landslides corresponds to annual precipitation. During heavy rainfall, the number of landslides increased significantly, while when the rainfall was slight, the increase in the number of landslides was not significant. By comparing the relationship between annual rainfall and landslide activity, we found that there were more active landslides in years with high rainfall. Figure 3 shows the rainfall distribution in the valley. Although rainfall was more abundant upstream, the impact of rainfall was unnoticeable owing to the shear strength of limestone. In the middle reaches near Jiudun-Yaodao, rainfall was heavy. This area consists of weak loess and weathered phyllite that can lead to landslides under the influence of rainfall. Before the earthquake, landslides were mainly distributed in the middle-lower reaches, and active landslides were concentrated in the vicinity of Jiudun-Yaodao. After the earthquake, with the weakening of the postearthquake effect, rainfall gradually enhanced landslide control in the study area. The new landslides gradually moved to the middle–lower reaches. When there was extreme rainfall in the valley in 2020, many reactivated and new landslides were concentrated in the middle–lower reaches. Therefore, in the study area, landslides triggered by rainfall were mainly concentrated in the middle–lower reaches and a few upstream regions. In addition, the rainfall in this region exhibited a trend of gradual increase, as is shown in Figure 15. Based on the five-year average rainfall of less than 400 mm in 1995–2000 to nearly 600 mm in 2015–2020, it can be inferred that rainfall will exhibit an upward trend in the future. An increase in rainfall may make the existing landslides more active, and more new landslides will occur.

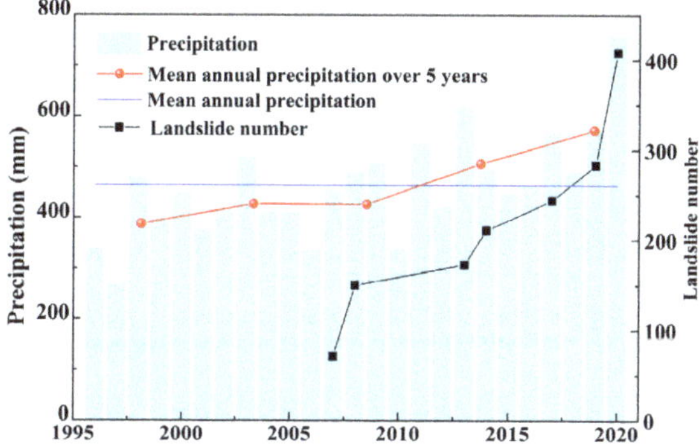

Figure 15. Precipitation and landslide number.

The land use has changed slightly since 2007 (Figure 16). Road engineering is the most important triggering factor of anthropogenic nature. It has been reported that roads in the upstream limestone area were built after the 2008 earthquake. Building roads through mountains provides the slope with good free-face conditions, which can easily trigger landslides. As is shown in Figure 8, there were some landslides along the road upstream, which is another reason for the increase in the number of landslides in this region after 2008. In addition, there has been road construction near Jiudun and Yaodao villages in recent years. Owing to the destruction of the original slopes caused by various engineering constructions, landslides are easily formed under seasonal rainfall. Human activities will amplify the likelihood of landslides triggered by seasonal rainfall. However, compared with the landslides triggered by rainfall and earthquake, landslides caused by human activities were relatively rare, with 29 landslides located on road slopes until 2020, accounting for 7% of the total number of landslides. Therefore, road construction is not the dominant driving mechanism of landslides in the study area. The landslides associated with human activity have a weak influence on the overall evolution pattern of the study area, so we do not conduct in-depth discussion.

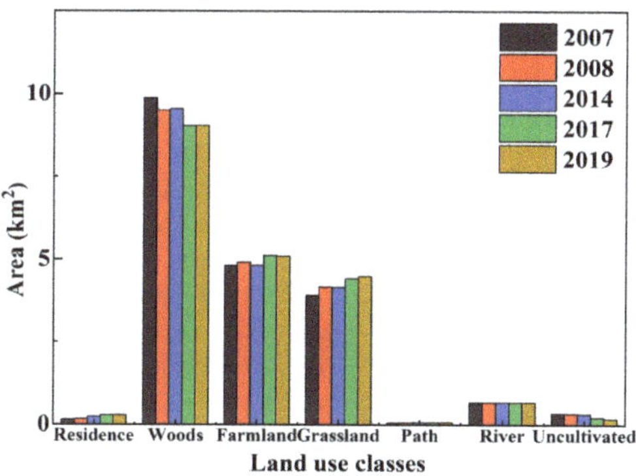

Figure 16. Land use in different years.

Based on the landslide evolution and driving mechanisms in the study area, we developed a conceptual model for the landslide evolution of the catchment (Figure 17). Before the earthquake, the landslides were primarily controlled by rainfall, most landslides were in the middle–lower section, and a few landslides were active (Figure 17a). In the upper reaches, much landslides and slope damages were caused by the earthquake. At the same time, the earthquake enhanced the activity of the existing landslides (Figure 17b). After the earthquake, the coseismic landslides gradually stabilized. Under the coupling effect of rainfall, the cracks upstream gradually developed into landslides. The spatial distribution of active landslides gradually shifted to the middle–lower reaches. The loose sediment of the landslides accumulated on the slope and channel and provided plentiful material for debris flow (Figure 17c). Currently, landslides are mainly driven by rainfall, and most of them are small slide landslides that have developed in the middle–lower sections. Abundant materials were deposited in the channel (Figure 17d).

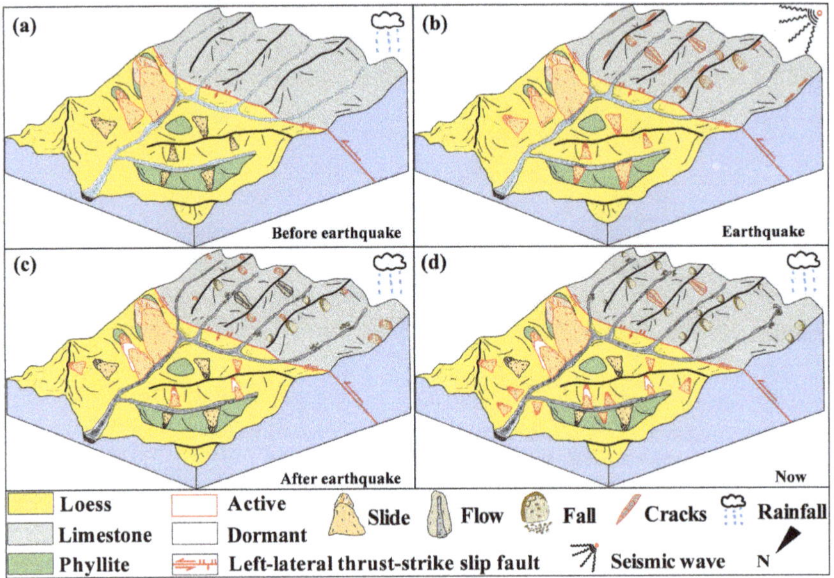

Figure 17. Conceptual model of landslide evolution in the study area. (**a**) Before earthquake; (**b**) during the earthquake; (**c**) after earthquake; (**d**) now.

5.2. Contribution of Landslide and Surface Erosion to Debris Flow

The study area is a typical high-frequency debris flow catchment, with debris flows occurring six times per year on average [89]. Three main factors control debris flow formation [90]. The volume of source material in a catchment is the most significant. Many landslides cause loose material to accumulate on hillslopes and channels, which increases the susceptibility of debris flows. To quantify the potential debris flow source provided by landslides, we divided the catchment into 15 sub-catchments and combined them with the landslide sediment supply capacity (ISl) to estimate the landslide potential material source [91]. The potential source quantity provided by landslide sediment for debris flows each year is shown in Figure 18. The variation trend of the potential source provided by landslides was generally consistent with that of the number of landslides, except for 2017. As the ISl index decreased in 2017, the volume also decreased slightly. According to the existing debris flow records and information gathered from local villagers, in recent years, the number of debris flows occurring in the basin was the highest in 2020, and these scales were more prominent than in previous years. Corresponding to the potential source, the volume in 2020 increased significantly compared to those in previous years. The additional landslides in this year were mainly small landslides. Small landslides induced by heavy rainfall can rapidly change the volume of channel material [92]. Therefore, it can be inferred that small landslides are an important source of debris flows.

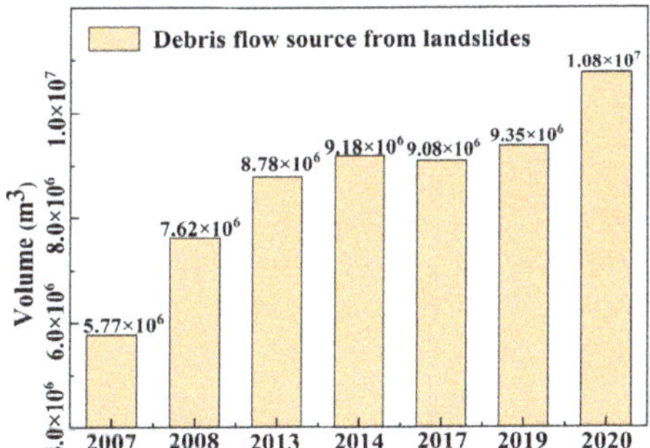

Figure 18. Contribution of landslide sediment to debris flow.

Furthermore, the surface erosion of slopes also significantly contributes to debris flow material sources. In a debris flow gully, erosion is distributed throughout the entire catchment; therefore, based on the SBAS–InSAR results, the obtained deformation data were used to estimate the volume of erosion occurring in the entire basin via statistical calculations performed based on the method of Cao et al. (2021) [90]. From 2014 to 2020, the catchment generated approximately 3.90×10^5 m^3 of slope erosion. As is shown in Figure 11c, owing to excellent water retention, the slope erosion intensity of the upper reaches with high vegetation cover was low. Erosion principally occurred on both sides of the gully in the middle–lower reaches. Thus, the erosion material of the slope originated mainly from the middle to the lower reaches. Overall, the greater the rainfall, the greater the amount of erosion. The annual amount of erosion from January to October from 2017 to 2020 is shown in Figure 19a. The amount of erosion in 2018 was the lowest, approximately 3.05×10^4 m^3, and the amount of erosion in 2020 was the highest, at approximately 8.31×10^4 m^3. Taking 2020 as an example, the time-series analysis of slope erosion within a year showed that the slope erosion in summer and autumn was greater than that in spring and winter (Figure 19b). Existing debris flow records show that debris flow occurred mainly from May to October, corresponding to slope erosion and landslide movement [51]. At the same time, the debris flow events that occurred in August 2020, corresponding to landslide activity and surface erosion at that time, were relatively large. Therefore, the intensification of landslides and surface erosion increased the material source of debris flow and the risk of a disaster chain.

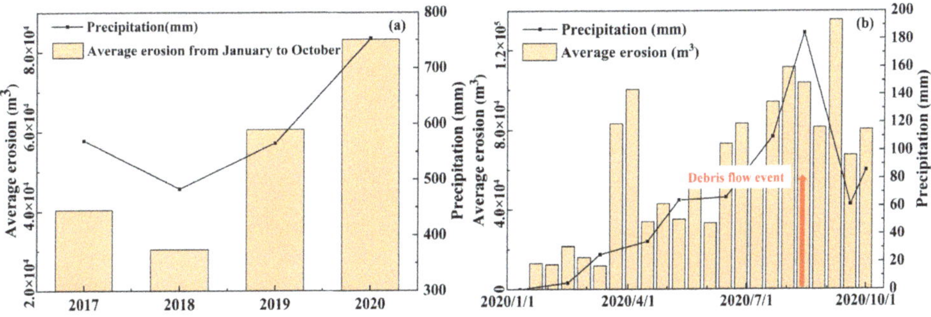

Figure 19. Changes in the surface erosion of slopes. (**a**) Relationship between the average erosion and precipitation from January to October; (**b**) relationship between time-series erosion and precipitation in 2020.

As this study was based only on remote-sensing data and field investigations, it is impossible to determine the relevant landslide process from stability to activity. Therefore, it is necessary to conduct further studies based on field experiments or physical models in the later stages to explore the physical mechanisms triggering landslides from a microscopic perspective. In the future work, we will conduct geophysical investigation to obtain more accurate parameters. The connected landslides are partitioned into individual landslides based on the methods of Fan et al. (2019) [93]. Thus, the accuracy of landslide volume calculation can be improved and we will estimate a more exact contribution to debris flow.

6. Conclusions

In this study, Goulinping valley, a typical medium–small catchment in the Bailong River basin, was used as the research area. A multitemporal inventory map and field investigation results revealed the spatiotemporal evolution pattern of landslides. We used multisource remote-sensing technology to evaluate landslide activity and analyze the driving mechanisms of landslides. Moreover, we quantified the potential contribution of landslides to debris flows in this catchment. The following conclusions were drawn from this study:

(1) The number of landslides increased nearly six times in 13 years (from 71 in 2007 to 408 in 2020), and the total volume of landslides approximately doubled (from 3.32×10^7 m^3 in 2007 to 5.87×10^7 m^3 in 2020). The growth was most significant in 2008 and 2020, and the volume of added landslides was less than 10^5 m^3.

(2) Landslides are mainly driven by rainfall and earthquake, and the responses of different lithologic strata to disturbances showed apparent differences. The evolution of landslides in the catchment can be divided into three stages. During the earthquake driving stage (2008), the coseismic landslides were mainly distributed in the limestone area, and the landslides in the catchment were primarily active. During the coupled driving stage of earthquake and rainfall (2008–2017), the damage of seismic rock mass in the limestone area developed into landslides, and the active landslides gradually concentrated in the loess–phyllite area. During the rainfall driving stage (2017–the present), rainfall triggered small landslides in the loess–phyllite region, the landslides in the limestone area were stable, and in the loess–phyllite area they were active. Human activities have a relatively small influence on landslides instead of the dominant control.

(3) Small landslides and mid-downstream slope erosion mainly determine the frequency and scale of debris flow. Many small landslides can rapidly provide abundant debris flow sources and reduce the threshold of debris flow, leading to an increase in the frequency and scale of debris flow. The upper reaches, with good vegetation cover, had a certain degree of soil and water conservation. Hence, vast slope erosion materials originated from the middle–lower reaches. When the rainfall intensity was high, the slope erosion intensified, making an outstanding contribution to debris flow.

Author Contributions: L.L.: methodology, software, validation, formal analysis, investigation, data curation, writing—original draft preparation, writing—review and editing, visualization; G.C.: conceptualization, formal analysis, investigation, resources, writing—review and editing, supervision, project administration, funding acquisition; W.S.: methodology, software, investigation; J.J.: investigation, data curation; J.W.: investigation, data curation; F.H.: investigation, data curation; Y.C.: investigation, data curation; Y.M.: investigation; Y.L.: investigation, resources; Y.Z.: software, resources. All authors have read and agreed to the published version of the manuscript.

Funding: This study was supported by the Second Tibetan Plateau Scientific Expedition and Research Program (STEP) (Grant No. 2021QZKK0201), National Key Research and Development Program of China (Grant No. 2017YFC1501005), Key Research and Development Program of Gansu Province (Grant No. 20YF8FA074), Science and Technology Major Project of Gansu Province (Grant No. 19ZD2FA002), the Fundamental Research Funds for the Central Universities (Grant No. lzujbky-2021-16), Geohazard prevention project of Gansu Province (Grant No. CNPC-B-FS2021012).

Data Availability Statement: The ENVISAT ASAR, Sentinel-1A, and precise orbits dates were provided by the European Space Agency. The SRTM3 DEM data were provided by NASA. We thank Japan Aerospace Exploration Agency for providing the 12.5 m DEM data and China Meteorological Data Network for providing the meteorological data.

Acknowledgments: We thank Yan Zhao, Wanyu Jiang, Shiqiang Bian, Yunpeng Yang, Hongxue Zhang, Yuanxi Li, Yiwen Liang and Wangcai Liu for their suggestions on software operation and data analysis. We are grateful to the editorial office and four anonymous reviewers for their constructive suggestions and comments to improve the paper.

References

1. Petley, D. Global patterns of loss of life from landslides. *Geology* **2012**, *40*, 927–930. [CrossRef]
2. Crozier, M.J. Deciphering the effect of climate change on landslide activity: A review. *Geomorphology* **2010**, *124*, 260–267. [CrossRef]
3. Haque, U.; Da Silva, P.F.; Devoli, G.; Pilz, J.; Zhao, B.; Khaloua, A.; Wilopo, W.; Andersen, P.; Lu, P.; Lee, J.; et al. The human cost of global warming: Deadly landslides and their triggers (1995–2014). *Sci. Total Environ.* **2019**, *682*, 673–684. [CrossRef] [PubMed]
4. Fan, X.; Scaringi, G.; Korup, O.; West, A.J.; Westen, C.J.; Tanyas, H.; Hovius, N.; Hales, T.C.; Jibson, R.W.; Allstadt, K.E.; et al. Earthquake-Induced Chains of Geologic Hazards: Patterns, Mechanisms, and Impacts. *Rev. Geophys.* **2019**, *57*, 421–503. [CrossRef]
5. Lin, Q.; Wang, Y. Spatial and temporal analysis of a fatal landslide inventory in China from 1950 to 2016. *Landslides* **2018**, *15*, 2357–2372. [CrossRef]
6. Qi, T.; Meng, X.; Qing, F.; Zhao, Y.; Shi, W.; Chen, G.; Zhang, Y.; Li, Y.; Yue, D.; Su, X.; et al. Distribution and characteristics of large landslides in a fault zone: A case study of the NE Qinghai-Tibet Plateau. *Geomorphology* **2021**, *379*, 107592. [CrossRef]
7. Yin, Y.; Wang, F.; Sun, P. Landslide hazards triggered by the 2008 Wenchuan earthquake, Sichuan, China. *Landslides* **2009**, *6*, 139–152. [CrossRef]
8. Qi, S.; Xu, Q.; Lan, H.; Zhang, B.; Liu, J. Spatial distribution analysis of landslides triggered by 2008.5.12 Wenchuan Earthquake, China. *Eng. Geol.* **2010**, *116*, 95–108. [CrossRef]
9. Bai, S.; Lu, P.; Thiebes, B. Comparing characteristics of rainfall- and earthquake-triggered landslides in the Upper Minjiang catchment, China. *Eng. Geol.* **2020**, *268*, 105518. [CrossRef]
10. Zhong, X.; Chen, W.; Hou, J.; Yuan, X. Distribution rules of geo-hazards induced by Wenchuan Earthquake. *Chin. J. Geotech. Eng.* **2011**, *33*, 356–360. (In Chinese)
11. Bai, S.; Cheng, C.; Wang, J.; Thiebes, B.; Zhang, Z. Regional scale rainfall- and earthquake-triggered landslide susceptibility assessment in Wudu County, China. *J. Mt. Sci. Engl.* **2013**, *10*, 743–753. [CrossRef]
12. Bai, S.; Wang, J.; Zhang, Z.; Cheng, C. Combined landslide susceptibility mapping after Wenchuan earthquake at the Zhouqu segment in the Bailongjiang Basin, China. *Catena* **2012**, *99*, 18–25. [CrossRef]
13. Liao, P.; Wu, W.; Zhe, X. Stability assessment and treatment for Hongtupo landslide in severe disaster area of southern Gansu induced by Wenchuan earthquake. *J. Eng. Geol.* **2012**, *20*, 204–212. (In Chinese) [CrossRef]
14. Wu, W.; Liao, P. An analysis of the treatment effect for Sanjiadi landslide in Wudu county, a heavy disaster area during the Wenchuan earthquake. *J. Lanzhou Univ. Nat. Sci.* **2014**, *50*, 633–638. (In Chinese) [CrossRef]
15. Guzzetti, F.; Mondini, A.C.; Cardinali, M.; Fiorucci, F.; Santangelo, M.; Chang, K. Landslide inventory maps: New tools for an old problem. *Earth-Sci. Rev.* **2012**, *112*, 42–66. [CrossRef]
16. Chen, M.; Tang, C.; Xiong, J.; Shi, Q.Y.; Li, N.; Gong, L.F.; Wang, X.D.; Tie, Y. The long-term evolution of landslide activity near the epicentral area of the 2008 Wenchuan earthquake in China. *Geomorphology* **2020**, *367*, 107317. [CrossRef]
17. Yang, H.; Yang, T.; Zhang, S.; Zhao, F.; Hu, K.; Jiang, Y. Rainfall-induced landslides and debris flows in Mengdong Town, Yunnan Province, China. *Landslides* **2020**, *17*, 931–941. [CrossRef]
18. Xu, C.; Xu, X.; Shyu, J.B.H. Database and spatial distribution of landslides triggered by the Lushan, China Mw 6.6 earthquake of 20 April 2013. *Geomorphology* **2015**, *248*, 77–92. [CrossRef]
19. Cevasco, A.; Pepe, G.; Brandolini, P. The influences of geological and land use settings on shallow landslides triggered by an intense rainfall event in a coastal terraced environment. *Bull. Eng. Geol. Environ.* **2014**, *73*, 859–875. [CrossRef]
20. Guzzetti, F.; Reichenbach, P.; Cardinali, M.; Galli, M.; Ardizzone, F. Probabilistic landslide hazard assessment at the basin scale. *Geomorphology* **2005**, *72*, 272–299. [CrossRef]
21. Galli, M.; Ardizzone, F.; Cardinali, M.; Guzzetti, F.; Reichenbach, P. Comparing landslide inventory maps. *Geomorphology* **2008**, *94*, 268–289. [CrossRef]
22. Bozzano, F.; Mazzanti, P.; Perissin, D.; Rocca, A.; De Pari, P.; Discenza, M. Basin Scale Assessment of Landslides Geomorphological Setting by Advanced InSAR Analysis. *Remote Sens.* **2017**, *9*, 267. [CrossRef]
23. Fan, X.; Juang, C.H.; Wasowski, J.; Huang, R.; Xu, Q.; Scaringi, G.; van Westen, C.J.; Havenith, H. What we have learned from the 2008 Wenchuan Earthquake and its aftermath: A decade of research and challenges. *Eng. Geol.* **2018**, *241*, 25–32. [CrossRef]
24. Tang, C.; Westen, C.; Tanyaş, H.; Jetten, V.G. Analysing post-earthquake landslide activity using multi-temporal landslide inventories near the epicentral area of the 2008 Wenchuan earthquake. *Nat. Hazard Earth Syst.* **2016**, *16*, 2641–2655. [CrossRef]

25. Yang, W.; Qi, W.; Wang, M.; Zhang, J.; Zhang, Y. Spatial and temporal analyses of post-seismic landslide changes near the epicentre of the Wenchuan earthquake. *Geomorphology* **2017**, *276*, 8–15. [CrossRef]
26. Fan, X.; Domènech, G.; Scaringi, G.; Huang, R.; Xu, Q.; Hales, T.C.; Dai, L.; Yang, Q.; Francis, O. Spatio-temporal evolution of mass wasting after the 2008 Mw 7.9 Wenchuan earthquake revealed by a detailed multi-temporal inventory. *Landslides* **2018**, *15*, 2325–2341. [CrossRef]
27. Zhang, S.; Zhang, L.; Lacasse, S.; Nadim, F. Evolution of Mass Movements near Epicentre of Wenchuan Earthquake, the First Eight Years. *Sci. Rep.* **2016**, *6*, 36154. [CrossRef]
28. Zhang, S.; Zhang, L.M. Impact of the 2008 Wenchuan earthquake in China on subsequent long-term debris flow activities in the epicentral area. *Geomorphology* **2017**, *276*, 86–103. [CrossRef]
29. Parise, M.; Wasowski, J.; Carrara, A.; Guzzetti, F. Landslide activity maps for landslide hazard evaluation; three case studies from southern Italy. *Nat. Hazards* **1999**, *20*, 159–183. [CrossRef]
30. Cigna, F.; Bianchini, S.; Casagli, N. How to assess landslide activity and intensity with Persistent Scatterer Interferometry (PSI): The PSI-based matrix approach. *Landslides* **2013**, *10*, 267–283. [CrossRef]
31. Stumpf, A.; Malet, J.; Delacourt, C. Correlation of satellite image time-series for the detection and monitoring of slow-moving landslides. *Remote Sens. Environ.* **2017**, *189*, 40–55. [CrossRef]
32. Comert, R.; Avdan, U.; Gorum, T.; Nefeslioglu, H.A. Mapping of shallow landslides with object-based image analysis from unmanned aerial vehicle data. *Eng. Geol.* **2019**, *260*, 105264. [CrossRef]
33. Strozzi, T.; Ambrosi, C.; Raetzo, H. Interpretation of Aerial Photographs and Satellite SAR Interferometry for the Inventory of Landslides. *Remote Sens.* **2013**, *5*, 2554–2570. [CrossRef]
34. Akbarimehr, M.; Motagh, M.; Haghshenas-Haghighi, M. Slope Stability Assessment of the Sarcheshmeh Landslide, Northeast Iran, Investigated Using InSAR and GPS Observations. *Remote Sens.* **2013**, *5*, 3681–3700. [CrossRef]
35. Borgomeo, E.; Hebditch, K.V.; Whittaker, A.C.; Lonergan, L. Characterising the spatial distribution, frequency and geomorphic controls on landslide occurrence, Molise, Italy. *Geomorphology* **2014**, *226*, 148–161. [CrossRef]
36. Gibson, A.D.; Culshaw, M.G.; Dashwood, C.; Pennington, C.V.L. Landslide management in the UK—the problem of managing hazards in a 'low-risk' environment. *Landslides* **2013**, *10*, 599–610. [CrossRef]
37. Schuster, R.L.; Nietothomas, A.S.; O'Rourke, T.D.; Crespo, E.; Plaza-Nieto, G. Mass wasting triggered by the 5 March 1987 ecuador earthquakes—ScienceDirect. *Eng. Geol.* **1996**, *42*, 1–23. [CrossRef]
38. Keeper, D.K. Landslides caused by earthquakes. *Bull. Geol. Soc. Am.* **1984**, *95*, 406. [CrossRef]
39. Crosta, G.B. Introduction to the special issue on rainfall-triggered landslides and debris flows. *Eng. Geol.* **2004**, *73*, 191–192. [CrossRef]
40. Gabet, E.J.; Burbank, D.W.; Putkonen, J.K.; Pratt-Sitaula, B.A.; Ojha, T. Rainfall thresholds for landsliding in the Himalayas of Nepal. *Geomorphology* **2004**, *63*, 131–143. [CrossRef]
41. Guzzetti, F.; Peruccacci, S.; Rossi, M.; Stark, C.P. Rainfall thresholds for the initiation of landslides in central and southern Europe. *Meteorol. Atmos. Phys.* **2007**, *98*, 239–267. [CrossRef]
42. Peruccacci, S.; Brunetti, M.T.; Luciani, S.; Vennari, C.; Guzzetti, F. Lithological and seasonal control on rainfall thresholds for the possible initiation of landslides in central Italy. *Geomorphology* **2012**, *139–140*, 79–90. [CrossRef]
43. Yang, H.; Wei, F.; Ma, Z.; Guo, H.; Su, P.; Zhang, S. Rainfall threshold for landslide activity in Dazhou, southwest China. *Landslides* **2020**, *17*, 61–77. [CrossRef]
44. Jiang, W.; Chen, G.; Meng, X.; Jin, J.; Zhao, Y.; Lin, L.; Li, Y.; Zhang, Y. Probabilistic rainfall threshold of landslides in Data-Scarce mountainous Areas: A case study of the Bailong River Basin, China. *Catena* **2022**, *213*, 106190. [CrossRef]
45. Lin, C.; Liu, S.; Lee, S.; Liu, C. Impacts of the Chi-Chi earthquake on subsequent rainfall-induced landslides in central Taiwan. *Eng. Geol.* **2006**, *86*, 87–101. [CrossRef]
46. Chiou, S.; Cheng, C.; Hsu, S.; Lin, Y.; Chi, S. Evaluating Landslides and Sediment Yields Induced by the Chi-Chi Earthquake and Followed Heavy Rainfalls along the Ta-Chia River. *J. GeoEng.* **2007**, *2*, 73–82. [CrossRef]
47. Weng, M.; Wu, M.; Ning, S.; Jou, Y. Evaluating triggering and causative factors of landslides in Lawnon River Basin, Taiwan. *Eng. Geol.* **2011**, *123*, 72–82. [CrossRef]
48. Bontemps, N.; Lacroix, P.; Larose, E.; Jara, J.; Taipe, E. Rain and small earthquakes maintain a slow-moving landslide in a persistent critical state. *Nat. Commun.* **2020**, *11*, 780. [CrossRef]
49. Zhang, S.; Zhang, L.M.; Glade, T. Characteristics of earthquake- and rain-induced landslides near the epicenter of Wenchuan earthquake. *Eng. Geol.* **2014**, *175*, 58–73. [CrossRef]
50. Chen, G.; Meng, X.; Qiao, L.; Zhang, Y.; Wang, S. Response of a loess landslide to rainfall: Observations from a field artificial rainfall experiment in Bailong River Basin, China. *Landslides* **2018**, *15*, 895–911. [CrossRef]
51. Li, Y.; Meng, X.; Guo, P.; Dijkstra, T.; Zhao, Y.; Chen, G.; Yue, D. Constructing rainfall thresholds for debris flow initiation based on critical discharge and S-hydrograph. *Eng. Geol.* **2021**, *280*, 105962. [CrossRef]
52. Xiong, M.; Meng, X.; Wang, S.; Guo, P.; Li, Y.; Chen, G.; Qing, F.; Cui, Z.; Zhao, Y. Effectiveness of debris flow mitigation strategies in mountainous regions. *Prog. Phys. Geog.* **2016**, *40*, 768–793. [CrossRef]
53. Li, Y.; Armitage, S.J.; Stevens, T.; Meng, X. Alluvial fan aggradation/incision history of the eastern Tibetan plateau margin and implications for debris flow/debris-charged flood hazard. *Geomorphology* **2018**, *318*, 203–216. [CrossRef]

54. Yuan, D.; Lei, Z.; Wang, A. Additional Textual Criticism of Southern Tianshui M8 Earthquake in Gansu Province in 1654. *China Earthq. Eng. J.* **2017**, *39*, 509–520. (In Chinese)

55. Yuan, D.; Lei, Z.; Yang, Q.; Wang, A.; Xie, H.; Su, Q. Seismic disaster features of the 1879 southern Wudu M8 earthquake in Gansu Province. *J. Lanzhou Univ. Nat. Sci.* **2014**, *50*, 611–621. (In Chinese) [CrossRef]

56. Yang, W.; Qi, W.; Zhou, J. Effects of precipitation and topography on vegetation recovery at landslide sites after the 2008 Wenchuan earthquake. *Land Degrad. Dev.* **2018**, *29*, 3355–3365. [CrossRef]

57. Li, X.; Cheng, X.; Chen, W.; Chen, G.; Liu, S. Identification of Forested Landslides Using LiDar Data, Object-based Image Analysis, and Machine Learning Algorithms. *Remote Sens.* **2015**, *7*, 9705–9726. [CrossRef]

58. CMONOC. Coseismic displacement fields of the 2008 Wenchuan Ms 8.0 earthquake measured by GPS. *Sci. Sin. Terrae.* **2008**, *38*, 12. (In Chinese)

59. Coe, J.A. Regional moisture balance control of landslide motion; implications for landslide forecasting in a changing climate. *Geology* **2012**, *40*, 323–326. [CrossRef]

60. Miller, C.V. *Photogeology*; Mac Graw Hill Book Company Inc.: London, UK, 1961. [CrossRef]

61. Allum, J.A.E. *Photogeology and Regional Mapping*; Pergamon Press: Oxford, UK, 1966.

62. Zuidam, R.A.V. *Aerial Photo-Interpretation in Terrain Analysis and Geomorphologic Mapping*; ITC. Smits Publishers: The Hague, The Netherlands, 1985.

63. Fiorucci, F.; Cardinali, M.; Carlà, R.; Rossi, M.; Mondini, A.C.; Santurri, L.; Ardizzone, F.; Guzzetti, F. Seasonal landslide mapping and estimation of landslide mobilization rates using aerial and satellite images. *Geomorphology* **2011**, *129*, 59–70. [CrossRef]

64. Du, J.; Glade, T.; Woldai, T.; Chai, B.; Zeng, B. Landslide susceptibility assessment based on an incomplete landslide inventory in the Jilong Valley, Tibet, Chinese Himalayas. *Eng. Geol.* **2020**, *270*, 105572. [CrossRef]

65. Shao, X.; Ma, S.; Xu, C.; Zhang, P.; Wen, B.; Tian, Y.; Zhou, Q.; Cui, Y. Planet Image-Based Inventorying and Machine Learning-Based Susceptibility Mapping for the Landslides Triggered by the 2018 Mw6.6 Tomakomai, Japan Earthquake. *Remote Sens.* **2019**, *11*, 978. [CrossRef]

66. Varnes, D.J. *Slope Movement Types and Processes*; Landslides, Analysis and Control, Special Report 176; Schuster, R.L., Krizek, R.J., Eds.; Transportation Research Board, National Academy of Sciences: Washington, DC, USA, 1978; pp. 11–33.

67. Hungr, O.; Leroueil, S.; Picarelli, L. The Varnes classification of landslide types, an update. *Landslides* **2014**, *11*, 167–194. [CrossRef]

68. Berardino, P.; Fornaro, G.; Lanari, R.; Sansosti, E. A new algorithm for surface deformation monitoring based on small baseline differential SAR interferograms. *IEEE Trans. Geosci. Remote* **2002**, *40*, 2375–2383. [CrossRef]

69. Tizzani, P.; Berardino, P.; Casu, F.; Euillades, P.; Manzo, M.; Ricciardi, G.; Zeni, G.; Lanari, R. Surface deformation of Long Valley caldera and Mono Basin, California, investigated with the SBAS-InSAR approach. *Remote Sens. Environ.* **2007**, *108*, 277–289. [CrossRef]

70. Shi, W.; Chen, G.; Meng, X.; Jiang, W.; Chong, Y.; Zhang, Y.; Dong, Y.; Zhang, M. Spatial-Temporal Evolution of Land Subsidence and Rebound over Xi'an in Western China Revealed by SBAS-InSAR Analysis. *Remote Sens.* **2020**, *12*, 3756. [CrossRef]

71. Guzzetti, F.; Ardizzone, F.; Cardinali, M.; Rossi, M.; Valigi, D. Landslide volumes and landslide mobilization rates in Umbria, central Italy. *Earth Planet. Sc. Lett.* **2009**, *279*, 222–229. [CrossRef]

72. Jin, J.; Chen, G.; Meng, X.; Zhang, Y.; Shi, W.; Li, Y.; Yang, Y.; Jiang, W. Prediction of river damming susceptibility by landslides based on a logistic regression model and InSAR techniques: A case study of the Bailong River Basin, China. *Eng. Geol.* **2022**, *299*, 106562. [CrossRef]

73. Zhang, Y.; Meng, X.M.; Dijkstra, T.A.; Jordan, C.J.; Chen, G.; Zeng, R.Q.; Novellino, A. Forecasting the magnitude of potential landslides based on InSAR techniques. *Remote Sens. Environ.* **2020**, *241*, 111738. [CrossRef]

74. Larsen, I.J.; Montgomery, D.R.; Korup, O. Landslide erosion controlled by hillslope material. *Nat. Geosci.* **2010**, *3*, 247–251. [CrossRef]

75. Jaboyedoff, M.; Carrea, D.; Derron, M.; Oppikofer, T.; Penna, I.M.; Rudaz, B. A review of methods used to estimate initial landslide failure surface depths and volumes. *Eng. Geol.* **2020**, *267*, 105478. [CrossRef]

76. Cascini, L.; Ciurleo, M.; Di Nocera, S. Soil depth reconstruction for the assessment of the susceptibility to shallow landslides in fine-grained slopes. *Landslides* **2017**, *14*, 459–471. [CrossRef]

77. Meier, C.; Jaboyedoff, M.; Derron, M.; Gerber, C. A method to assess the probability of thickness and volume estimates of small and shallow initial landslide ruptures based on surface area. *Landslides* **2020**, *17*, 975–982. [CrossRef]

78. Zhao, C.; Lu, Z.; Zhang, Q.; de la Fuente, J. Large-area landslide detection and monitoring with ALOS/PALSAR imagery data over Northern California and Southern Oregon, USA. *Remote Sens. Environ.* **2012**, *124*, 348–359. [CrossRef]

79. Wli, U. A suggested method for describing the activity of a landslide. *B Int. Assoc. Eng. Geol.* **1993**, *47*, 53–57. [CrossRef]

80. Cruden, D.; Varnes, D. *Landslide Types and Processes*; Special Report; National Research Council; Transportation Research Board: Ottawa, ON, Canada, 1996; pp. 36–75.

81. Wasowski, J.; Bovenga, F. Investigating landslides and unstable slopes with satellite Multi Temporal Interferometry: Current issues and future perspectives. *Eng. Geol.* **2014**, *174*, 103–138. [CrossRef]

82. Novellino, A.; Cesarano, M.; Cappelletti, P.; Martire, D.D.; Calcaterra, D. Slow-moving landslide risk assessment combining Machine Learning and InSAR techniques. *Catena* **2021**, *203*, 105317. [CrossRef]

83. Notti, D.; Davalillo, J.C.; Herrera, G.; Mora, O. Assessment of the performance of X-band satellite radar data for landslide mapping and monitoring: Upper Tena Valley case study. *Nat. Hazard. Earth Syst.* **2010**, *10*, 1865–1875. [CrossRef]

84. Bayer, B.; Simoni, A.; Mulas, M.; Corsini, A.; Schmidt, D. Deformation responses of slow moving landslides to seasonal rainfall in the Northern Apennines, measured by InSAR. *Geomorphology* **2018**, *308*, 293–306. [CrossRef]

85. Tang, C.; Zhu, J.; Qi, X.; Ding, J. Landslides induced by the Wenchuan earthquake and the subsequent strong rainfall event: A case study in the Beichuan area of China. *Eng. Geol.* **2011**, *122*, 22–33. [CrossRef]

86. Luo, Y.; Fan, X.; Huang, R.; Wang, Y.; Yunus, A.P.; Havenith, H.B. Topographic and near-surface stratigraphic amplification of the seismic response of a mountain slope revealed by field monitoring and numerical simulations. *Eng. Geol.* **2020**, *271*, 105607. [CrossRef]

87. Huang, R.; Pei, X.; Fan, X.; Zhang, W.; Li, S.; Li, B. The characteristics and failure mechanism of the largest landslide triggered by the Wenchuan earthquake, May 12, 2008, China. *Landslides* **2012**, *9*, 131–142. [CrossRef]

88. Karpouza, M.; Chousianitis, K.; Bathrellos, G.D.; Skilodimou, H.D.; Kaviris, G.; Antonarakou, A. Hazard zonation mapping of earthquake-induced secondary effects using spatial multi-criteria analysis. *Nat. Hazards* **2021**, *109*, 637–669. [CrossRef]

89. Zhao, Y.; Meng, X.; Qi, T.; Li, Y.; Chen, G.; Yue, D.; Qing, F. AI-based rainfall prediction model for debris flows. *Eng. Geol.* **2022**, *296*, 106456. [CrossRef]

90. Cao, C.; Zhang, W.; Chen, J.; Shan, B.; Song, S.; Zhan, J. Quantitative estimation of debris flow source materials by integrating multi-source data: A case study. *Eng. Geol.* **2021**, *291*, 106222. [CrossRef]

91. Xiong, J.; Tang, C.; Chen, M.; Gong, L.; Li, N.; Zhang, X.; Shi, Q. Long-term changes in the landslide sediment supply capacity for debris flow occurrence in Wenchuan County, China. *Catena* **2021**, *203*, 105340. [CrossRef]

92. Fan, L.; Lehmann, P.; Zheng, C.; Or, D. The Lasting Signatures of Past Landslides on Soil Stripping From Landscapes. *Water Resour. Res.* **2021**, *57*, e2021WR030375. [CrossRef]

93. Fan, R.L.; Zhang, L.M.; Shen, P. Evaluating volume of coseismic landslide clusters by flow direction-based partitioning. *Eng. Geol.* **2019**, *260*, 105238. [CrossRef]

Article

A Landslide Numerical Factor Derived from CHIRPS for Shallow Rainfall Triggered Landslides in Colombia

Cheila Avalon Cullen [1,*], Rafea Al Suhili [2] and Edier Aristizabal [3]

[1] CUNY-Remote Sensing Earth System Institute (CUNY-CREST Institute), The City University of New York, New York, NY 10453, USA

[2] Department of Civil Engineering, The City College of New York, New York, NY 10031, USA; ralsuhili@ccny.cuny.edu

[3] Departamento de Geociencias y Medio Ambiente, Universidad Nacional de Colombia, Sede Medellín 050034, Colombia; evaristizabalg@unal.edu.co

* Correspondence: ccullen@gradcenter.cuny.edu

Abstract: Despite great advances in remote sensing technologies, accurate satellite information is sometimes challenged in tropical regions where dense vegetation prevents the instruments from retrieving reliable readings. In this work, we introduce a satellite-based landslide rainfall threshold for the country of Colombia by studying 4 years of rainfall measurements from The Climate Hazards Group Infrared Precipitation with Stations (CHIRPS) for 346 rainfall-triggered landslide events (the dataset). We isolate the two successive rainy/dry periods leading to each landslide to create variables that simulate the dynamics of antecedent wetness and dryness. We test the performance of the derived variables (Rainfall Period 1 (PR1), Rainfall Sum 1 (RS1), Rainfall Period 2 (PR2), Rainfall Sum 2 (RS2), and Dry Period (DT)) in a logistic regression that includes three (3) static parameters (Soil Type (ST), Landcover (LC), and Slope angle). Results from the logistic model describe the influence of each variable in landslide occurrence with an accuracy of 73%. Subsequently, we use these dynamic variables to model a landslide threshold that, in the absence of satellite antecedent soil moisture data, helps describe the interactions between the dynamic variables and the slope angle. We name it the Landslide Triggering Factor—LTF. Subsequently, with a training dataset (65%) and one for testing (35%) we evaluate the LTF threshold performance and compare it to the well-known event duration (E-D) threshold. Results demonstrate that The LTF performs better than the E-D threshold for the training and testing datasets at 71% and 81% respectively.

Keywords: rainfall-triggered landslides; tropics; statistical analysis; CHIRPS

Citation: Cullen, C.A.; Al Suhili, R.; Aristizabal, E. A Landslide Numerical Factor Derived from CHIRPS for Shallow Rainfall Triggered Landslides in Colombia. *Remote Sens.* **2022**, *14*, 2239. https://doi.org/10.3390/rs14092239

Academic Editor: Francesca Ardizzone

Received: 25 April 2022
Accepted: 4 May 2022
Published: 7 May 2022

Publisher's Note: MDPI stays neutral with regard to jurisdictional claims in published maps and institutional affiliations.

1. Introduction

Landslides are a physical hazard that frequently result in devastating human and economic losses around the world [1–3]. There are various underlying geological, lithological, and morphological characteristics that make an area prone to these hazards. Nonetheless, landslides can happen as a result of anthropogenic activities or can be triggered by natural forces such as earthquakes, melting snow, or extreme precipitation [4–6].

Landslides that are triggered by rainfall are common phenomena in mountainous tropical regions. These landslides are associated with long-term, high-intensity periods of precipitation that have dangerous potential to initiate mass soil movement due to changes in pore pressure and seepage forces in the soil [7–9]. Rainfall-triggered landslides are usually shallow (0.3–2 m) and often driven by two different mechanisms. In the first, the hydraulic conductivity of the weathering profile decreases, creating a perched water flow that is parallel to the slope. This results in a reduction of the shear strength of the soil, which leads to slope failure. In the second mechanism, water from the surface advances on the slope while it is still unsaturated and, in this case, low suction results in a rigid mass slope failure [8,10].

Over the years, scholars have tried to define statistical or empirical correlations between slope failures and rainfall intensity and duration. These relationships are often defined mathematically as rain thresholds that attempt to define the rainfall curve in between the slope's stability and failure zones [11]. Since their inception in Cane (1980) [12], precipitation thresholds have been established for Rainfall Intensity-Duration (I-D), Cumulative Rainfall Event-Duration (E-D), Cumulative Rainfall Event Intensity (E-I), Rainfall Cumulative (R), and other relationships between intraday rain and antecedent rainfall [13].

Inevitably, these thresholds are highly influenced by temporal and spatial factors such as the location, range of the study area, and the instruments (rain gauges or remote sensors) used to calculate them. To a large extent, in-situ sensors (gauges) have been used to derive rainfall thresholds in various areas of the world. In Indian's Himalayan region, for example, several authors have used in-situ-based data for the definition of rainfall thresholds. These scholars combined intensity-duration thresholds based on the daily rainfall and antecedent rain by aggregating several days in different combinations, such as 2, 3, 5, and 20 days [14–16].

Nonetheless, recent advancements in satellite technologies have been a promising and reliable source of data to map and model susceptibility, hazard, risk, and landslide impacts in various areas of the world. Satellite products such as the Global Precipitation Measurement Mission (GPM), for example, provide rainfall estimates that can help evaluate rain as a landslide trigger at large scales [9,17]. Satellite soil moisture products have also been successfully adapted in various shallow landslide studies. Ray et al. (2007), for example, used moisture settings from the Advanced Microwave Scanning Radiometer-Earth Observing System (AMSR-E) to demonstrate the correlation between moisture conditions, rainfall patterns observed from the Tropical Rainfall Measuring Mission (TRMM), and landslide occurrence. Brocca et al. (2012) used the soil water index (SWI) value derived from the Advanced SCATterometer (ASCAT) to obtain soil moisture indicators that can help predict landslide occurrence. Cullen et al. (2016) developed a shallow landslide index (SLI) derived from the Soil Moisture Active Passive mission (SMAP) and GPM that can be used as a dynamic indicator of the total amount of antecedent moisture and rainfall needed to trigger a shallow landslide in North America.

Various studies have used remote sensors, or a combination of remote sensors and gauges, to derive rainfall landslide thresholds. Brunetti et al. (2021), for example, used GPM, SM2RAIN (Soil Moisture to Rain)—ASCAT rainfall products and daily rain gauge observations from the Indian Meteorological Department to study 197 rainfall-induced landslides. In this instance, results demonstrated that the satellite products outperformed the in-situ sensors due to the better satellite spatial and temporal resolutions [18]. Contrary to these results, M Rossi et al. (2017) described three statistical procedures for defining satellite and gauge threshold methods in central Italy. In this case, the results indicated that the thresholds derived from satellite data were lower than those obtained from gauges as the satellite products underestimated the "ground" rainfall measured by the gauges [19].

Despite these developments, accurate satellite information is sometimes challenged by the area's physical characteristics. This is the case in tropical regions where dense vegetation prevents the instruments from retrieving reliable readings. Complex and heavily vegetated tropical areas usually pose a significant challenge for remote earth observations. For example, exploratory analysis of the expected association between rainfall and soil moisture is not observed when looking at data retrieved from NASA's GPM and SMAP missions in Colombia, South America. Methods such as those described in Cullen et al. (2016) perceive the connection between remotely sensed precipitation and soil moisture content but are useful only for less complex and less vegetated terrains.

Perhaps for this reason, various physical, and not satellite-based, rainfall thresholds have been determined for the Colombian region. Marin et al. (2021) for example, applied a physically based model to define rainfall intensity-duration thresholds and predict areas susceptible to shallow landslides in tropical mountain basins of the Colombian Andes [20].

Nonetheless, to the knowledge of the authors, as of the time of this writing, satellite-based landslide rainfall thresholds for this area are not available.

Framework

This work proposes a framework for the development of a rainfall-triggered landslide threshold derived from a system that incorporates satellite observations and physical ground instrumentation at regional and global scales. As previously stated, remotely sensed antecedent soil moisture conditions for the range of the study area are not available. Therefore, we derive soil wetness conditions using a four-year (2016–2019) rainfall time series from The Climate Hazards Group Infrared Precipitation with Stations (CHIRPS).

First, we establish a relationship between the two successive rainfall episodes and the dry period in between for the entire series. Subsequently, we test the performance of these parameters in conjunction with static factors in a logistic regression. Here we leverage the information provided in inventories, the expert opinion of specialists in the region, and the various heuristic, statistical, and deterministic analyses in C.J. van Western (2008) [21] to determine the static factors that should be incorporated into the analysis.

Dividing the dataset into training and testing sets, we then formulate a relationship between the slope angle and the new dynamic parameters expressed as a threshold that once exceeded will trigger a landslide. Later, we compare the performance between the proposed threshold and the well-known event-duration (E-D) method for the training and testing sets. Finally, we represent the proposed threshold values in a hazard map of the study area.

2. Data

2.1. Geological and Climatological Settings

The country of Colombia is located in the northwest part of South America. It lies between latitudes 4°S and 12°N and longitudes 67°W and 79°W bordering Panama to the northwest; Brazil and Venezuela to the east; and Ecuador and Peru to the south. It is surrounded by the Caribbean Sea to the east and the Pacific Ocean to the west. According to the latest Colombian census of 2018, there are approximately 45,500,000 million residents in the country, many of whom inhabit the interior mountain ranges [22].

The Colombian Mountain ranges, or the Colombian Andes, are the result of subduction-accretion in the triple-plate junction of the Nazca, Caribbean, and South American plates, where tough terrains with steep slopes dominate the landscape. Moreover, these hillslopes are overlain by thick weathering profiles that consist of residual soils, saprolites, and weathered rock horizons. They are divided into three mountain ranges, known as the Western, Central, and Eastern mountains. Geomorphologically, it is a diverse country divided into five distinct natural regions: The Andean Mountain range, the Caribbean Sea coastal region, the Pacific Ocean coastal region, the lowlands of the Amazon, and the Orinoco region. The Andean Mountains have only 33% of the landmass, but 78% of the national population. This region also presents 92.5% of the total landslide reports, where 92% are triggered by rainfall [23].

Colombia has a tropical climate that exhibits a highly intermittent rainfall behavior in space and time. The mean annual precipitation over the whole country is 2830 mm [24]. Along the Andean region, the mean annual precipitation ranges from 1000 to 3000 mm, where strong topographic features produce local atmospheric circulations and convective rainstorm events, which commonly trigger flash floods and landslides [25]. Additionally, the double migration of the Inter-Tropical Convergence Zone (ITCZ) strongly controls the bimodal regime of rainy periods. This pattern is mainly observed in central and western Colombia where rainfall peaks are prominent in the months of March-April-May (MAM) and September-October-November (SON) [26]. The interannual rainfall variability is controlled mainly by El Niño/Southern Oscillation (ENSO). During El Niño, there is a decrease in precipitation and mean monthly flow of Colombia's rivers as well as a decrease in soil moisture, whereas La Niña is generally associated with positive precipitation anomalies [27].

The Colombian territory is a complex tectonic setting under extreme hydro-meteorological conditions. These conditions create a multi-hazard landscape where geomorphological phenomena such as landslides, debris flows, earthquakes, and flooding are frequent. The Andes mountains that cross Colombia from south to north, for example, make the country extremely vulnerable to landslides. On Saturday, 1 April 2017, the Mocoa landslide is just one illustration of a heavy rainfall-triggered event in Colombia. More than 300 people were killed, 400 were injured, and the city was destroyed [28]. Later in the same month, on 20 April 2017, another landslide, in a different region, resulted in at least 17 deaths and dozens injured [29]. Although significant, these are just a few examples of the vast occurrence of landslide events in the region; Aristizábal and Sanchez (2020) recorded 32,022 landslides occurring in the 116 years between 1900 and 2016. Out of these events, 93% occurred in the Colombian Andes region, and 92% were landslides triggered by rainfall.

2.2. Landslide Inventory

El Servicio Geológico Colombiano (SGC) (https://simma.sgc.gov.co, accessed on 5 May 2020) and the National University of Colombia (https://geohazards.com.co/, accessed on 5 May 2020) keep landslide records in inventories for the Colombian territory. Both these registers are used in this work in such a way that one helps complement the other for the basis of evaluating static parameters. There were 396 landslides between the years 2016 and 2019 in Colombia, 346 triggered by rain, some of which resulted in numerous fatalities. For this work, each landslide event listed in the inventories is converted into a Geo Information System (GIS) point format for geo-location and corresponding retrieval of parameters.

Figure 1 below shows landslide events occurring between 2016 and 2019 as listed in these inventories. A bimodal annual pattern of landslide occurrence is observed in the data, with two maximum peaks in May and October-November, reflecting the strong influence of rainfall on landslide occurrence.

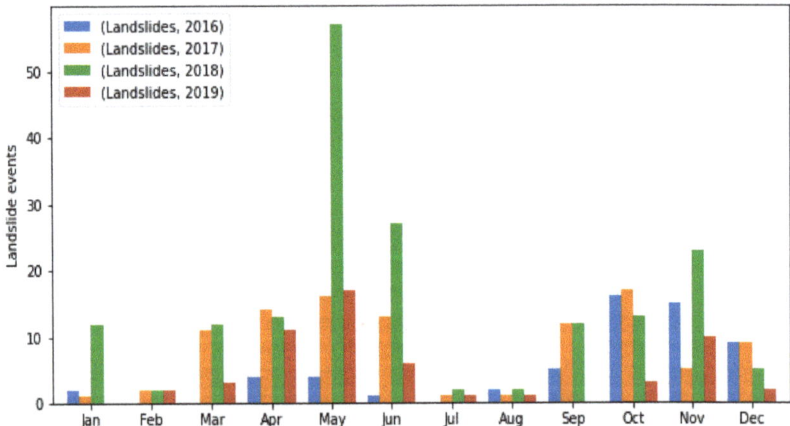

Figure 1. Landslide events in Colombia between 2016 and 2019.

2.3. Static Parameters

Slope gradient is invariably one of the most significant physical parameters as shown by the various heuristic, statistical, and deterministic analyses in C.J. van Western (2008). In this work, slope angles are derived from NASA's Shuttle Radar Topography Mission (SRTM) V3 [30]. SRTM provides digital elevation information at a 30 m resolution. We obtain the slope angle measurements using Google Earth Engine and extract slope angle values for each of the landslide event points.

Landcover of the study area was attained from the Copernicus Global Land Cover Layers: CGLS-LC100 Collection 3 (CGLS). The Copernicus information compiles a global landcover map at 100 m spatial resolution based on a database of landcover training sites and ancillary datasets. Copernicus released a landcover map for each year in the 2015–2019 period. Copernicus provides up to 80% accuracy over all the years of analysis [31]. Landcover information was also retrieved using Google Earth Engine and similarly corresponding event values are set aside in the landslide inventory database.

Major soils distribution for the study area was acquired from the United States Department of Agriculture (USDA), the National Cooperative Soil Survey (NCSS) [32] in GeoTiff format, and the Colombian Instituto Geográfico Agustín Codazzi (IGAC) [33] in shapefile format. These maps provide information at scale: 1:5,000,000 and 1:100,000, respectively. We use both these datasets to corroborate and complement soil characteristics but select major categories as described in the USDA format. We use ArcGIS 10.7 to extract values corresponding to each one of the landslide events.

2.4. Dynamic Parameters

It is well established that rainfall intensity and duration can trigger shallow landslide activity [8,34]. Tropical areas that climatologically are more exposed to extreme rainfall events are usually located at the top of mountainous catchments. Not surprisingly, civil infrastructure and human settlements are very common along the gentle surfaces of these mountainous areas, which are composed of alluvial sediments in the lower catchments close to the mouth.

Although various satellite products are available to retrieve rainfall information globally, accurate measurements in tropical regions are usually hindered by dense vegetation that prevents the instruments from retrieving reliable readings. For this reason, data that are from a hybrid between satellite and ground instrumentation present an opportunity to remotely study rainfall dynamics in these areas.

The Climate Hazards Group Infrared Precipitation with Stations (CHIRPS) effort was developed by the United States Geological Survey (USGS) and the Climate Hazard Group (CHG) at the University of California Santa Barbara (USCB) to support the United States Agency for International Development Famine Early Warning System Network (FEWS NET). CHIRPS builds on various thermal infrared (TIR) precipitation products, calibrates global cold cloud duration rainfall estimates with the Tropical Rainfall Measuring Mission Multi-Satellite Precipitation Analysis version 7 (TMPA 3B42 v7), and uses various interpolated gauge products. CHIRPS data are available globally from 6-hour to 3-month aggregates at a 0.050×0.050-degree spatial resolution [35]. CHIRPS has been validated for Colombia. At least 338 rain gauges from the Colombian Instituto de Hidrología, Meteorología y Estudios Ambientales (IDEAM) were used to compare CHIRPS results, and correlations of R = 0.97 were reported accompanied by the benefit of relatively low latencies [35].

For this work, CHIRPS Daily: Version 2.0 Final was retrieved using Google Earth Engine. Daily precipitation between 2016 and 2019 was obtained for each latitude and longitude landslide point in the inventory. Table 1 below describes all the static and dynamic datasets used in this study, and Figure 2 shows them in the study area.

Table 1. Datasets used in this work.

Data Type	Dataset	Resolution/Accuracy	Extent	Source
Slope	SRTM	30 m	Global	NASA/USGS/JPL-Caltech
Landcover	Copernicus	100 m	Global	Copernicus
Soils	USDA	1:5,000,000	Global	USDA
Rainfall	CHIRPS	$0.05° \times 0.05°$	Global	UCSB/CHG
Landslide inventory	Universidad Nacional De Colombia/SGC	Various mapping scales and survey types	National	Universidad Nacional De Colombia/SGC

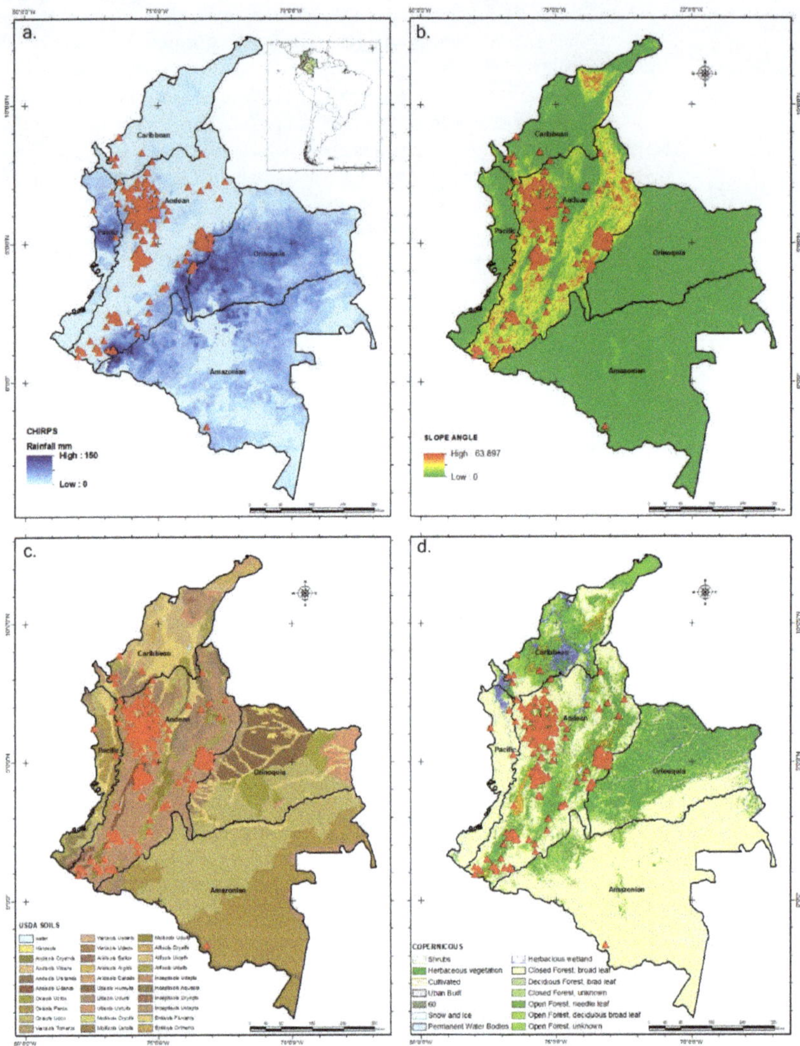

Figure 2. Datasets used for the study area and corresponding landslide events happening between 2016 and 2019: (**a**) CHIRPS rainfall (mm)—used in the logistic model and LTF method; (**b**) slope angle derived from SRTM V3—used in the logistic model and LTF method. (**c**) USDA Soils Classification—used in the logistic model; (**d**) Copernicus Landcover 2019—used in the logistic model.

The landslide inventories used in this work are maintained by the SGC and the Universidad Nacional and are two of the most complete in Colombia. For the 2016–2019 period, they list 346 rainfall triggered landslides. Both inventories are complemented with field campaigns that provide detailed descriptions of the events. However, landslide inventories are usually not free of inaccuracies that challenge the use of statistical methods to determine relationships between factors that can lead to a landslide event. This process becomes more intricate at the regional level because the related spatial and temporal data are extracted from remote sensors. Figure 3 below shows rain triggered landslide occurrence as listed in the inventory in the study area and the respective average cumulative rainfall for those events derived from CHIRPS. A remarkable peak in the number of landslides and rainfall in May 2018 is observed. This period corresponds to a strong La Niña ENSO event as reported in the Multivariate ENSO Index (MEI) [26].

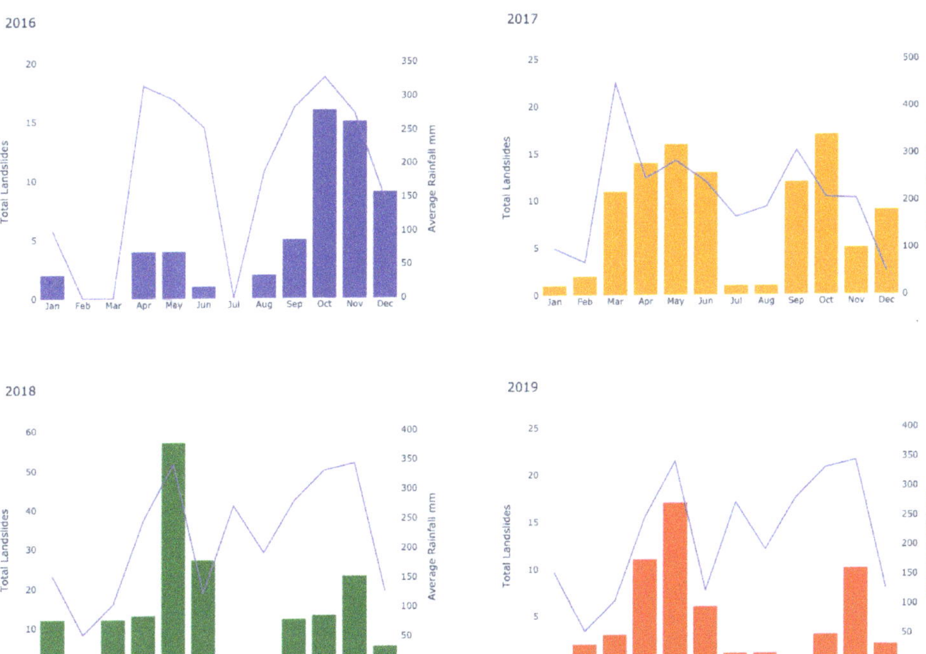

Figure 3. Monthly rainfall-triggered landslide events during the study period 2016–1019 and monthly cumulative rainfall data for the same timeframe from CHIRPS.

3. Methods

3.1. Dynamic Factors Modeling—Soil Moisture and Rainfall

For the basis of this analysis, we assume that soil moisture content for a specific location is dependent on the amount and duration of the rainfall that occurs before the landslide event and in the non-rain (dry) periods between rainfall episodes. By following the successive rainfall/dry periods, we can characterize how much moisture is left in the soil directly before the landslide. A rainfall series for each landslide event location for 4 years (starting on 1 January 2016 and ending on 31 December 2019) is used to find precipitation characteristics for every two successive rainfall episodes as described in Table 2 and illustrated in Figure 4 below.

Table 2. Variables created to track rainfall days, dry days, and rainfall amounts leading to a landslide event in the inventory.

Variable Name	Represents
PR1	Total number of days of Precedent Rainfall event
RS1	Rainfall Sum during PR1 in mm
PR2	Total number of days of rainfall event following PR1
RS2	Rainfall Sum during PR2 in mm
DT	Non-rainfall day period between two consecutive rainfall events

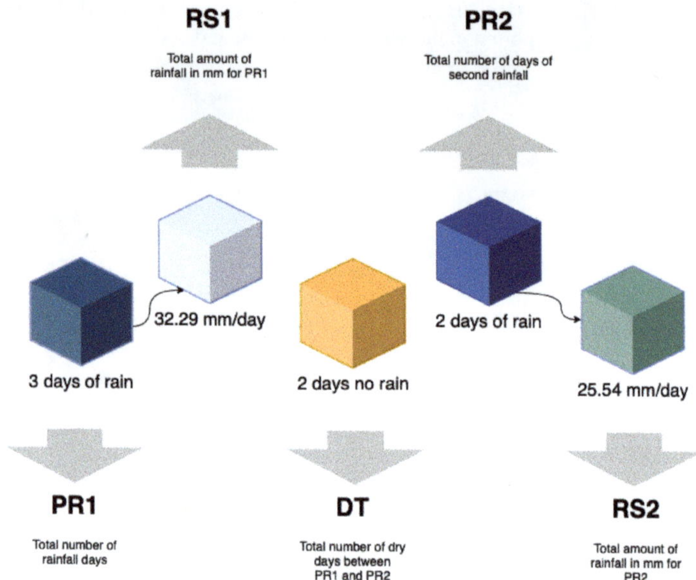

Figure 4. An example of rainfall characteristics for two successive rainfall/dry periods at one landslide location. PR1 and RS1 represent the very first rainfall recorded at one location: (PR1) number of days that the rainfall (intermittently lasted) and (RS1) the total amount of rainfall in that period. Then, there was no rain for 2 days (DT), but later it rained for 2 extra days (PR2) amounting to the triggering rainfall (RS2). These rainfall/dry pairs start at the beginning of the rainfall time series (01/01/2016) until the time of the landslide as listed in the inventory. Each pair is stored as a "running" data line where PR2 and RS2 become PR1 and RS1 of the following line.

As we investigate the rainfall/dry periods for each landslide location throughout the 4 years, we obtain approximately 125,901 data lines of RS1, RS2, PR1, PR2, and DT pairs. Table 3 below shows an example of the successive rainfall/dry periods for one location.

Table 3. Example of rainfall/dry periods for one location from the beginning of the time series until the moment of the event—one landslide.

PR1 (Days)	RS1 (mm/Day)	DT (Days)	PR2 (Days)	RS2 (mm/Day)
1	34.30	2	1	34.30
1	34.30	1	3	52.31
3	52.31	27	1	11.87
1	11.87	3	1	6.00
1	6.00	2	1	8.17
...	...	...	...	...

Consequently, we look for the two successive rainfall occurrences that lead to each one of the landslide events in the inventory. We investigate the rainy and dry periods right before the event and generalize them in Equation (1) as Pre-event Moisture Content.

$$PMC = \frac{RS1}{PR1 * DT} \tag{1}$$

where PMC is an indicator factor of the moisture content in the soil before the landslide event and it is deemed as a first landslide triggering condition. Theoretically, any high values of RS1, associated with low PR1 and DT values, will increase the probability of a landslide event and therefore increase the PMC value.

The second condition is the triggering rainfall episode when the landslide happened, expressed in Equation (2) as Rainfall Sum Trigger.

$$\text{RST} = \frac{\text{RS2}}{\text{PR2}*\text{DT}} \tag{2}$$

where RS2 and PR2 are as described above. Mathematically, the RST value will be higher for higher RS2 associated with a low PR2 value. With a higher RST value, the soil will receive a higher amount of rainfall in a shorter period, therefore increasing the possibility of a landslide event.

3.2. Logistic Modeling—Dynamic and Static Factors

In this study, we define rainfall-triggered landslide susceptibility as the probability that a downslope movement would occur with rainfall as a trigger, but we also consider the pre-existing factors that underlie the physical characteristics of the region. We employ a stochastic approach in the form of logistic regression to understand the relationships between the newly derived dynamic variables, (PR1, RS1, PR2, RS2, and DT), some of the most important static factors (Slope, Landcover, and Soil type), and the landslide probability for the region. We use a logistic regression algorithm because it is a straightforward method for when some of the predictor variables are categorical [36]. The logistic regression method differs from the linear one in that the dependent variable is binary and represents a probability of the outcome. In our case, 1—landslide and 0—no landslide. This dichotomy makes the logistic regression approach a simple and helpful process to delineate landslide susceptibility as demonstrated by the numerous applications in various settings [9,36–46].

The logistic model can be, in its simplest form, as indicated in Equation (3), where the "log-odds" is denoted by Z, which depends linearly on the dynamic and static variables. The coefficients of this dependency are estimated in Section 4.1, using maximum likelihood estimation (MLE).

$$\text{Probability (Landslide} = 1) = \frac{1}{1 + e^{-z}} \tag{3}$$

In Equation (3), as Z increases, the probability of a shallow landslide event increases and vice-versa.

Four-years of rainfall data on each landslide location resulted in several (125,901) sets of dynamic values as explained above in Section 3.1. However, many of these values did not lead to the landslide event. For the proper setting of logistic regression, there were too many no-landslide to landslide values; the ratio was 93:7. Performing a logistic regression under these conditions would result in an erroneous predictive accuracy; therefore, we employed an under-sampling technique known as Synthetic Minority Under-Sampling Technique (SMOTE). SMOTE is an under-sampling method where the larger class is under-sampled by randomly selecting examples [47]. The resulting data for modeling had a 1:1 ratio of 1—landslide to 0—no-landslide. We use a Python subroutine that calculates the logistic regression for all dynamic and static variables.

Using training data for model assessment is not acceptable for evaluating the model's performance in machine learning processes. Hence, we divide the landslide and no-landslide datasets into two groups: one for training, with approximately 70% of the data, and one for testing with the remaining 30%, as shown in Table 4 below.

Table 4. Training and testing data—under sampling SMOTE.

Event	Cases	Under SMOTE	Cases Training	Cases Testing
1	346	346	241	105
0	125,901	346	238	108
Percentage	100%	100%	~70%	~30%

The receiver-operating characteristic plot (ROC) is a solid logistic regression model performance evaluation, as it classifies the sensitivity (true positive rate) and the specificity (true negative rate) of the model. A model with high discrimination ability possesses high sensitivity and specificity simultaneously. Therefore, the area-under-ROC (AUC) grants the evaluation of thresholds in the model.

3.3. Landslide Thresholds

3.3.1. Landslide Triggering Factor—LTF

We integrate the two activating conditions PMC and RST into a landslide triggering factor LTF and normalize it by the slope angle parameter under the premise that as the slope increases, the amount of rainfall necessary to trigger a landslide event will decrease and vice-versa. In this manner, the LTF simply describes the relationship between the expected antecedent moisture content in the soil PMC and the rainfall RST that, according to the landslide inventory, triggered the event. The LTF can be seen in its equation form in Equation (4) below:

$$\text{LTF} = \frac{\text{PMC} + \text{RST}}{\text{Slope}} \tag{4}$$

where PMC and RST are as described above.

Equations (1), (2), and (4) were developed in this work to satisfy the mathematical probability of a landslide occurring. With the LTF, when the dynamic variables PMC and RST for a given slope angle increase, the probability of a landslide event increases as well. In contrast, as the dry period between the two successive rainfall occurrences (DT) increases, the probability of a landslide decreases.

3.3.2. Cumulative Rainfall Event-Duration (E-D) Threshold

The intensity or duration of rainfall periods can be used to establish statistical or empirical correlations to shallow landslide occurrence. These relationships are often expressed as rainfall thresholds that once exceeded, will cause a landslide [48]. The intensity-duration (I-D) threshold, for example, relates these quantities in a power law ($I = \alpha D^{-b}$, where I is the rainfall mean intensity in mm/h; D is the duration of the rainfall event expressed hourly or daily; α is the scaling constant; and b represents the slope) [9].

Contrary to the I-D method, the event-duration (E-D) threshold does not consider the intensity, but instead, the total or cumulative rain. The cumulative precipitation event E is expressed as the total rain from beginning to end that triggered the landslide event. In the literature, various authors have expressed the E-D threshold in a similar power function to the I-D as shown in Equation (5a), where E is measured in (mm) for the full precipitation period; D is the duration of the rainfall episode, and it is expressed hourly or daily; α is the rainfall depth; and γ is the threshold inclination generated from the power function regression [13]. Similarly, the E-D threshold linear form was introduced by Valenzuela et al. (2019) as per Equation (5b) below:

$$E = \alpha D^{-\gamma} \tag{5a}$$

$$E + \alpha D = C \tag{5b}$$

where E is the cumulative rain in (mm); D is the rainfall period expressed hourly or daily; α is the slope of the fitted line; and where C represents the y-intercept [49].

In this study, we evaluate both the power and linear functions but implement the E-D threshold in its linear form as a baseline to evaluate the LTF threshold performance. As previously noted, the landslide inventories used in this work do not provide the time, but the date of the event. In the lack of rainfall intensity information, the E-D threshold is most useful in this application. We also divide the dataset into a training group of approximately 65% and a testing group of approximately 35% for evaluating each of these methods.

3.4. Assumptions

Studying rainfall-triggered landslides at large scales using remote sensors and databases presents various challenges due to the non-structured nature of the information. In this study, it is important to highlight the following:

1. Both the logistic regression model and the LTF threshold are data driven approaches.
2. We assume that pore pressure increases due to liquefaction of the material.
3. We suppose that soil moisture content for a specific location is dependent on the amount and duration of the rainfall that occurs before the landslide event and on the non-rain (dry) period between the two events. We do not incorporate root uptake or evapotranspiration information.
4. Daily rainfall temporal resolution is used because the landslide inventory lists a date, not a timestamp of when the event occurred.
5. It is understood that a landslide changes the physical characteristics of the area. It may flatten the slope and remove the weak soil layer, which in return may change the landcover. Under these circumstances, the calculated LTF for that location no longer applies because conditions have changed.

4. Results

4.1. Logistic Regression—Dynamic and Static Factors

The estimated coefficients for each factor affecting the "log-odds" using the maximum likelihood estimate (MLE) in the logistic model is presented in the Z-factor for Equation (3) above:

$$Z = -(0.33PR1 - 0.01RS1 + 1.87DT + 0.33PR2 - 0.01RS2 + 0.20slope + 0.90SoilType) \tag{6}$$

In Equation (3), P tends to 1 as Z in Equation (6) increases. As Z increases, the probability of a shallow landslide event tends towards 1 (landslide). In contrast, as Z decreases, the probability tends to 0 (no-landslide). The relationship between the coefficients and the probability is expressed as positive (landslide) or negative (no-landslide).

Landcover and soil type are categorical variables with six and five categories, respectively, for the study area and are described in depth in H. Eswaran (2016) and Smets (2020), respectively. Because this is a data-driven model, we do not assign any weight to any soil or landcover type. Instead, we create dummy variables for each category at each location. After segregating meaningful types of these categories using the RFE process, we are left with those that either influence the event to occur or not. Those with positive coefficients have a positive relationship to event causation and vice versa.

Validation results of the logistic regression demonstrate that the model can correctly predict 73% of the cases using the newly created dynamic variables. Figure 5a shows the ROC curve that helps summarize the model predictability based on the area under the curve (AUC). The AUC reflects the probability that a randomly chosen actual landslide incident will have a high chance of classification as being an actual event. The model has an AUC of 0.73, suggesting good data-driven predictability for landslide events. Figure 5b shows the confusion matrix for the model. These values help explain the precision (true positive)/(true positive + false positive), recall (true positive)/(true positive + false negative), and F1-measure, which combines the precision and recall. These measurements can be seen in Table 5 below.

Table 5. Precision, recall and F-measure.

Class	Precision	Recall	F1-Score
0	0.71	0.79	0.75
1	0.75	0.67	0.71

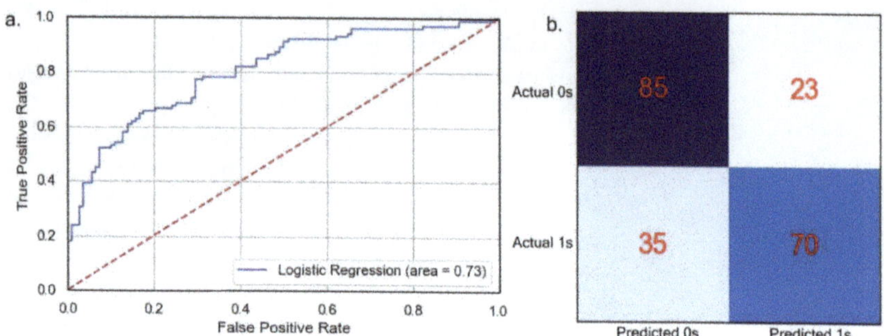

Figure 5. (**a**). ROC performance; (**b**). model confusion matrix.

The odds ratio (OR) demonstrates how a one-unit increase or decrease in a variable affects the odds of initiating a landslide event. In Table 6 below we see that for one unit increase in PR1, we expect that there is 0.718 times increase in the odds of a landslide happening. The other independent variables can be interpreted the same way.

Table 6. Odds Ratio.

Variable	Coefficients	OR
PR1	−0.33	0.718
RS1	0.01	1.013
DT	−1.87	0.153
PR2	−0.33	0.715
RS2	0.01	1.011
Slope	−0.20	0.851
Soil Type	−0.90	0.404

4.2. Landslide Triggering Factor (LTF) Thresholds—Dynamic Factors and Slope

The LTF value for the training (65%) and testing (35%) datasets is assessed using Equation (4) for every two rainfall/dry periods at each landslide location from 1 January 2016 to 31 December 2019. LTF values that are associated with an actual landslide event are set as thresholds for the corresponding slope angles. Equation (7) and Figure 6 below show the inverse function that relates the LTF and the Slope angle.

$$\text{LTFThreshold} = 42.91\,\text{Slope}^{-1.068} \tag{7}$$

where the determination coefficient (R^2) for the LTF threshold-Slope relationship as per Equation (7) is $R^2 = 0.836$. Figure 6 shows that the LTF-Slope angle relation rapidly changes in smaller slope angles, whereas it barely fluctuates in larger ones. Slopes greater than 25° show an asymptote average threshold value of 1.227 with a standard deviation of 0.104.

The LTF-slope relationship is congruent with the physical mechanisms that drive rainfall-triggered landslides. Physically, a rainfall-triggered landslide develops as the moisture content of the soil and its pore pressure increase. The slope fails when the driving force along the slip failure surface is greater than the shear strength of the material and its cohesion [50]. In steep-slope angles, the weight of the soil along the slope surface is already significant. In this case, a small amount of additional water weight is likely to initiate a failure. This explains the low variation of the LTF in the high slope ranges. Alternatively, for small slope angles, the soil's weight component along the slip surface that contributes to the slope failure is relatively small, and therefore, a substantial additional water weight component is needed to initiate a failure. Hence, for small slope ranges, the LTF exhibits more heightened variations.

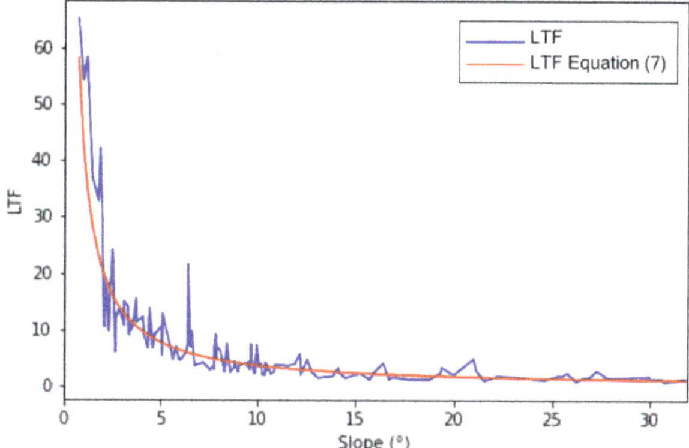

Figure 6. Calculated LTF threshold values (**blue**) fitted with Equation (7) (**red**) for the training dataset.

Landslide Triggering Factor Error—False Positive Rate (FPR)

The false positive rate is defined as the probability of falsely rejecting the null hypothesis. In our case, it represents the negative cases in the data that were mistakenly reported as positive or where the LTF threshold was exceeded but there was no landslide event. Consequently, we use the FPR concept to check the adequacy of the LTF threshold value as per Equation (8) below:

$$\text{FPRLTF} = \frac{\text{LTFOver}}{\text{TotalPeriods} - 1} \tag{8}$$

where LTFOver are the times in the rainfall series where the LTF value exceeded the established LTF threshold. TotalPeriods are the total number of rainfall/dry periods from the first day of the rainfall series to the actual landslide event. In this case, the maximum observed FPR value for all training cases was 0.271, demonstrating a 73% overall accuracy. Similarly, the testing dataset, 35% of the cases, presented a maximum FPR of 0.274 showing an overall performance of 72.6%.

4.3. Accumulated Rainfall Duration (E-D) Threshold—Dynamic Factors and Slope

The E-D values in the power and linear forms for the training dataset are shown in Figure 7a,b below. In these Figures, the E-D threshold values are fitted with the corresponding E-D linear and power forms as per Equations (5a) and (5b) above.

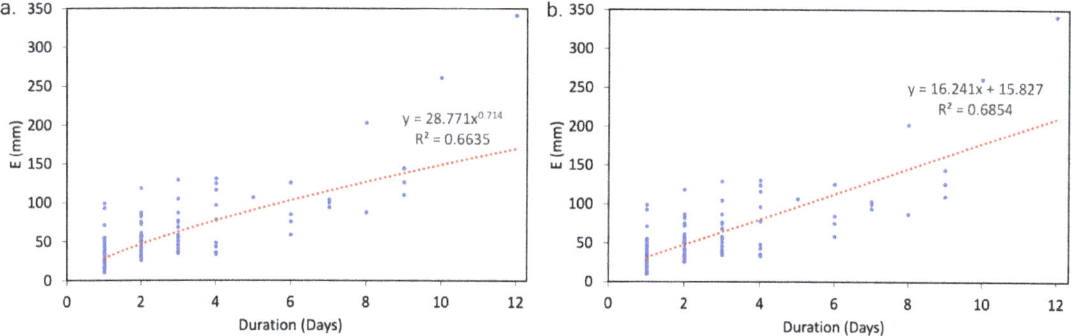

Figure 7. E-D threshold values for the training dataset with corresponding equations and determination coefficients: (**a**) Equation (5a)—power form and (**b**) Equation (5b)—linear form.

For each instance, the corresponding curve becomes as per Equations (9a) and (9b).

$$EDThreshold = 28.77\ D^{0.714} \tag{9a}$$

$$EDThreshold = 15.83 + 16.24\ D \tag{9b}$$

Equations (9a) and (9b) define whether a landslide event happens or not. Here, for an observed accumulated rainfall (E) for a time duration (D), if E is greater than the threshold, a landslide should be expected. However, in this case, the linear form exhibits a higher determination coefficient (R^2 = 0.68), thus, we use this form.

Accumulated Rainfall Duration (E-D) Thresholds Error—False Positive Rate (FPR)

We calculate the E-D threshold false positive rate (FPR) using the same approach as for the LTF FPR. In Equation (10) below, we substitute LTFOver for EDOver from Equation (8) above.

$$FPRED = \frac{EDOver}{TotalPeriods - 1} \tag{10}$$

where EDOver are the times in the rainfall series where the E value exceeded the established E-D threshold, and TotalPeriods are the total number of rainfall/dry periods from the first day of the rainfall series to the actual landslide event. In this case, the maximum observed FPR value for all training cases was 0.60.

4.4. LTF Threshold vs. E-D Threshold

It is well known that the antecedent moisture conditions of the soil before a landslide event are critical for landslide initiation. Regardless of the intensity and duration of a rainfall episode, shallow landslides are directly affected by soil moisture conditions [51,52]. Various physically based analyses have demonstrated that slope instability does not only depend on the intensity of the rain or its duration. It is the case that extensive precipitation within a dry period can trigger a landslide as much as a low-intensity rainfall during a wet period [53]. Similarly, pre-existing wet conditions can cause large debris flow during or following a downpour [54].

The E-D method uses the duration and accumulation of the triggering rainfall to establish a threshold value that, once exceeded, will lead to a landslide event. The LTF method, instead, not only considers the triggering rainfall but also evaluates the effects of the preceding precipitation before the triggering rain and the dry period in between the two rainfalls. The maximum observed FPR for the E-D method for the training dataset was 0.60. Conversely, the LTF FPR was 0.271 as noted above.

Figure 8a,b below show the E-D and LTF threshold FPR values for the training and testing datasets. In both cases, the LTF threshold performs better than the E-D threshold for 71% and 81% of the cases, respectively.

The difference in performance between the two thresholds can potentially be explained by the introduction of parameters that simulate the state of the soil before the landslide event and by relating them to the slope inclination. The LTF method considers five dynamic variables (RS1, PR1, RS2, PR2, and DT) that precariously simulate the wet state of the soil before the landslide event affecting the probability of a landslide. This assumption is possible under the notion that the slope angle inclination is inversely proportional to the amount of rainfall necessary to trigger a landslide. Conversely, by design, the E-D threshold does not consider any information about the soil wetness pre-event, therefore limiting its performance.

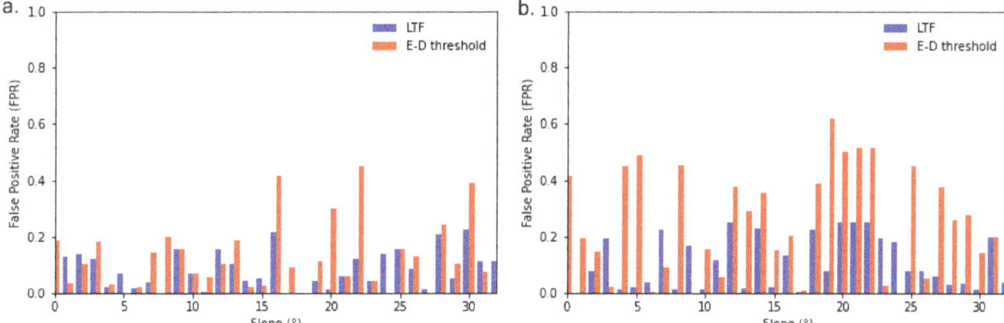

Figure 8. (**a**) E-D and LTF threshold values with corresponding false positive rate for the training dataset (65%) and (**b**) for the testing dataset (35%).

4.5. Landslide Triggering Factor—(LTF) Thresholds Hazard Map

A landslide hazard map that shows the probability of where and when an event would happen, as defined by Guzzetti et al. (2005) [55], can be derived by applying the LTF threshold concept to the slope angle distribution in the study area. From Equation (4) above, we can now derive a dynamic value that can be mapped for all areas where the quantities of both rainfall periods (RS1, RS2), their duration (PR1, PR2), and the dry period in between (DT) exceeds the LTF threshold as follows:

$$\text{DynamicMap (DM)}; \; \text{LTF} > \text{LTFThreshold} \tag{11}$$

By defining DM as in Equation (11):

$$\text{DM} = \frac{\text{RS1}}{\text{PR1}*\text{DT}} + \frac{\text{RS2}}{\text{PR2}*\text{DT}} \tag{12}$$

And then substituting Equations (4), (7), and (10) into Equation (11) and simplifying:

$$\text{DM} > 42.91*\text{Slope}^{-0.068} \tag{13}$$

The DynamicMap (DM) quantities represent the average rainfall in mm/day of the two rainfalls divided by the dry period between the two rainfalls in days. In Figure 9 below, it is anticipated that when the DM quantity exceeds the LTF threshold multiplied by the slope, a landslide should be expected. Areas located in the mountainous Andes region show relative low DM (Equation (12)) values necessary to trigger a landslide. This is evident, as the area is characterized by high slopes. Conversely, areas located in the Caribbean Sea coastal region, the Pacific Ocean coastal region, and the lowlands of the Amazon and Orinoco regions with gentle or very low slopes require higher DM values.

The DM values map highlights the Andes region as the area that is at most risk of landslides with lower DM threshold values. From a risk management perspective, new DM values can be derived from weather forecasts where a landslide should be expected if the new calculated DM values overpass the DM thresholds presented here.

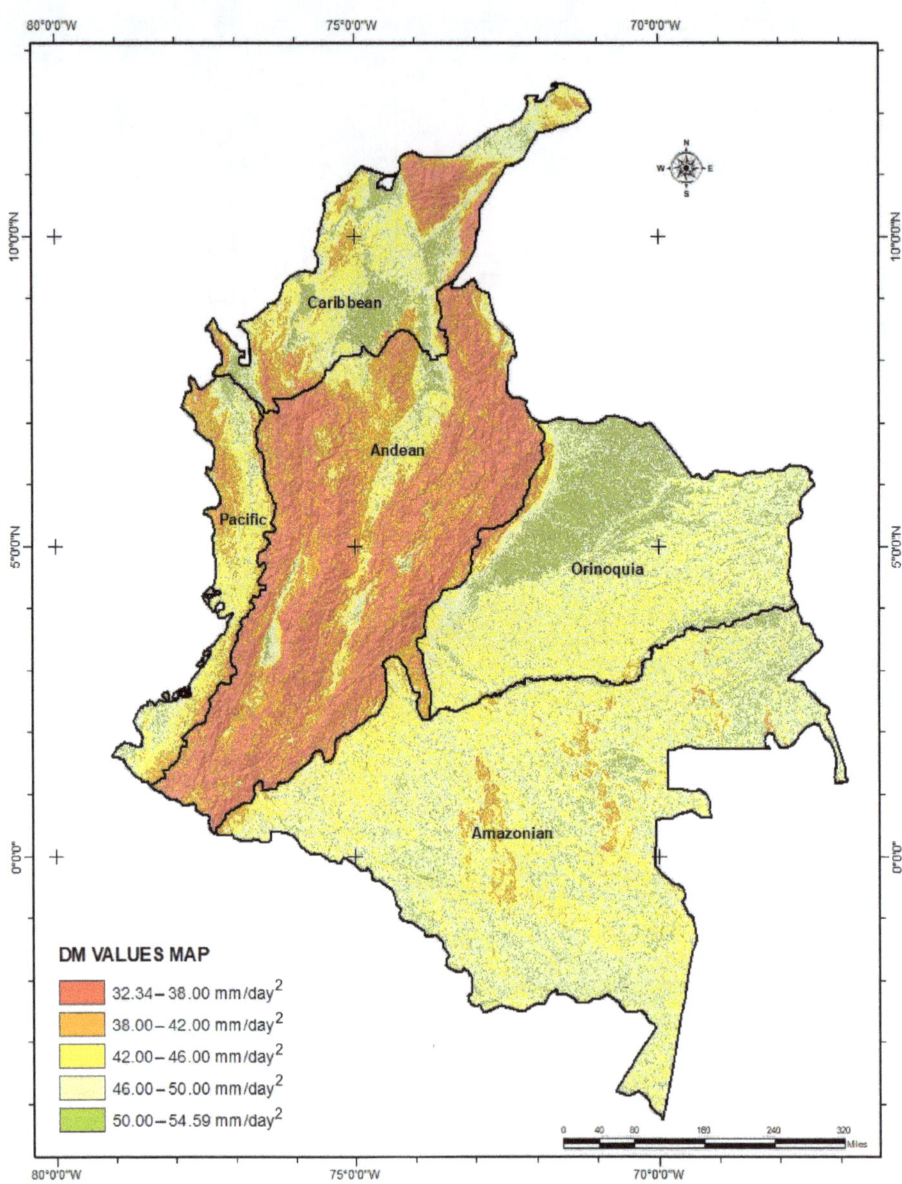

Figure 9. DynamicMap quantities between two rainfall episodes that will trigger a landslide once exceeded at that specific slope angle. Days are expressed as (day^2) as per Equation (12) where the two rainfalls average is measured in mm/day and then are divided by the dry period between the two rainfalls, also measured in days.

4.6. Challenges and Limitations

It is important to highlight the limitations of this work. First, the landslide records on which this work was based do not provide a timestamp of the event. For this reason, a daily average of rainfall estimates is used to develop both the logistic model and the LTF thresholds. Information that reflects the exact time of the event could have a significant impact on the LTF and DM thresholds. CHIRPS, for example, provides rainfall information every 6 h, and having a landslide inventory that gives a time of event could be useful to improve performance.

Second, because of the lack of satellite-based antecedent soil moisture information for the study area, wet and dry periods are used to simulate the effect that antecedent soil wetness would have in landslide initiation. This point itself is a significant assumption, as all the intrinsic physical dynamics of soil moisture conditions are not accounted for. Furthermore, this notion is based on rainfall estimates, and although these approximations have demonstrated high correlations with in-situ gauges, they have their limitations and uncertainties. Consequently, they have a direct effect on the LTF and DM thresholds.

Finally, although the DM map could serve as guidance for the landslide rainfall threshold, it is essential to note that the DM map is based on a 30 m resolution, but many landslides occur at smaller scales.

5. Conclusions

Rainfall-triggered landslides are a significant and constant hazard for the Andes region. This danger, coupled with the lack of availability of on-site instrumentation and the reliability of remotely sensed information, opens the need for more imaginative ways to unravel the problem. We present a data-driven solution in the form of dynamic variables derived from a satellite infrared precipitation and station data, CHIRPS, to simulate soil moisture variations and build a landslide triggering threshold in the Colombian region.

With the assumptions detailed above, we study four years of daily rainfall at 346 landslide events in the region. We focus on the two consecutive rainfall occurrences and corresponding dry periods leading to each one of the landslide events in the inventory. We then use them as dynamic variables. We first investigate the relationship of these rainy/dry periods in a logistic algorithm where results demonstrate acceptable performance of 73%.

Consequently, we take the rainy and dry periods right before the landslide event and simplify them as the PMC value. This value serves as an indicator for the moisture content in the soil before the landslide event. The triggering rainfall episode is then expressed as the RST value. Accordingly, these two factors are normalized with the effect of the slope angle giving rise to the LTF concept.

The LTF model allows for the allocation of threshold values associated with slope angles, and we see that as the slope increases the LTF decreases. The LTF also serves as a guidance for landslide hazards in the region, as areas with lower threshold values and high slopes are at a higher risk of landslides and vice-versa.

Although the simulated LTF lacks details about the complex processes that drive soil moisture mechanics, it attempts to simulate them by including the triggering rain, the antecedent rainfall, and the dry period in between. When the LTF is compared to the E-D threshold, the LTF performs better for 81% of the testing cases.

Although various physically based landslide rainfall thresholds have been developed for the study area, no satellite-based thresholds currently exist. The LTF threshold is one of the first satellite-based thresholds for the Colombian region that attempts to simulate the effect of this parameter in the area.

DM values for the Andes region range between 32 and 46 mm/day^2 mark this as the region with the least precipitation (in two rainfall episodes) necessary to trigger a landslide. And although the DM map could serve as a guide for vulnerability and risk, several challenges should be resolved to "fine-tune" the thresholds. These include introducing a "time of event" parameter and physical or reliable satellite-based antecedent soil moisture information when it becomes available.

Author Contributions: Conceptualization, R.A.S. and C.A.C.; methodology, R.A.S. and C.A.C.; software, C.A.C. and R.A.S.; validation, R.A.S., C.A.C. and E.A.; formal analysis, R.A.S., C.A.C. and E.A.; investigation, R.A.S., C.A.C. and E.A.; resources, C.A.C.; data curation, R.A.S. and C.A.C.; writing—original draft preparation, C.A.C.; writing—review and editing, C.A.C., R.A.S. and E.A.;

visualization, C.A.C. and E.A.; project administration, C.A.C.; funding acquisition, C.A.C. All authors have read and agreed to the published version of the manuscript.

Funding: This publication was made possible by the support of CUNY-Remote Sensing Earth (CUNY-CREST) Institute. Its contents are solely the responsibility of the authors and do not necessarily represent the official views of the CREST Institute and the City University of New York.

Data Availability Statement: The data presented in this study are available on request from the corresponding author.

Acknowledgments: The global soils regions map was obtained courtesy of USDA-NRCS, Soil and Plant Science Division National Agricultural Library.

Conflicts of Interest: The authors declare no conflict of interest. The funders had no role in the design of the study; in the collection, analyses, or interpretation of data; in the writing of the manuscript, or in the decision to publish the results.

References

1. Girty, G.H. Perilous Earth: Understanding Processes behind Natural Disasters, ver. 1.0 Chapter 8 Landslides. 2009. Available online: http://www.sci.sdsu.edu/visualgeology/naturaldisasters/ (accessed on 1 July 2021).
2. Petley, D. Global patterns of loss of life from landslides. *Geology* **2012**, *40*, 927–930. [CrossRef]
3. Kim, H.G.; Lee, D.K.; Park, C. Assessing the cost of damage and effect of adaptation to landslides considering climate change. *Sustainability* **2018**, *10*, 1628. [CrossRef]
4. Cruden, D.J.; Varnes, D.M. Landslides: Investigation and Mitigation. *Transp. Res. Board Spec. Rep.* **1996**, *247*, 36–75.
5. Sidle, R.C.; Ochiai, H. *Landslides: Processes, Prediction, and Land Use*; American Geophysical Union: Washington, DC, USA, 2006; Volume 18. [CrossRef]
6. Kirschbaum, D.B.; Adler, R.; Hong, Y.; Lerner-Lam, A. Evaluation of a preliminary satellite-based landslide hazard algorithm using global landslide inventories. *Nat. Hazards Earth Syst. Sci.* **2009**, *9*, 673–686. [CrossRef]
7. Aristizábal, E.; García, E.; Martínez, C. Susceptibility assessment of shallow landslides triggered by rainfall in tropical basins and mountainous terrains. *Nat. Hazards* **2015**, *78*, 621–634. [CrossRef]
8. Aristizábal, E.; Velez, J.; Martínez, C.; Jaboyedoff, M. SHIA_Landslide: A distributed conceptual and physically based model to forecast the temporal and spatial occurrence of shallow landslides triggered by rainfall in tropical and mountainous basins. *Landslides* **2015**, *13*, 497–517. [CrossRef]
9. Cullen, C.A.; Al-Suhili, R.; Khanbilvardi, R. Guidance index for shallow landslide hazard analysis. *Remote Sens.* **2016**, *8*, 866. [CrossRef]
10. Collins, B.D.; Znidarcic, D. Stability Analyses of Rainfall Induced Landslides. *J. Geotech. Geoenviron. Eng.* **2004**, *130*, 362. [CrossRef]
11. Glade, T.; Anderson, M.; Crozier, M. *Landslide Hazard and Risk*; John Wiley & Sons, Ltd.: Hoboken, NJ, USA, 2004; Available online: https://books.google.com/books?id=UFQk0I4EUiwC&printsec=frontcover&source=gbs_ge_summary_r&cad=0#v=onepage&q&f=false (accessed on 22 March 2022).
12. Caine, N. The rainfall intensity-duration control of shallow landslides and debris flows. *Geogr. Ann. Ser. A Phys. Geogr.* **1980**, *62*, 23–27.
13. Maturidi, A.M.A.M.; Kasim, N.; Taib, K.A.; Azahar, W.N.A.W. Rainfall-Induced Landslide Thresholds Development by Considering Different Rainfall Parameters: A Review. *J. Ecol. Eng.* **2021**, *22*, 85–97. [CrossRef]
14. Dikshit, A.; Satyam, N.; Pradhan, B.; Kushal, S. Estimating rainfall threshold and temporal probability for landslide occurrences in Darjeeling Himalayas. *Geosci. J.* **2020**, *24*, 225–233. [CrossRef]
15. Naidu, S.; Sajinkumar, K.S.; Oommen, T.; Anuja, V.J.; Samuel, R.A.; Muraleedharan, C. Early warning system for shallow landslides using rainfall threshold and slope stability analysis. *Geosci. Front.* **2018**, *9*, 1871–1882. [CrossRef]
16. Mandal, P.; Sarkar, S. Estimation of rainfall threshold for the early warning of shallow landslides along National Highway-10 in Darjeeling Himalayas. *Nat. Hazards* **2021**, *105*, 2455–2480. [CrossRef]
17. Kirschbaum, D.B.; Stanley, T.; Simmons, J. A dynamic landslide hazard assessment system for Central America and Hispaniola. *Nat. Hazards Earth Syst. Sci.* **2015**, *15*, 2257–2272. [CrossRef]
18. Brunetti, M.T.; Melillo, M.; Gariano, S.L.; Ciabatta, L.; Brocca, L.; Amarnath, G.; Peruccacci, S. Satellite rainfall products outperform ground observations for landslide prediction in India. *Hydrol. Earth Syst. Sci.* **2021**, *25*, 3267–3279. [CrossRef]
19. Rossi, M.; Luciani, S.; Valigi, D.; Kirschbaum, D.; Brunetti, M.T.; Peruccacci, S.; Guzzetti, F. Statistical approaches for the definition of landslide rainfall thresholds and their uncertainty using rain gauge and satellite data. *Geomorphology* **2017**, *285*, 16–27. [CrossRef]
20. Marin, R.J.; Velásquez, M.F.; García, E.F.; Alvioli, M.; Aristizábal, E. Assessing two methods of defining rainfall intensity and duration thresholds for shallow landslides in data-scarce catchments of the Colombian Andean Mountains. *Catena* **2021**, *206*, 105563. [CrossRef]
21. van Westen, C.J.; Castellanos, E.; Kuriakose, S.L. Spatial data for landslide susceptibility, hazard, and vulnerability assessment: An overview. *Eng. Geol.* **2008**, *102*, 112–131. [CrossRef]

22. Vallejo-Zamudio, L.E. El incierto crecimiento económico colombiano. *Apuntes Cenes* **2017**, *36*, 9–10. [CrossRef]
23. Aristizábal, E.; Sánchez, O. Spatial and temporal patterns and the socioeconomic impacts of landslides in the tropical and mountainous Colombian Andes. *Disasters* **2020**, *44*, 596–618. [CrossRef] [PubMed]
24. Poveda, G.; Vélez, J.I.; Mesa, O.J.; Cuartas, A.; Barco, J.; Mantilla, R.I.; Mejía, J.F.; Hoyos, C.D.; Ramírez, J.M.; Ceballos, L.I.; et al. Linking Long-Term Water Balances and Statistical Scaling to Estimate River Flows along the Drainage Network of Colombia. *J. Hydrol. Eng.* **2007**, *12*, 4–13. [CrossRef]
25. Álvarez-Villa, O.D.; Vélez, J.I.; Poveda, G. Improved long-term mean annual rainfall fields for Colombia. *Int. J. Climatol.* **2011**, *31*, 2194–2212. [CrossRef]
26. NOAA—Physical Science Laboratory. Multivariate ENSO Index Version 2 (MEI.v2). NOAA ENSO. 2022. Available online: https://psl.noaa.gov/enso/mei/ (accessed on 25 February 2022).
27. Poveda, G. Diagnóstico del Ciclo Anual y Efectos del ENSO Sobre la Intensidad Máxima de Lluvias de Duración Entre 1 y 24 Horas en los Andes de Colombia. *Meteorol. Colomb.* **2002**, *5*, 67–74.
28. El Espectador. Avalancha en Mocoa, una de las Peores Tragedias de 2017. 2017. Available online: https://www.elespectador. com/noticias/nacional/avalancha-en-mocoa-una-de-las-peores-tragedias-de-2017/ (accessed on 19 October 2020).
29. Benfield, A. Global Catastrophe Recap. 2019. Available online: http://thoughtleadership.aonbenfield.com/Documents/20190508 -analytics-if-april-global-recap.pdf (accessed on 4 July 2020).
30. Farr, T.G.; Rosen, P.A.; Caro, E.; Crippen, R.; Duren, R.; Hensley, S.; Kobrick, M.; Paller, M.; Rodriguez, E.; Roth, L.; et al. The shuttle radar topography mission. *Rev. Geophys.* **2007**, *45*, 2. [CrossRef]
31. Buchhorn, M.; Bertels, L.; Smets, B.; De Roo, B.; Lesiv, M.; Tsendbazar, N.E.; Masiliunas, D.; Linlin, L. *Copernicus Global Land Service: Land Cover 100m: Version 3 Globe 2015–2019: Algorithm Theoretical Basis Document*; Zenodo: Geneve, Switzerland, 2020. [CrossRef]
32. Eswaran, H.; Reich, P.; Padmanabhan, E. World soil resources opportunities and challenges. In *World Soil Resources and Food Security*; CRC Press, Taylor and Francis Group: Boca Raton, FL, USA, 2016; pp. 29–52.
33. Instituto Geográfico Agustín Codazzi- Subdirección de Agrología—Grupo Interno de Trabajo Geomática. Mapas de Suelos del Territorio Colombiano a Escala 1:100.000. 31 December 2017. Available online: http://metadatos.igac.gov.co/geonetwork/srv/ spa/catalog.search#/metadata/b857e651-b8d2-4bf2-9e03-41a038c7206a (accessed on 14 August 2020).
34. Lehmann, P.; Gambazzi, F.; Suski, B.; Baron, L.; Askarinejad, A.; Springman, S.M.; Holliger, K.; Or, D. Evolution of soil wetting patterns preceding a hydrologically induced landslide inferred from electrical resistivity survey and point measurements of volumetric water content and pore water pressure. *Water Resour. Res.* **2013**, *49*, 7992–8004. [CrossRef]
35. Funk, C.; Peterson, P.; Landsfeld, M.; Pedreros, D.; Verdin, J.; Shukla, S.; Husak, G.; Rowland, J.; Harrison, L.; Hoell, A.; et al. The climate hazards infrared precipitation with stations—A new environmental record for monitoring extremes. *Sci. Data* **2015**, *2*, 150066. [CrossRef] [PubMed]
36. Gorsevski, P.V.; Gessler, P.E.; Foltz, R.B.; Elliot, W.J. Spatial prediction of landslide hazard using logistic regression and ROC analysis. *Trans. GIS* **2006**, *10*, 395–415. [CrossRef]
37. Guns, M.; Vanacker, V. Logistic regression applied to natural hazards: Rare event logistic regression with replications. *Nat. Hazards Earth Syst. Sci.* **2012**, *12*, 1937–1947. [CrossRef]
38. Thomas, D.R.; Zumbo, B.D.; Dutta, S. On Measuring the Relative Importance of Explanatory Variables in a Logistic Regression. *J. Mod. Appl. Stat. Methods* **2008**, *7*, 4. [CrossRef]
39. Zhu, L.; Huang, J. GIS-based logistic regression method for landslide susceptibility mapping in regional scale. *J. Zhejiang Univ. Sci. A* **2006**, *7*, 2007–2017. [CrossRef]
40. Akbari, A.; Bin, F.; Yahaya, M.; Azamirad, M.; Fanodi, M. Landslide Susceptibility Mapping Using Logistic Regression Analysis and GIS Tools. *Electron. J. Geotech. Eng.* **2014**, *19*, 1687–1696.
41. Regmi, N.R.; Giardino, J.R.; McDonald, E.V.; Vitek, J.D. A comparison of logistic regression-based models of susceptibility to landslides in western Colorado, USA. *Landslides* **2014**, *11*, 247–262. [CrossRef]
42. Lee, S. Cross-verification of spatial logistic regression for landslide susceptibility analysis: A case study of Korea. In Proceedings of the 31st International Symposium on Remote Sensing of Environment, ISRSE 2005: Global Monitoring for Sustainability and Security, St. Petersburg, Russia, 20–24 June 2005; Available online: http://www.scopus.com/inward/record.url?eid=2-s2.0-84879 728712&partnerID=tZOtx3y1 (accessed on 4 March 2020).
43. Kavzoglu, T.; Sahin, E.K.; Colkesen, I. Landslide susceptibility mapping using GIS-based multi-criteria decision analysis, support vector machines, and logistic regression. *Landslides* **2013**, *11*, 425–439. [CrossRef]
44. Pourghasemi, H.R.; Moradi, H.R.; Aghda, S.M.F. Landslide susceptibility mapping by binary logistic regression, analytical hierarchy process, and statistical index models and assessment of their performances. *Nat. Hazards* **2013**, *69*, 749–779. [CrossRef]
45. Shahabi, H.; Khezri, S.; Ahmad, B.B.; Hashim, M. Landslide susceptibility mapping at central Zab basin, Iran: A comparison between analytical hierarchy process, frequency ratio and logistic regression models. *Catena* **2014**, *115*, 55–70. [CrossRef]
46. Ayalew, L.; Yamagishi, H. The application of GIS-based logistic regression for landslide susceptibility mapping in the Kakuda-Yahiko Mountains, Central Japan. *Geomorphology* **2005**, *65*, 15–31. [CrossRef]
47. Chawla, N.V.; Bowyer, K.W.; Hall, L.O.; Kegelmeyer, W.P. SMOTE: Synthetic Minority Over-sampling Technique. *J. Artif. Intell. Res.* **2002**, *16*, 321–357. [CrossRef]

48. Segoni, S.; Rossi, G.; Rosi, A.; Catani, F. Landslides triggered by rainfall: A semi-automated procedure to define consistent intensity–duration thresholds. *Comput. Geosci.* **2014**, *63*, 123–131. [CrossRef]

49. Valenzuela, P.; Zêzere, J.L.; Domínguez-Cuesta, M.J.; García, M.A.M. Empirical rainfall thresholds for the triggering of landslides in Asturias (NW Spain). *Landslides* **2019**, *16*, 1285–1300. [CrossRef]

50. Mathew, J.; Babu, D.G.; Kundu, S.; Kumar, K.V.; Pant, C.C. Integrating intensity-duration-based rainfall threshold and antecedent rainfall-based probability estimate towards generating early warning for rainfall-induced landslides in parts of the Garhwal Himalaya, India. *Landslides* **2014**, *11*, 575–588. [CrossRef]

51. Glade, T.; Crozier, M.; Smith, P. Applying probability determination to refine landslide-triggering rainfall thresholds using an empirical 'Antecedent Daily Rainfall Model. *Pure Appl. Geophys.* **2000**, *157*, 1059–1079. Available online: http://link.springer.com/article/10.1007/s000240050017 (accessed on 14 August 2014). [CrossRef]

52. Liao, Z.; Hong, Y.; Wang, J.; Fukuoka, H.; Sassa, K.; Karnawati, D.; Fathani, F. Prototyping an experimental early warning system for rainfall-induced landslides in Indonesia using satellite remote sensing and geospatial datasets. *Landslides* **2010**, *7*, 317–324. [CrossRef]

53. Godt, J.W.; Baum, R.L.; Chleborad, A.F. Rainfall characteristics for shallow landsliding in Seattle, Washington, USA. *Earth Surf. Processes Landf.* **2006**, *31*, 97–110. [CrossRef]

54. Baum, R.L.; Godt, J.W. Early warning of rainfall-induced shallow landslides and debris flows in the USA. *Landslides* **2009**, *7*, 259–272. [CrossRef]

55. Guzzetti, F.; Stark, C.P.; Salvati, P. Evaluation of flood and landslide risk to the population of Italy. *Environ. Manag.* **2005**, *36*, 15–36. [CrossRef] [PubMed]

remote sensing

Article

Targeted Rock Slope Assessment Using Voxels and Object-Oriented Classification

Ioannis Farmakis [1,*], David Bonneau [1], D. Jean Hutchinson [1] and Nicholas Vlachopoulos [2]

1   Department of Geological Sciences and Geological Engineering, Queen's University,
    Kingston, ON K7L 3N6, Canada; david.bonneau@queensu.ca (D.B.); hutchinj@queensu.ca (D.J.H.)
2   Department of Civil Engineering, Royal Military College of Canada, Kingston, ON K7K 7B4, Canada;
    vlachopoulos-n@rmc.ca
*   Correspondence: i.farmakis@queensu.ca; Tel.: +30-698-410-1585

Abstract: Reality capture technologies, also known as close-range sensing, have been increasingly popular within the field of engineering geology and particularly rock slope management. Such technologies provide accurate and high-resolution n-dimensional spatial representations of our physical world, known as 3D point clouds, that are mainly used for visualization and monitoring purposes. To extract knowledge from point clouds and inform decision-making within rock slope management systems, semantic injection through automated processes is necessary. In this paper, we propose a model that utilizes a segmentation procedure which delivers segments ready to classify and be retained or rejected according to complementary knowledge-based filter criteria. First, we provide relevant voxel-based features based on the local dimensionality, orientation, and topology and partition them in an assembly of homogenous segments. Subsequently, we build a decision tree that utilizes geometrical, topological, and contextual information and enables the classification of a multi-hazard railway rock slope section in British Columbia, Canada into classes involved in landslide risk management. Finally, the approach is compared to machine learning integrating recent featuring strategies for rock slope classification with limited training data (which is usually the case). This alternative to machine learning semantic segmentation approaches reduces substantially the model size and complexity and provides an adaptable framework for tailored decision-making systems leveraging rock slope semantics.

Keywords: 3D point cloud; voxels; supervoxels; rock slope management; classification; knowledge extraction; semantics; object-oriented; change detection

Citation: Farmakis, I.; Bonneau, D.; Hutchinson, D.J.; Vlachopoulos, N. Targeted Rock Slope Assessment Using Voxels and Object-Oriented Classification. *Remote Sens.* **2021**, *13*, 1354. https://doi.org/10.3390/rs13071354

Academic Editor: Daniele Giordan

Received: 22 January 2021
Accepted: 25 March 2021
Published: 1 April 2021

Publisher's Note: MDPI stays neutral with regard to jurisdictional claims in published maps and institutional affiliations.

1. Introduction

Reality capture technologies, such as both terrestrial and aerial laser scanning and dense stereo-matching (also known as close-range sensing [1]), have become increasingly popular for visualization and monitoring purposes for geohazard assessment along rock slopes. Such technologies provide accurate and high-resolution (a few centimeters) n-dimensional spatial representations of our physical world, known as 3D point clouds. Many advances in the utilization of 3D point clouds have occurred over the last fifteen years [2,3]. These advances have predominantly focused on survey planning and optimization, pre-processing, registration, and change detection among periodically acquired datasets providing critical geometrical and geographical information of rock slopes. Such information includes: the exact change positioning [4–6], change volume (loss/gain) and shape [7,8], and motion kinematics [9,10]. These derivatives have proven quite useful for enriching the failure inventory building pipelines with valuable additional information with regard to landslide risk management. In addition, a significant amount of the workflow from data acquisition to processing and change detection has been automated [11]. There are, however, still efficiencies in automation that can be realized.

As CS discipline has been advancing rapidly, geoscientists have been interested in automated semantic labelling of 3D point clouds depicting natural scenes of geological and geotechnical engineering interest. As such, there has been a push to work within a multidisciplinary framework with a view to incorporate such computer advances within the geological realm. Although semantic segmentation is a well-investigated field in 2D image analysis [12–14] (including rasterized 2.5D representations such as DEMs), sound preliminary results for slope-scale landslide analysis purposes are now being published [15]. However, 3D point cloud semantic labelling is a relatively unexplored research area.

In rock slope management, only minor interest has been shown in knowledge extraction from 3D point clouds. A common yet very time-consuming practice includes the manual annotation of the content of a 3D point cloud. Masking processes have been adopted for semantic injection to point cloud time-series [8]. However, due to the amount of filtering and editing required, coupled with the users' subjectivity, and the dynamic nature of a certain setting it might not always be a practical solution. Moreover, in datasets that do not include colour information the task becomes even more challenging. As a result, there is a great need to automate such processes to speed up current analysis frameworks and allow practitioners to interact with the computer more efficiently and make interpretations more quickly. This requires the integration of semantic segmentation approaches to extract the appropriate information for a certain task. However, the semantic concepts that are attached to case studies can vary depending on site-specific reasoning (e.g., considering a rock outcrop as the object, or its discontinuity planes, overhanging blocks, eroded areas.), and thus a single model cannot directly satisfy all the objectives. The engagement of semantic meaning into virtual 3D natural scenes including different geomaterials (rock, talus, soil, vegetation), landslide elements (i.e., scarp, deposition zone, toe, overhanging blocks), and other geomorphological features such as rock outcrops and debris channels is critical for an enhanced landslide risk assessment and management. The missing parts required for the completion of this enhanced management framework based on object identification are: interoperability, integration, and automation.

In this paper, an object-oriented knowledge-based semantic segmentation framework utilizing a voxel-based point cloud clustering approach [16] is proposed. The goal is to extract segments of the voxel grid (supervoxels or objects) in a way that the information is both transferable and reproducible, and one that permits flexible usage to benefit different application objectives. The approach aims to extract meaningful semantic information from a rock slope by replicating expert reasoning and knowledge. The final objective is to integrate spatio-semantic reasoning to landslide risk management frameworks by means of a geo-database that would directly link GIS concepts to 3D point clouds. The knowledge extraction phase includes the partitioning of the dataset into semantic objects and their characterization by means of geometrical, topological, and contextual descriptors, followed by a site-specific set of informed sequences, filters and rules used for classification. At this point, the lack of ground-truth reference datasets for training and validation discourages supervised learning in rock slope point cloud semantic segmentation. Object-oriented analysis is expected to bridge the gap towards the development of semantically rich point clouds of rock slope settings. It proposes a framework for the extraction of informative features and the incorporation of simple knowledge-based rules without overcomplicating the task. This study provides insights into the reliability challenges in rock slope semantic segmentation due to the absent of solid methodologies for generating and ensuring the quality of reference datasets and the suitability of traditional performance assessment protocols.

2. Current State and Related Works

Due to the increasing use of 3D point clouds in geosciences research, a wide range of very-high-resolution information has become available, including geographical, spectral, intensity, and full waveform data sets. However, point cloud analyses for geo-engineering purposes typically aim to exploit geographical information (XYZ) by means of geometric and topographic signatures using low-level local descriptors at user-defined point neighbourhoods. To semantically label complex natural scenes, local descriptors are usually assigned to the neighbourhoods' origin points which are then fed to ML classifiers. Low-level descriptors refer to features that can be understood by an end-user without any previous processing knowledge, but do not carry any semantic meaning. Such descriptors (also known as features or attributes) are typically computed based on PCA by encoding the XYZ point sets into a 3D structure tensor (3×3 covariance matrix), and calculating its eigenvalues (λ_1, λ_2, λ_3) and eigenvectors [17]. Other features, such as intensity and/or colour, can be used for specific analyses but are rarely applied since intensity requires a series of non-trivial corrections to be applied in order to be consistently reliable and the latter is not a usable discriminator except where contrast results from different rock or soil formations, the products of rock weathering or vegetation cover.

The geometric and topographic low-level local descriptors are products of the eigenvalues and eigenvectors derived from PCA applied on the XYZ space of the point neighbourhoods. Such descriptors have been commonly accepted by the geosciences community and the expeditious calculation of them has been recently enabled within the popular open-source 3D point cloud processing software CloudCompare v.2.10. These geometric descriptors are: normalized eigenvalues (p_1, p_2, p_3), omnivariance, linearity, planarity, sphericity, anisotropy, eigenentropy, and sum of eigenvalues (definitions of these parameters can be found in [18]). Additional eigen-based descriptors are the parameters of slope and aspect that are computed based on the orientation of the eigenvector corresponding to the lowest eigenvalue (normal vector). The subsequent ML-based classification of the 3D point cloud can be conducted based on low-level local descriptors calculated within one of the three types of point neighbourhoods (Figure 1) listed below:

1. Single-sized point neighbourhoods;
2. Multi-sized point neighbourhoods; and,
3. Adjusted-sized point neighbourhoods.

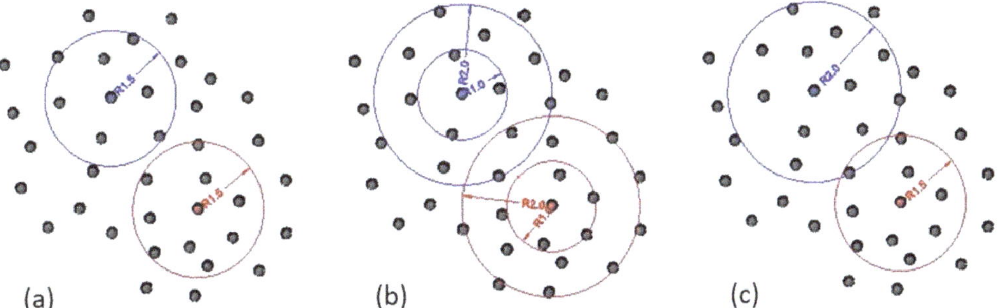

Figure 1. Schematic representation of different types of point neighbourhoods used in point cloud featuring approaches: (**a**) single-sized neighbourhoods calculate features within a fixed volume around individual points; (**b**) multi-sized neighbourhoods calculate features within multiple volumes, around individual points; (**c**) adjusted-sized neighbourhoods calculate features within a volume adjusted to the local geometry of individual points.

In geosciences research involving scenes of geo-engineering interest, only the first two types of point neighbourhoods, as shown in Figure 1a,b, have been considered so far in the literature. In particular, [19] proposed the CANUPO methodology that includes a SVM [20] classification based on the dimensionality (normalized eigenvalues) of single-sized point neighbourhoods. The size of the point neighbourhood is determined based on the analysis of the balanced accuracy of the SVM classifier over a range of different sizes. The CANUPO approach has had high prediction scores among classes such as defining ground and vegetation as well as fine- and coarse-grained debris areas [21].

More recently, [22] employed a RF [23] classification based on a large number of low-level local descriptors derived from multi-sized point neighbourhoods (Figure 1c) utilizing an approach adopted in point cloud processing (e.g., [24–27]). In particular, they examined the suitability of geometric, topographic, intensity, and difference descriptors to the classification of areas of rock, talus, vegetation, and snow on rock cuts adjacent to highways in Colorado, USA. The descriptors used were calculated within point neighbourhoods of ten different sizes between 0.1 m and 5 m. Their analysis of the aforementioned four classes (bedrock, talus, vegetation, and snow) found that a combination of geometric and topographic descriptors generates high discriminating power between bedrock and talus. In detail, the descriptors used are: normalized eigenvalues (p_1, p_2, p_3) as geometric descriptors, as well as the mean, standard deviation, skewness, and kurtosis of slope as topographic descriptors, resulting in a total of 70 feature vectors. The calculation of the above slope statistics requires the normal vector calculation for individual points in each neighbourhood, based on another smaller neighbourhood. The performance of that model was compared to the SVM-based CANUPO model showing 0.07, 0.1, and 0.14 higher F_1-score for vegetation, talus, and bedrock classes, respectively. F1-score is a weighted average of precision and recall and ranges up to 1, with higher values indicating better performance. The formulation of the different evaluation metrics is provided in 5.1.

The third type of point neighbourhoods (adjusted-sized: Figure 1c) has been used by [18] in an urban scene classification problem. This approach incorporates the estimation of the optimal neighbourhood size for each point. The optimization is achieved by minimizing the Shannon entropy (a measure of unpredictability) [28] within each point's neighbourhood through the evaluation of a range of sizes. This approach inspired the voxel size selection process incorporated in the proposed method and is also considered as a featuring strategy for the ML model in the comparison section.

An alternative approach published by [29] includes the application of an object-oriented framework in the field of landslide geomorphology. They investigated a rotational landslide through object-based erosion monitoring on a soil slope, utilizing geometrical information derived from TLS point clouds. The process was initiated with a primary over-segmentation of the scene by means of k-means clustering at the point level, followed by a seeded region growing algorithm with randomly selected seed points. In addition to the clustering process, a 2D (x, y coordinates) maximum distance to the seed point threshold was applied in order to keep the segments (objects) small. The subsequent object classification is performed based on a RF classifier using 43 object descriptors as well as expert-based topological refinement rules for corrections, achieving an overall F1-score of 0.82.

The work by [30] on the development of a point cloud based rockfall hazard assessment methodology also includes a morphology classification component. In their study, the authors proposed an unsupervised rule-based rock slope classification scheme based on a decision tree applied to a grid structure projected to the best-fit plane of the entire point cloud. The scheme includes classes such as intact, closely- and widely- spaced, and fragmented rock, as well as talus and overhangs based on each cell's (0.05 m) slope angle and roughness (i.e., low roughness cells are classified as either intact rock or talus based on a slope threshold). However, the classification result was only qualitatively evaluated and thus no performance metrics are available.

3. Materials and Methods

The proposed object-oriented framework is focused on addressing the inherent lack of structure and semantic meaning of 3D point clouds, as pointed out by [29]. The primary objective of the model is to replicate the human perception. This is achieved by homogenizing the raw 3D point cloud to delineate perceptually meaningful segments or semantic object primitives (homogenous and meaningful areas with respect to the geometry and topography of the whole scene) using an unsupervised segmentation process. These segments aim at balancing the conflicting goals of reducing the complexity of a scene while avoiding under-segmentation and can be essentially extracted by any segmentation algorithm. Semantic objects can be combined in application-dependent classes within the subsequent knowledge-based classification task substituting the initial mapping unit (e.g., spherical point neighbourhoods or voxels). Many methods for object recognition rely on the organization of the scene into semantic objects since they are better aligned with edges than a sphere or cube. This is essentially the principal of the object-oriented model, leveraging the new properties that these newly formed object primitives yield. The methodology can be summarized in three steps: voxelization, scene homogenization, and classification. A detailed schematic representation of the entire proposed semantic segmentation workflow is provided in Figure 2.

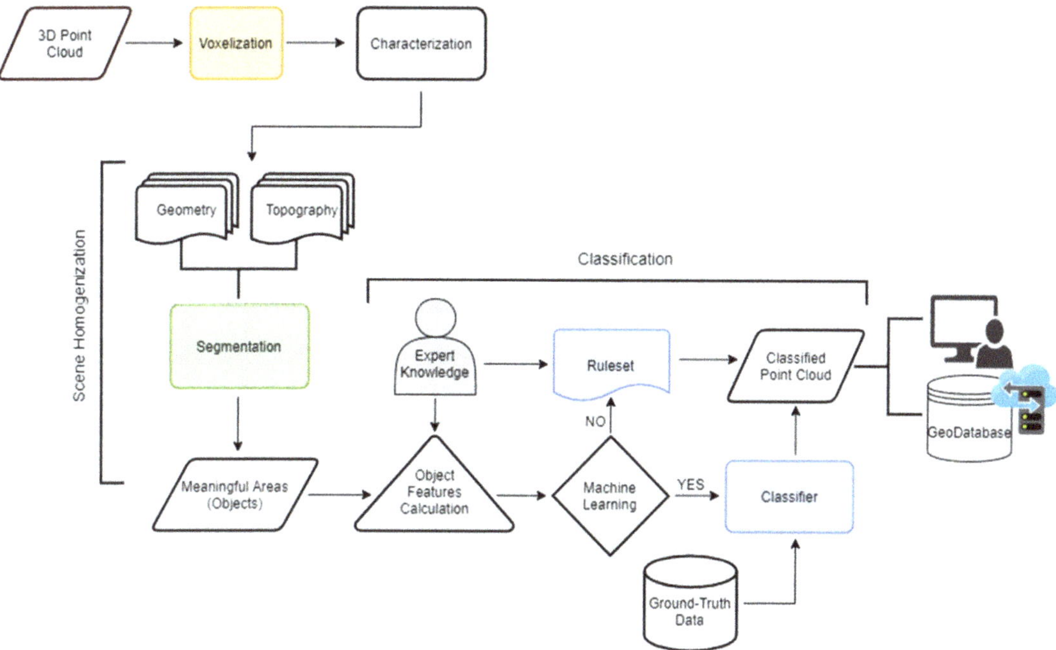

Figure 2. The flowchart representing the proposed semantic segmentation framework. The three main processes of the workflow are: voxelization (yellow), scene homogenization (green), and classification (blue), as discussed in Sections 3.2–3.4, respectively.

3.1. Study Dataset

A section of a steep natural slope in the White Canyon, British Columbia, Canada as shown in Figure 3 is investigated for the purposes of this study. The terrain mainly consists of large rock exposures and debris accumulations on a 500 m high slope above the railway line running adjacent to the Thompson River. The setting, without vegetation, and with a clear line-of-sight for the scanner, is an ideal case to demonstrate the identification of the two major material classes and the associated geomorphological features observed along natural rock slopes. Large and long debris channels are present between the rock outcrops which generate regular rockfalls. Both rockfalls and debris flows can impact the railway line, which is protected by rocksheds, ditches, and slide detector fences as shown in Figure 3. As part of the Canadian Railway Ground Hazard Research Program (RGHRP), the site has been scanned multiple times each year since 2012 [31]. Each scanning campaign includes multiple vantage points, and the different point clouds are aligned based on the ICP algorithm [32] and the fine registration routine in RiScan Pro software following a manual coarse registration. The final dataset is then resampled using a space-based resampling algorithm resulting in an approximately evenly distributed point spacing of 5 cm. The latter aims to provide a standardized distribution of the quality of the extracted local information within the dataset.

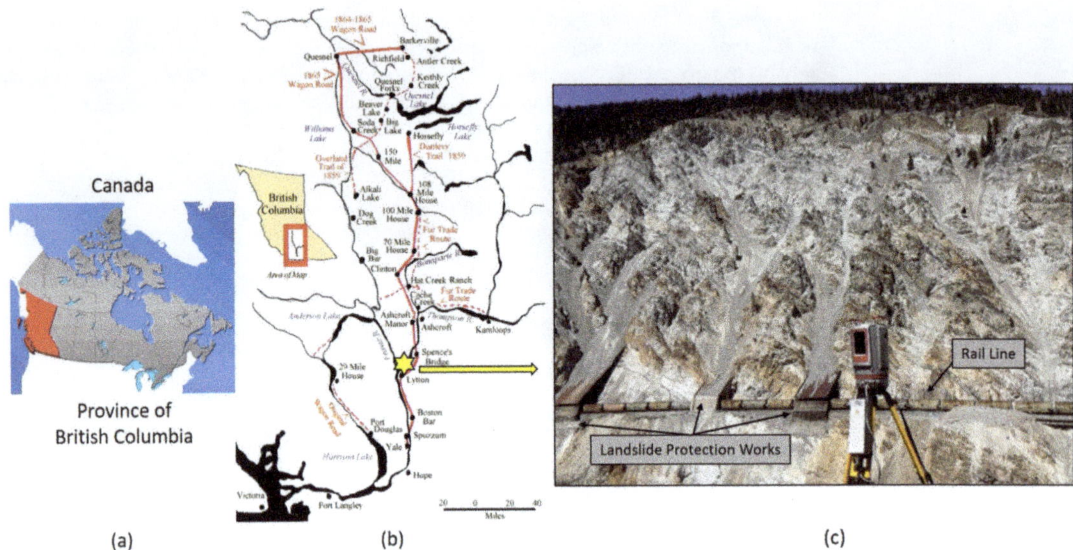

Figure 3. Study area (**a**) Province of British Columbia (BC); (**b**) White Canyon Site Location within BC; (**c**) Picture of the Scheme 2019.

3.2. Voxelization

The point cloud is first stored into a voxel grid, using an octree as shown in Figure 4. The octree is a 3D data structure within which the bounding box of the dataset (root node) is recursively subdivided into eight child voxels. The size of the initial root node is technically defined by the bounding box of the input point cloud. From the resulting child voxels of each subdivision level (known as depth level), only those containing points proceed to further sub-division, while the rest are rejected. The sub-division continues until a termination criterion is met. There are several termination criteria that can be used such as: the voxel size, the depth level, or the minimum number of points per voxel. The advantage of voxelization is that it provides the dataset with an organization and structure which enables neighbourhood searches by means of adjacency graph representations. Voxel adjacency in the 3D space can be modeled in two ways: (i) shared facets (6-connectivity), and (ii) shared facets, edges, and vertices (26-connectivity; Figure 5).

In the methodology proposed in this paper, the termination criterion is based on the voxel size and is defined by minimizing the mean voxel eigenentropy via the Shannon entropy [28], similar to the definition of optimal point neighbourhoods used by [18]. This defines the resolution of the extracted information which technically means that structures smaller than this size are not "visible". This optimization procedure aims to increase the probability that the XYZ information is distributed along the voxel grid in a way that local geometric differences are highlighted with the minimum resolution cost. However, user interpretation regarding the desired level of detail is still a key factor that controls the selection of the voxel size range to be optimized.

3.3. Characterization

As in the case of an image where each pixel is characterized by the RGB values, this methodology aims to extract robust descriptors to characterize each leaf voxel within the generated 3D structure. The information can be propagated in the same way to different levels of the octree (Figure 4). The descriptors used in this method do not contain spectral information but rather are focused on utilizing exclusively XYZ-based geometric and topographic information. In order to provide a framework dedicated to the analysis of natural terrain of geo-engineering interest, the local geometry is expressed by means of dimensionality, as used previously in natural terrain classifications [19] and the topography through orientation, which is key information in any geological investigation.

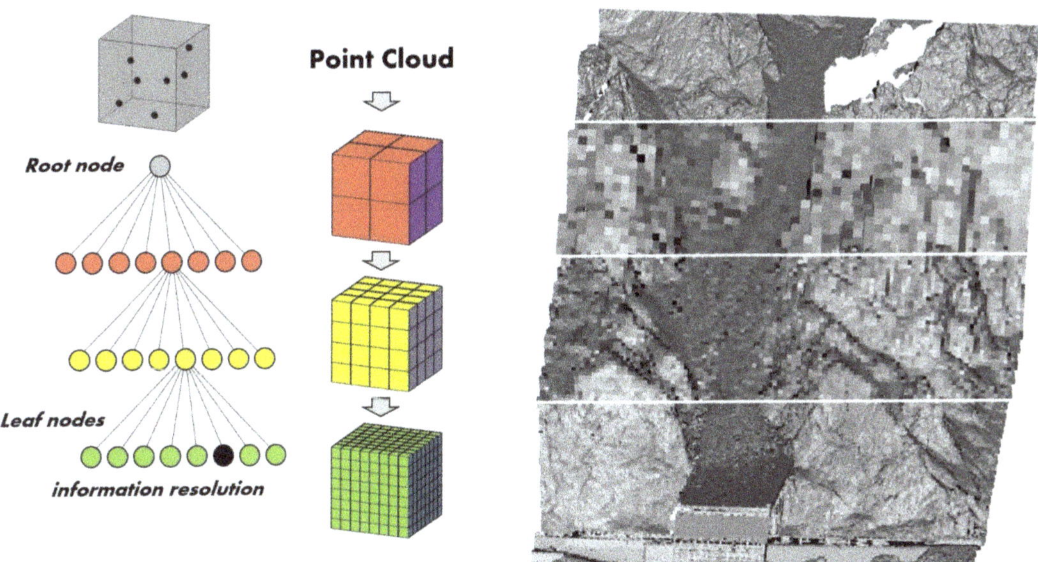

Figure 4. Representation of the octree data structure and the effect of resolution on the spatial information. The size of the leaf nodes (green) represents the resolution of the final voxel grid created based on the octree data structure.

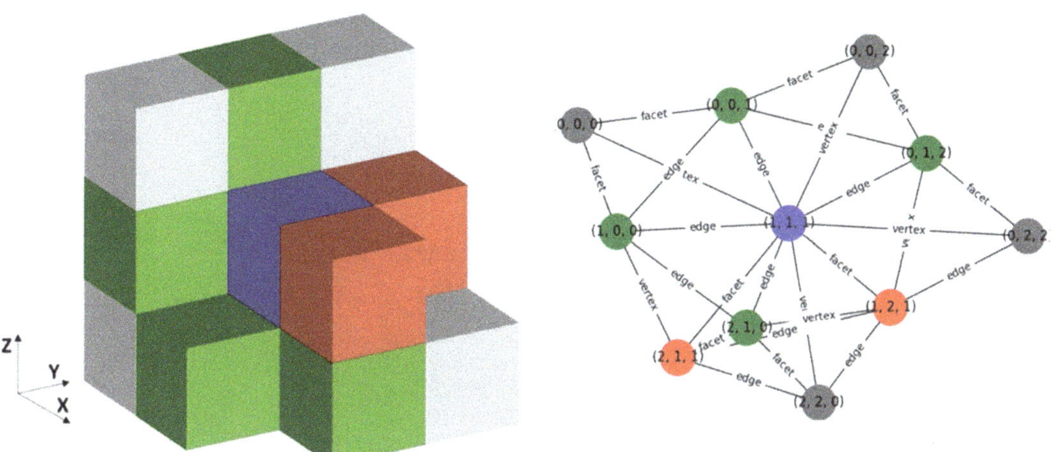

Figure 5. Voxel adjacency graph representation of a given voxel (blue). The different adjacency types of a voxel in 3D space are shown using different colours: shared facet (red), edge (green), vertex (grey).

The normalized eigenvalues (p_1, p_2, p_3) and the normal vector ($\vec{n}_x$, $\vec{n}_y$, $\vec{n}_z$) are calculated for each voxel describing its dimensionality and orientation, respectively. This process includes the application of SVD and PCA to the covariance matrix (Q) calculated at the point level as follows:

$$Q = \frac{\sum_{i=1}^{n}\left(X_i - \overline{X}\right)\left(X_i - \overline{X}\right)^{T}}{n-1},\tag{1}$$

Figure 6 provides a schematic representation of the descriptors used to express the geometry and topography of the scene. The relative proportions of the eigenvalues indicate whether the content of a voxel is closer to being 1-, 2-, or 3-dimensional thereby defining the dimensionality. The orientation of the point set included within a voxel can be represented by the slope and aspect using an equal area stereonet plot. The dataset is therefore characterized by means of geometry and topography and provided with structure. The feature vectors are then normalized in a [0:1] range using the Min-Max normalization method to prevent outweighing and to equalize their contribution to the following object partitioning process:

$$X_i = \frac{X_i - \min(X)}{\max(X) - \min(X)},\tag{2}$$

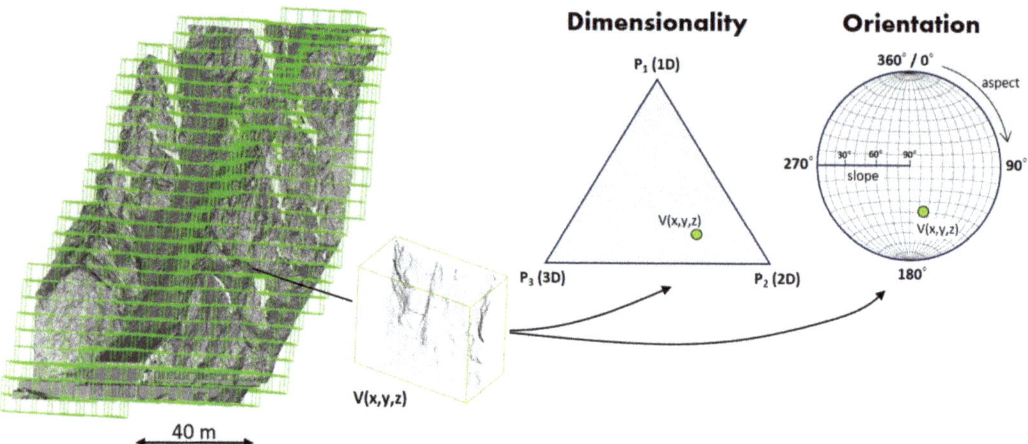

Figure 6. Illustration of the proposed descriptor extraction process for point cloud characterization by means of local geometry and topography. Non-empty voxels allow for the calculation of the desired descriptors at the point level.

3.4. Scene Homogenization

The clustering of a scene into homogenous segments (segmentation) is a fundamental process in image understanding with a variety of algorithms proposed for this specific task (i.e., chessboard, region merging, model-fitting, k-means, hierarchical, graph cut), mainly for pixel grids. In this study, the objective was to integrate data-driven segmentation into the object-oriented model. For this reason, we employed a non-seeded region merging procedure which starts with the initial voxels and proceeds to a pairwise merging of connected voxels at each iteration, forming an assembly of semantic objects, the so-called supervoxels (Figure 7). This step aims to provide an initial over-segmentation while simultaneously generating new properties to support the subsequent classification step. The main difference when compared to point neighbourhoods and voxels, is that the size and shape of supervoxels are formed according to the content of the scene, and they are unique for each object.

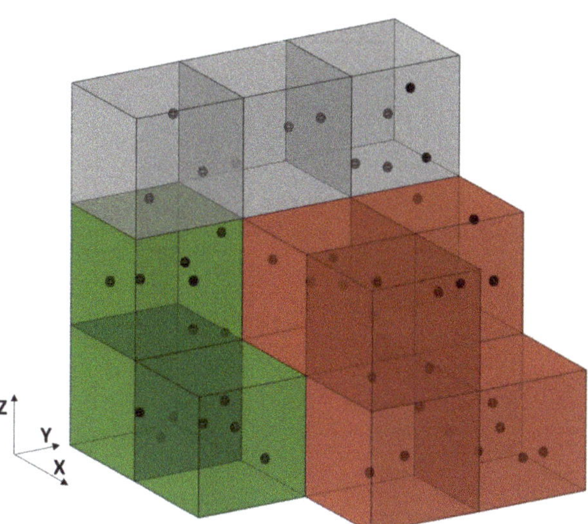

Figure 7. Schematic representation of supervoxel assembly. Different colours (red, green, and grey) represent the supervoxels, cubes represent voxels, and black dots are the points within each voxel.

To ensure that the process is reproducible, and to avoid using seed points, local mutual best-fitting heuristics are employed [33]. Local mutual best-fitting assures that each merge is the best possible in the local vicinity of any object. The merging criterion, similar to [34], evaluates the balanced sum of the overall dimensionality and orientation variation between neighbouring objects and the merge only happens with the neighbour where the minimum score is achieved. This score represents the degree of homogenization of a certain object and is technically a factor that defines the object size. A task-adjustable parameterization methodology for region merging based segmentation of 2D images has been proposed [35]. After each iteration, and when all of the objects have been examined, the overall LV of the scene is calculated and compared to the previous iteration LV. The process terminates when the LV is lower or equal to the previous. In detail, it is assumed that as long as the LV is increasing the homogenization is progressing up to the point where the peak LV value is recorded.

New graphs are thus constructed in order to retain inter-relationships among the objects, as well as link the object with the voxel level (Figure 8). Such representation allows each object to know the spatio-semantic relationships with both its neighbours and its previous level content by means of the initial characterization. This semantic information enables classification of complex scenes by assigning labels to the graph edges and extracting subgraphs via connected components hierarchically based on specific ontologies.

3.5. Classification

After the dataset has been partitioned into semantic objects, they can be classified. The classification can be conducted in two ways within the current object-oriented model. It can be either ML- or knowledge-based. For ML-based classification, the collection and preparation of training, validation, and test datasets is required prior to the application of the model. In contrast, knowledge-based classification is built on domain knowledge applied directly to the extracted objects. In the current knowledge-based classification scheme, object features are first selected based on domain conceptualization and by visualizing each feature layer as a scalar field through an experimental process. These features can vary from the previously described low-level local descriptors to higher-level descriptors such as shape and size measures, neighbourhood relations, or even statistics on finer levels. In contrast to both point neighbourhoods and voxels, the latter is possible within the current methodology since each generated object has a unique shape and size and also retains topological knowledge via the graph representation (Figure 5). Therefore, objects can be characterized based on their geometry, topology, content, and also their semantic relation to their neighbourhood. The use of voxel-based objects is also favoured over point neighbourhoods because it is a time-consuming and computationally expensive process to either perform distance searches within the cloud or manually segment out and correct misclassified points.

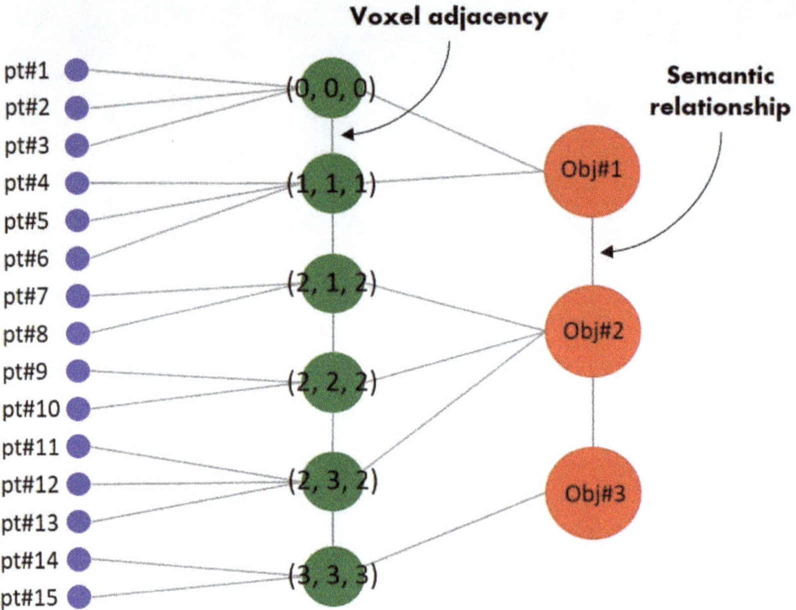

Figure 8. Graph representation of the semantic network with interrelationships within different levels of organization of the 3D point cloud (blue: point cloud, green: voxel grid, red: object assembly).

The object descriptors proposed for the current knowledge-based classification and the reasoning behind their selection are listed in Table 1. The rules are applied hierarchically, and the graphs are updated at each step providing additional semantic information to be used in the establishment of the next step rule. A semantic network is thus constructed in order to represent the scene at multiple levels of detail.

This process allows the model developer to design a domain-knowledge-driven classification schema by means of ontologies specifically focused on the needs of the project. In Figure 9, an idealized schematic representation of a steep railway rock slope ontology is depicted. It includes knowledge formalization regarding different materials, geomorphological features, and constructed structures involved in landslide risk management. The current classification procedure incorporates such an ontology-based conceptualization of railway rock slopes. Connecting ontologies to classification schemas allows information to be generalized more easily. The different aspects of knowledge generated through the process can either be used in rule establishment, be output as knowledge representation at the corresponding level, or both. The latter depends on the desired level of detail and the scope of each project.

Table 1. Presentation and description of both the lower- and higher-level descriptors used within the proposed object-oriented framework.

Descriptor Name	Description	Purpose/Expert Reasoning
Slope	Dip angle between 0–90° to the horizontal plane. Is used to describe slope variations associated with the objects of interest.	Eroded areas such as debris channels usually retain lower slope angles than the surrounding rock mass. Constructed infrastructure is usually vertical or horizontal.
Aspect	Angle between 0–360° (North). Is used to describe the alignment of the various objects based on the azimuth.	Features such as rock benches formed from the bedrock structure are usually oriented perpendicular to the main slope dip direction.
Linearity	The difference of the two major axes of a 3D shape divided by the longest.	Transportation corridor structures appear to be more elongated than geological structures.
2D compactness	Expresses how close to a square is a 3D shape projected to its best-fitting plane.	Constructed infrastructure appears to be more squared-shaped and platy than geological structures.
Relative adjacent class	Expresses the object based to its adjacent classes.	Topological rules used for refinement.
Relative elevation to class	Object's position along Z-axis relative to other class objects.	Topological rules used for refinement.

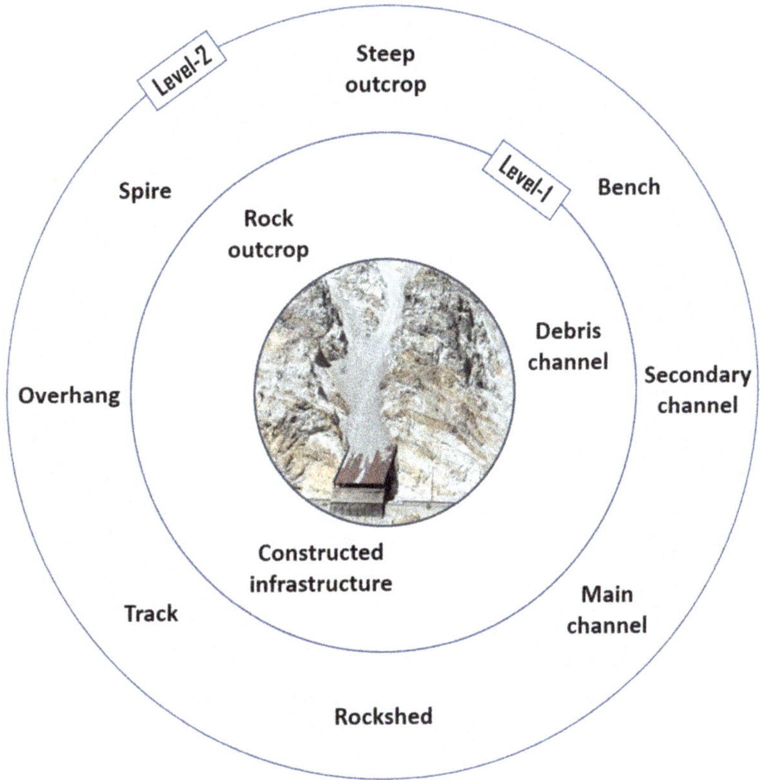

Figure 9. Railway rock slope ontology representation of the White Canyon slopes. Semantic representation of the geometrical knowledge permits definition of materials, geomorphological structures, and anthropogenic objects.

The slope value used for the primary detection of the "debris channel" objects is based on the range of angles of repose related to such structures, linearity is required for the extraction of the elongated rail lines and barrier walls, and the 2D compactness is used to identify the rockshed parts that are more square-shaped and platy than other natural objects. Furthermore, all the classification steps are followed by semantic refinement rules based on spatio-semantic relationships among objects through graph-based clustering and connected components. At a finer level of detail, rock benches within the rock outcrop class are detected using the aspect at the object level. Since such features are usually formed along planes of weakness (i.e., foliation, shear, or fault zones) traversing the slope, they are likely to be oriented sub-perpendicular to the main slope orientation compared to the rest of rock outcrop objects.

4. Analysis and Results

The developed algorithms were used for the semantic segmentation of a section of the railway rock slope described in Section 3.1. The aim was to automatically identify the areas that represent different landslide hazard source zones while ensuring that the model was as simple as possible. The goal was to highlight the potential of simple and explainable automated knowledge-based workflows for effective and efficient multi-hazard monitoring and management. The results of the segmentation process, without using training data, were compared to different ML models trained with limited data from a separate portion of the same site. Selected areas of interest within the slopes include the debris channels and rock outcrops which generate debris flows and rockfalls, respectively, under the appropriate conditions. Such landslide hazards may impact the constructed infrastructure directly during an event when the main channel flushes or an event may fill a secondary channel which in time will contribute debris to the main channel. Therefore, secondary channels or rock benches comprise a lower-granularity class of interest within the semantic network of this site. Constructed infrastructure is another important class since it represents the main element at risk in almost any risk scenario in the study area.

4.1. Metrics

Semantic segmentation results are typically evaluated based on valid labeled data repositories depicting the classes of interest. However, in engineering geology, which deals with natural environments, the availability of such repositories is very limited or non-existent. Although some attempts have been made in regional scale mapping (i.e., SLIDO) utilizing ALS point clouds and satellite imagery, there is no commonly accepted slope-scale annotated 3D point cloud repository at this time. For this reason, the authors suggest that at this point, performing a visual interpretation and assessment together with providing the reader with a detailed representation of the area of interest for clarity, is the most pragmatic approach and better reflects the current state of geoscience practice, as in [30].

A quantitative assessment of the performance of the proposed algorithms was also conducted for completeness. The reference dataset was generated by manual mapping of the study dataset based on gigapixel imagery and point cloud interpretation, as well as field observations. The assessment was conducted at the point level, in order to also account for the voxelization impact and scene homogenization quality. This was based on the following metrics extracted from a confusion matrix which provides visualization of the performance of a classification algorithm by plotting the predicted against the true instances of the four classes (rock outcrop, debris channel, benches/secondary channels, and constructed infrastructure):

$$Recall = \frac{TP}{TP + FN}, \tag{3}$$

$$Precision = \frac{TP}{TP + FP}, \tag{4}$$

$$F1 - score = \frac{2TP}{2TP + FP + FN}, \tag{5}$$

where,

TP: Instances correctly predicted to be positive.
TN: Instances correctly predicted to be negative.
FP: Instances erroneously predicted to be positive.
FN: Instances erroneously predicted to be negative.

Accuracy is the most intuitive performance measure, reflected by the ratio of correctly predicted instances divided by the total. However, accuracy provides solid estimates only for symmetric datasets where values of FP and FN are almost the same. Therefore, other metrics should complement the performance assessment of a model. Precision reflects the ability of the model to avoid FP, recall is the ability of the model to predict the positive instances, and F1-score is essentially a weighted average of precision and recall and is usually preferred over accuracy for uneven class distribution datasets.

4.2. Evaluation

4.2.1. Predictions

In this Section, the final semantic segmentation result of the examined White Canyon section, based on the proposed methodology, is presented and evaluated both qualitatively and quantitively. A hierarchical classification of the generated objects was implemented based on domain knowledge and according to the site-specific conceptualization depicted in Figure 9. Two different ML models were also trained based on one half of the study site (split symmetrically) and used to evaluate differences of the semantic segmentation results compared to the proposed knowledge-based model.

Figure 10 provides a step-by-step demonstration of the classification of the objects generated by the homogenization process based on the following simple rules:

4. "Debris channel" candidate segments are labeled based on their slope. Debris channels represent features generated by erosion and usually dip at lower angles relative to the rest of the slope.
5. Non-debris channel candidate segments surrounded by "debris channel" candidates are aggregated and refined. This includes potential large rock boulders within the main channel.
6. A refined object is then tagged "debris channel" if its Z-axis range at the point level is higher than a length threshold.
7. A linear or compact object located lower than the "debris channel" is tagged "constructed infrastructure". Rail lines and barrier walls along the track are aggregated as linear elements while rockshed components are defined as compact.
8. Remaining objects surrounded by "constructed infrastructure" are incorporated and refined. The remaining objects are tagged "rock outcrop". This rule is applied to prevent potential misclassifications of parts of the infrastructure as natural features.
9. At a lower-granularity level, lower-slope "rock outcrop" objects dipping at an aspect (sub)-perpendicular to the average rock slope aspect are tagged "rock bench/ secondary channel".

In Figure 11, the results of the proposed semantic segmentation are presented together with a RGB photo of the site, the raw point cloud representation, and the ground truth dataset for qualitative visual assessment. In addition, both the confusion matrix in Figure 12 and Table 2 provide a quantitative assessment of the performance of the proposed semantic segmentation over the White Canyon dataset and a per-class classification report, respectively. The analysis shows very high-performance scores (e.g., >94% F1-score) for steep outcrop, debris channel, and constructed infrastructure with the minimum observed value (64%) associated with the rock bench/secondary channel class (Table 2). The latter also highlights the effect of occlusion due to the orientation of the rock benches and secondary channels with respect to the scanner location that potentially leads to misrepresentation of these specific features. Please note that, although the dataset includes only one debris channel, the fact that almost its entire extent is well-separated from the rest of the outcrop identifies as a unique semantic feature (no matter its non-consistent geometry). This finding is of particular importance for the subsequent multi-temporal analyses of the site due to the local geometric changes caused by mass wasting processes in time. Misclassified "debris channel" or "rock outcrop" areas would lead to false assessments regarding movement of the debris within the channel or rockfall activity, respectively, within an automated monitoring workflow similar to the method discussed in Section 4.3.

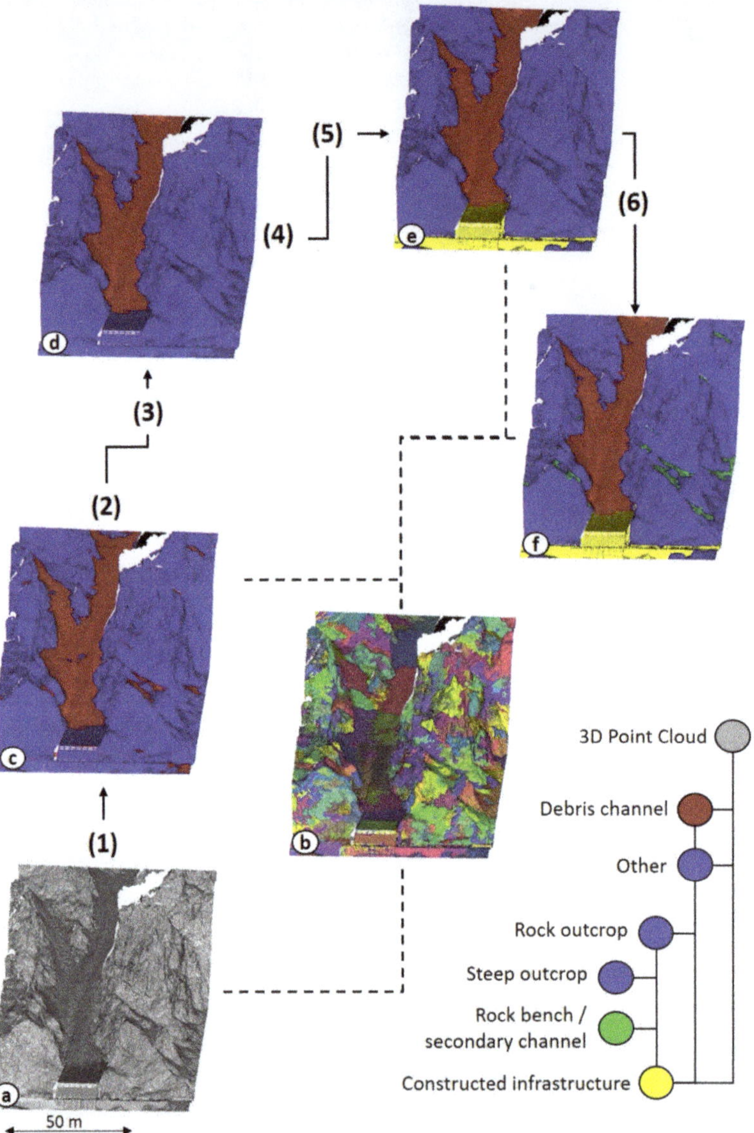

Figure 10. Object-oriented classification results of the study site. (**a**) 3D point cloud; (**b**) initial segmentation (objects are assigned random RGB values); (**c**) detection of "debris channel" candidates; (**d**) "debris channel" tagging; (**e**) extraction of "constructed infrastructure"; (**f**) extraction of guest rock outcrop features (rock benches) at a lower ontology level. The numbers (1–6) correspond to the classification rules, respectively.

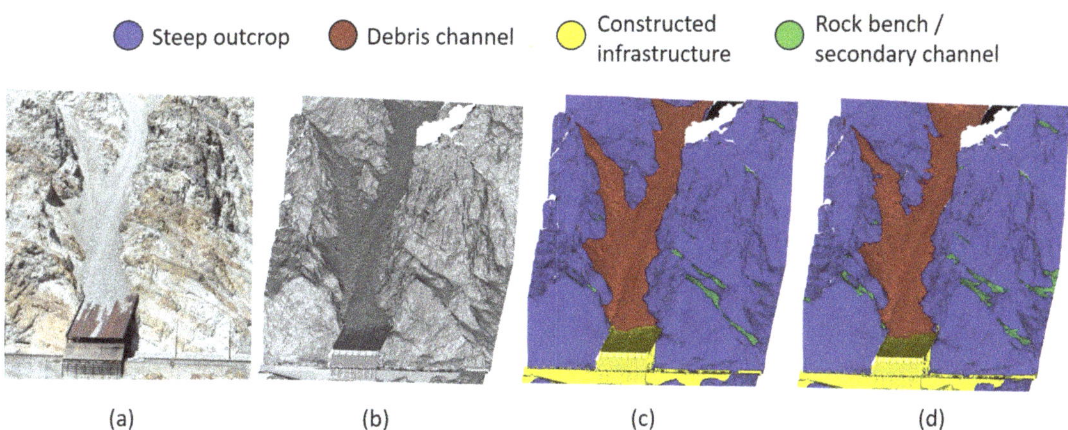

Figure 11. Results of the proposed object-oriented semantic segmentation on a White Canyon section. (**a**) RGB photo; (**b**) Raw point cloud; (**c**) Ground Truth; (**d**) Prediction using the process shown in Figure 10.

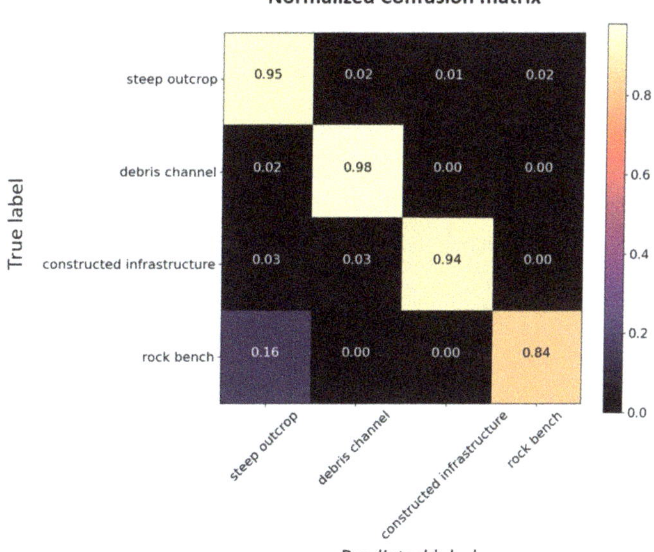

Figure 12. Normalized confusion matrix of the proposed unsupervised object-based classification approach over the study White Canyon section.

Table 2. Per-class classification report of the proposed semantic segmentation on the study dataset.

	Precision	Recall	F_1-Score
Steep outcrop	0.99	0.95	0.97
Debris channel	0.91	0.98	0.95
Constructed infrastructure	0.95	0.94	0.94
Rock bench/secondary channel	0.51	0.84	0.64

4.2.2. Comparison to Point-Based Machine Learning

The knowledge-based semantic segmentation model is also assessed against ML using the limited available training data, which is typically the case with rock slope point clouds. In particular, a RF classifier trained with both multi- and adjusted-sized point neighbourhood descriptors was employed. The examined point neighbourhood sizes were picked from existing methodologies in rock slope classification. In detail, both a wide point neighbourhood size range (0.1 to 5 m) and the optimal point neighbourhoods discussed in [18] are evaluated together in a geological environment. The tested descriptors consist of the eigen-based dimensionality features as well as slope statistics as discussed in [22].

For the purposes of the analysis, the study dataset was split proportionally into training and validation sets (Figure 13). The split was vertical to represent the variations in down-slope geometry and topography in both the training and validation processes. The models were trained and implemented using the open-source ML Python package named scikit-learn. Although the multi-sized neighbourhood model performs significantly better than the adjusted-sized, there are still important misclassifications observed (Figure 13). For instance, constructed infrastructure is predicted to be rock outcrop sections and vice versa, which might cause issues and lead to incorrect decision-making if adopted in the design of a future intelligent management system of the site.

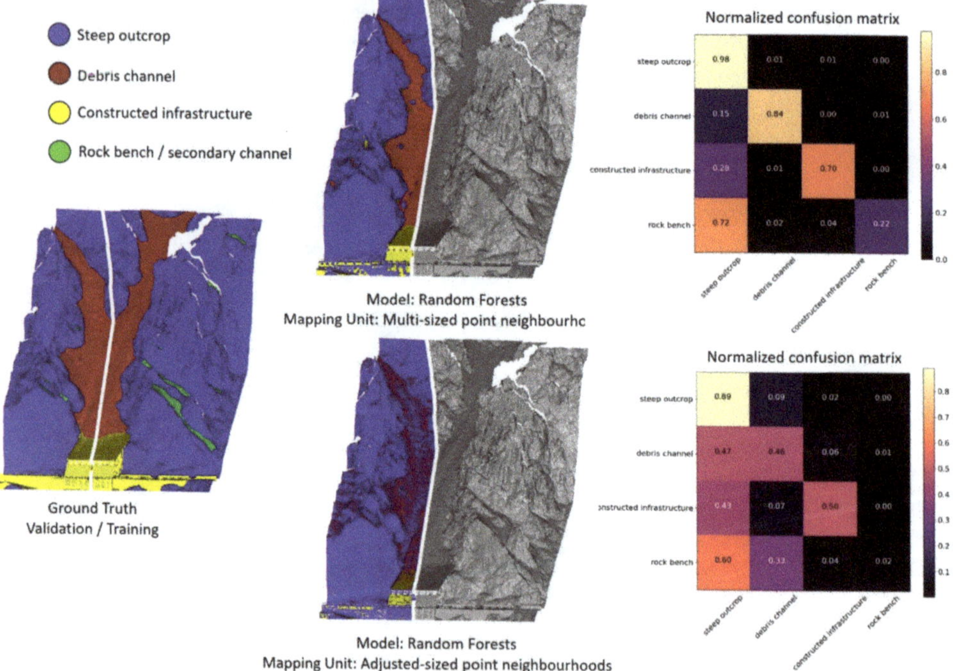

Figure 13. Semantic segmentation of the study site based on Machine Learning using multi- and adjusted-sized point neighbourhood features and a RF classifier.

Table 3 provides a quantitative per-class assessment of two of the ML models as compared to the knowledge-based semantic segmentation approach based on the F_1-score. The analysis shows that the multi-sized neighbourhood RF model performs almost equally as well for almost all the classes regardless of the limited training data. For the rock outcrop and debris channel classes, a 3% and 6% difference were observed, respectively, while the difference is wider for the constructed infrastructure (13%) and rock bench/secondary channel (41%) as can be seen in Figure 13. It is, however, important to note that the significantly higher difference in the "rock bench/secondary channel" class may be due to occlusion bias. This is due to the vantage point of the LiDAR scanner located across the valley, reducing the amount of data collected from horizontal surfaces located above the elevation of the scanner. Additional training data may be needed in order to compare the classification performances more confidently for this particular class.

Table 3. Quantitative F_1-score-based per-class comparison of proposed Machine Learning models to the proposed unsupervised semantic segmentation.

Model	Steep OutCrop	Debris Channel	Constructed Infrastructure	Rock Bench/ Secondary Channel
Multi-sized RF	0.94	0.89	0.81	0.23
Adjusted-sized RF	0.83	0.52	0.58	0.03
Knowledge-based	0.97	0.95	0.94	0.64

Therefore, it is clear that the integration of multiple point neighbourhood sizes better characterizes the scene, however, the object-oriented conceptualization provides significant advantages. Apart from the pure arithmetic assessment, it is also important to see the location of the errors in both methods. Constructed structures, predicted as rock outcrop exist within the actual constructed infrastructure class (Figure 13). The same applies more or less to all the four classes. In contrast, in the object-oriented approach the errors are only observed on the boundaries, especially between rock outcrop and debris channels. The scene homogenization component of this method provides strong support to knowledge-based feature engineering by adding structure to the data and yielding new properties. The scene complexity is minimized, and the dataset can be modeled as a graph where the nodes represent the different objects and the edges the semantic relationships (rules). This supports the conceptualization, prevents mixed-class features, and restricts the errors to the boundaries.

4.3. Application

The current object-based semantic segmentation model was used for the example analysis presented in Figure 14. It includes the change detection between point clouds of the study area captured in June 2018 and June 2019, respectively. Employing the proposed model, the debris channel areas are automatically extracted without any user intervention and the changes are subsequently detected through M3C2 algorithm [36] by filtering the negative values. The output provides a direct estimate and visualization of the eroded channel material volume which can be used for further filing and interpretations.

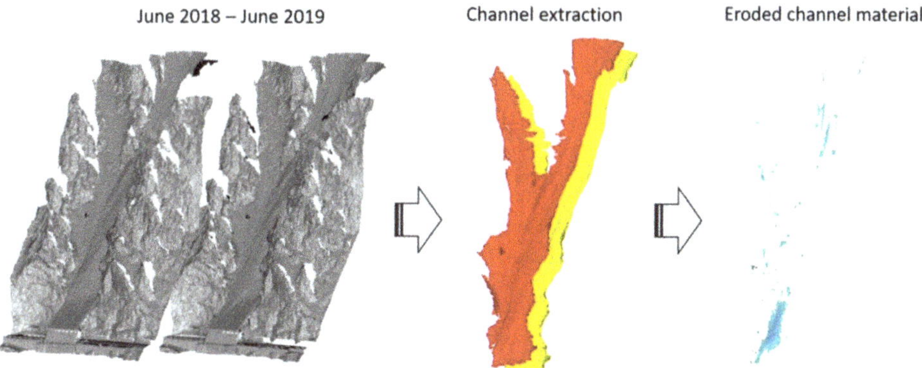

Figure 14. Application of the proposed semantic segmentation for automated change detection with semantics. The red and yellow information in the channel extraction panel display debris channel point cloud data taken from different times. The difference between the two instances is the change that records the erosion and deposition processes on the slope, Figure 5.

5. Discussion

To extract knowledge from 3D point clouds and inform decision-making within rock slope management systems, semantic injection though automated processes is being developed. Advances in CV have provided promising and novel approaches for semantic injection into 3D urban scenes and every-day objects [37–40] but the adaptation of such techniques for engineering geology purposes seems to be facing some challenges. DL/ML models typically require large volumes of diverse training data to be able to generalize well and produce reproducible and reliable results. To date, publicly available ground-truth rock slope point cloud data are not available. In contrast, research on anthropogenic objects of interest (urban scenes), such as constructed infrastructure, for instance, benefits from numerous publicly available ground-truth datasets and the explicitly defined outlines and geometric relationships of such objects within anthropogenic ontologies like houses (e.g., Semantic3D [41], S3DIS [42]). It is clear that ground-truthing natural scenes involves a high degree of subjectivity and it is quite a challenge to put together a commonly accepted dataset. The lack of implicit quantitative definitions regarding the classes involved in rock slope processes makes the ground-truthing of such scenes a challenge, considering the high degree of subjectivity among different researchers. The latter also relies on the statement made by [14] saying that: "the term 'agreement' instead of 'accuracy' is better fitted in landslide mapping procedures". Due to the above reason, researchers tend to evaluate the performance of their algorithms based on their own hand-mapped datasets. According to [34] (p. 4), "as segmentation procedures are used for automation, they are replacing the activity of visual digitizing. No segmentation result-even if quantitatively proofed-will convince if it does not satisfy the human eye". In addition, due to the dynamic and varying nature of the geologic environments, the ability to model site-specific concepts seems critical in rock slope semantic segmentation.

Although some researchers have proposed strategies for integrating engineering geological knowledge into 3D point clouds for further reasoning (refer to ML-based methods in Figure 13), little has been done in the data structuration and knowledge organization. Effective segmentation methods which are able to partition the point cloud into semantically meaningful areas, are very helpful to model a scene [42]. To accomplish this, the point cloud needs to be structured retaining both spatial and relational information. The proposed method provides semantic structure to the 3D point cloud (Figures 4 and 5), which promotes the development of a multi-level graph representation. Thus, robust object-based descriptors can be incorporated in the workflow following tailored conceptualizations in order to address the substantial complexity observed in rock slopes.

The main limitation that such an unsupervised semantic segmentation model has to overcome is the fact that, in engineering geology, ontologies are not explicitly defined. As a result, the design of such models is based on the modeler's conceptualization and has to be adjusted to the specific environment based on expert characterization, potential geological and geomorphological processes and the ensuing reasoning each time. However, expert mapping does not always provide coherent validation due to the bias associated with expert perception subjectivity. Therefore, researchers might be discouraged to investigate the potential of incorporating automatic semantic segmentation into engineering geology workflows due to the need to prepare their own reference datasets for validation and/or training. Without defined ontologies, and a collective effort to develop annotations, semantic segmentation results may remain undeveloped and satisfactory levels of interoperability may not be achieved. However, and previous from this study, the conceptual framework for integrating formalized knowledge within rock slope point cloud processing was also largely absent.

6. Summary and Conclusions

This paper proposes a rock slope assessment methodology based on LiDAR point cloud processing. Specifically, it includes a knowledge-based object-oriented framework for tailored rock slope semantic segmentation using raw XYZ information (Figure 2). The investigated site is a steep railway rock slope in British Columbia, Canada, where different types of mass wasting mechanisms occur periodically (Figure 3). The proposed methodology, which has been developed based on a data-driven homogenization procedure generating meaningful object primitives, supports the extraction of informative descriptors able to replicate expert perception. The developed rule-base leverages geometric, topographic, and contextual features to extract specific classes of the investigated scheme by employing regulation through tailored rules (Table 1). This methodology provides an alternative to existing supervised ML-based semantic segmentation approaches [22] in cases where sufficient amounts of annotated reference data are not available or higher-level conceptualization is needed (apart from texture identification). For this reason, a comparison with two different ML models was performed to validate the capabilities of the proposed model (Figure 13 and Table 3). The results demonstrate that one can efficiently segment a rock slope point cloud, in a knowledge-based fashion, by employing only a few informative features, without overcomplicating the task. The classification was performed based on an experts' conceptualization scheme of the site (Figure 9). The application of the method to temporally different instances of the study site, showcases its incorporation of it into automatic multi-temporal analysis workflows for targeted rock slope assessment (Figure 14).

This study shows that in engineering geology, satisfactory tailored semantic segmentation can be achieved even without the employment of ML classifiers and the associated largely time-consuming need for training data collection. This, however, requires proper conceptualization for the selection of object features and the definition of the ruleset. The model demonstrates almost equal performance to ML, with over 90% scores for three of the four classes examined. However, although ML approaches using multi-sized point neighbourhood descriptors performed very well in identifying different materials/textures, they cannot accommodate expert conceptualization. It has been proven that bedrock and talus can be identified as different materials using ML, but the general concept is missed (e.g., is the talus accumulation detected to rest on a channel or bench?). Knowledge-based object-oriented models value quality over quantity in descriptor selection and thus complexity is reduced and explainability increased. This analytic knowledge can then be efficiently represented through specific data structures. In contrast, point-based procedures analyze every point in the context of its spherical neighbourhood(s), without being able to leverage spatial or contextual information related to the point of interest since the point cloud data inherently lacks structure and semantic meaning.

The paper shows the potential of object-oriented models to be used in rock slope assessment and highlights new aspects of further research in this direction and especially in knowledge formalization through ontologies. Considerations for future work include the extension of the object-oriented model to deeper levels of the examined rock slope conceptualization, as well as the calibration to other rock slope sites within different geomorphologic settings and concepts to test the sensitivity of the rulesets. In addition, comparison between point- and object-based ML models using rich and varying training data would provide interesting insights. Colour information from photogrammetric point clouds will also be considered in future tests together with the potential of other segmentation methods to be integrated for the supervoxel generation.

Another interesting future challenge is the investigation of the processes modelling potential, utilizing semantic relationships within multi-temporal semantically-rich point clouds of a rock slope. The latter consideration aims to provide deeper insights regarding future potential hazardous zones based on a certain slope dynamic behaviour and eventually contributes to the establishment of quantitative definitions of different rock slope elements. The development of computer-aided methodologies that can accommodate expert-reasoning is considered essential in such a site-specific domain as landslide research, and knowledge formalization is necessary towards this direction.

Author Contributions: Conceptualization, I.F.; methodology, I.F.; software, I.F. and D.B.; formal analysis, I.F.; investigation, I.F. and D.B.; resources, N.V. and D.J.H.; data curation, I.F.; writing—original draft preparation, I.F.; writing—review and editing, D.B., D.J.H., and N.V.; visualization, I.F.; supervision, D.J.H. and N.V. All authors have read and agreed to the published version of the manuscript.

Funding: This research is supported by the RMC Green Team and the Canadian Department of National Defense as well as the Canadian Railway Ground Hazard Research Program (CN Rail, CP Rail, Transport Canada, Geological Survey of Canada) which is funded through a Natural Sciences and Engineering Research Council of Canada (NSERC) Alliance grant.

Acknowledgments: The authors would like to thank the three anonymous reviewers for helping to improve the quality of the manuscript with their comments and suggestions.

Conflicts of Interest: The authors declare no conflict of interest. The funders had no role in the design of the study; in the collection, analyses, or interpretation of data; in the writing of the manuscript, or in the decision to publish the results.

Abbreviations

2, 2.3, 3D	2-, 2.5, 3-dimensional;
CS	computer science;
CV	computer vision;
DEM	digital elevation model;
GIS	geographic information system;
ML	machine learning;
PCA	principal component analysis;
SVM	support vector machine;
RF	random forests;
TLS	terrestrial laser scanner;
SCV	singular value decomposition;
LV	local variance;
ICP	iterative closest point;
SLIDO	statewide landslide information database for Oregon;
TP	true positive;
TN	true negative;
FP	false positive;
FN	false negative;
RGB	red-green-blue;
M3C2	Multi-scale model to model cloud comparison;
DL	deep learning.

References

1. Rutzinger, M.; Höfle, B.; Lindenbergh, R.; Oude Elberink, S.; Pirotti, F.; Sailer, R.; Scaioni, M.; Stötter, J.; Wujanz, D. Close-Range Sensing Techniques in Alpine Terrain. *ISPRS Ann. Photogramm. Remote Sens. Spat. Inf. Sci.* **2016**, *III–6*, 15–22. [CrossRef]
2. Scaioni, M.; Longoni, L.; Melillo, V.; Papini, M. Remote sensing for landslide investigations: An overview of recent achievements and perspectives. *Remote Sens.* **2014**, *6*, 9600–9652. [CrossRef]
3. Hutchinson, D.J.; Lato, M.; Gauthier, D.; Kromer, R.; Ondercin, M.; Van Veen, M.; Harrap, R. Applications of remote sensing techniques to managing rock slope instability risk. In Proceedings of the 68th Canadian Geotechnical Conference-GEOQuébec 2015, Quebec City, QC, Canada, 21–23 September 2015; p. 10.
4. Abellán, A.; Jaboyedoff, M.; Oppikofer, T.; Vilaplana, J.M. Detection of millimetric deformation using a terrestrial laser scanner: Experiment and application to a rockfall event. *Nat. Hazards Earth Syst. Sci.* **2009**, *9*, 365–372. [CrossRef]
5. Kromer, R.A.; Hutchinson, D.J.; Lato, M.J.; Gauthier, D.; Edwards, T. Identifying rock slope failure precursors using LiDAR for transportation corridor hazard management. *Eng. Geol.* **2015**, *195*, 93–103. [CrossRef]
6. Lato, M.; Diederichs, M.S.; Hutchinson, D.J.; Harrap, R. Optimization of LiDAR scanning and processing for automated structural evaluation of discontinuities in rockmasses. *Int. J. Rock Mech. Min. Sci.* **2009**, *46*, 194–199. [CrossRef]
7. Bonneau, D.A.; Jean Hutchinson, D.; DiFrancesco, P.M.; Coombs, M.; Sala, Z. Three-dimensional rockfall shape back analysis: Methods and implications. *Nat. Hazards Earth Syst. Sci.* **2019**, *19*, 2745–2765. [CrossRef]
8. Bonneau, D.; DiFrancesco, P.M.; Jean Hutchinson, D. Surface reconstruction for three-dimensional rockfall volumetric analysis. *ISPRS Int. J. Geo-Inf.* **2019**, *8*, 548. [CrossRef]

9. Rowe, E.; Hutchinson, D.J.; Kromer, R.A. An analysis of failure mechanism constraints on pre-failure rock block deformation using TLS and roto-translation methods. *Landslides* **2018**, *15*, 409–421. [CrossRef]

10. Oppikofer, T.; Jaboyedoff, M.; Blikra, L.; Derron, M.H.; Metzger, R. Characterization and monitoring of the Åknes rockslide using terrestrial laser scanning. *Nat. Hazards Earth Syst. Sci.* **2009**, *9*, 1003–1019. [CrossRef]

11. Williams, J.G.; Rosser, N.J.; Hardy, R.J.; Brain, M.J.; Afana, A.A. Optimising 4-D surface change detection: An approach for capturing rockfall magnitude-frequency. *Earth Surf. Dyn.* **2018**, *6*, 101–119. [CrossRef]

12. Martha, T.R.; Kerle, N.; Jetten, V.; van Westen, C.J.; Kumar, K.V. Characterising spectral, spatial and morphometric properties of landslides for semi-automatic detection using object-oriented methods. *Geomorphology* **2010**, *116*, 24–36. [CrossRef]

13. Feizizadeh, B.; Blaschke, T. A semi-automated object based image analysis approach for landslide delineation. In Proceedings of the 2013 European Space Agency Living Planet Symposium, Edinburgh, UK, 9–13 September 2013; pp. 9–13. [CrossRef]

14. Hölbling, D.; Eisank, C.; Albrecht, F.; Vecchiotti, F.; Friedl, B.; Weinke, E.; Kociu, A. Comparing manual and semi-automated landslide mapping based on optical satellite images from different sensors. *Geosciences* **2017**, *7*, 37. [CrossRef]

15. Karantanellis, E.; Marinos, V.; Vassilakis, E.; Christaras, B. Object-based analysis using unmanned aerial vehicles (UAVs) for site-specific landslide assessment. *Remote Sens.* **2020**, *12*, 1711. [CrossRef]

16. Farmakis, I.; Bonneau, D.; Hutchinson, D.J.; Vlachopoulos, N. Supervoxel-Based Multi-Scale Point Cloud Segmentation Using FNEA for Object-Oriented Rock Slope Classification Using TLS. *Int. Arch. Photogramm. Remote Sens. Spat. Inf. Sci.* **2020**, *43*, 1049–1056. [CrossRef]

17. Pauly, M.; Keiser, R.; Gross, M. Multi-scale Feature Extraction on Point-Sampled Surfaces. *Comput. Graph. Forum* **2003**, *22*, 281–289. [CrossRef]

18. Weinmann, M.; Jutzi, B.; Hinz, S.; Mallet, C. Semantic point cloud interpretation based on optimal neighborhoods, relevant features and efficient classifiers. *ISPRS J. Photogramm. Remote Sens.* **2015**, *105*, 286–304. [CrossRef]

19. Brodu, N.; Lague, D. 3D terrestrial lidar data classification of complex natural scenes using a multi-scale dimensionality criterion: Applications in geomorphology. *ISPRS J. Photogramm. Remote Sens.* **2012**, *68*, 121–134. [CrossRef]

20. Bosser, B.; Guyon, I.; Vapnik, V. A Training Algorithm for Optimal Margin Classifiers. In Proceedings of the Fifth Annual Workshop on Computational Learning Theory, New York, NY, USA, 27–29 July 1992; pp. 144–152.

21. Bonneau, D.A.; Hutchinson, D.J. The use of terrestrial laser scanning for the characterization of a cliff-talus system in the Thompson River Valley, British Columbia, Canada. *Geomorphology* **2019**, *327*, 598–609. [CrossRef]

22. Weidner, L.; Walton, G.; Kromer, R. Classification methods for point clouds in rock slope monitoring: A novel machine learning approach and comparative analysis. *Eng. Geol.* **2019**, *263*, 105326. [CrossRef]

23. Breiman, L. Random Forests. *Mach. Learn.* **2001**, *45*, 5–32. [CrossRef]

24. Feng, C.C.; Guo, Z. Automating parameter learning for classifying terrestrial LiDAR point cloud using 2D land cover maps. *Remote Sens.* **2018**, *10*, 1192. [CrossRef]

25. Blomley, R.; Weinmann, M.; Leitloff, J.; Jutzi, B. Shape distribution features for point cloud analysis—A geometric histogram approach on multiple scales. *ISPRS Ann. Photogramm. Remote Sens. Spat. Inf. Sci.* **2014**, *II-3*, 9–16. [CrossRef]

26. Thomas, H.; Goulette, F.; Deschaud, J.E.; Marcotegui, B.; Gall, Y. Le Semantic classification of 3d point clouds with multiscale spherical neighborhoods. In Proceedings of the 2018 International conference on 3D vision (3DV), Verona, Italy, 5–8 September 2018; pp. 390–398. [CrossRef]

27. Ioannou, Y.; Taati, B.; Harrap, R.; Greenspan, M. Difference of normals as a multi-scale operator in unorganized point clouds. In Proceedings of the 2012 Second International Conference on 3D Imaging, Modeling, Processing, Visualization & Transmission, Zurich, Switzerland, 13–15 October 2012; pp. 501–508. [CrossRef]

28. Shannon, C.E. A Mathematical Theory of Communication. *Bell Syst. Tech. J.* **1948**, *27*, 623–656. [CrossRef]

29. Mayr, A.; Rutzinger, M.; Bremer, M.; Oude Elberink, S.; Stumpf, F.; Geitner, C. Object-based classification of terrestrial laser scanning point clouds for landslide monitoring. *Photogramm. Rec.* **2017**, *32*, 377–397. [CrossRef]

30. Dunham, L.; Wartman, J.; Olsen, M.J.; O'Banion, M.; Cunningham, K. Rockfall Activity Index (RAI): A lidar-derived, morphology-based method for hazard assessment. *Eng. Geol.* **2017**, *221*, 184–192. [CrossRef]

31. Hutchinson, D.J.; Lato, M.; Gauthier, D.; Kromer, R.; Ondercin, M.; MacGowan, T.; Edwards, T. Rock Slope Monitoring and Risk Management for Railway Infrastructure in the White Canyon, British Columbia, Canada. In *Engineering Geology for Society and Territory*; Springer International Publishing: Cham, Switzerland, 2015; Volume 2, pp. 435–439.

32. Besl, P.J.; McKay, N.D. A method for registration of 3-D shapes. *IEEE Trans. Pattern Anal. Mach. Intell.* **1992**, *14*, 239–256. [CrossRef]

33. Wu, L.; Wang, Y.; Long, J.; Liu, Z. *Image and Graphics*; Zhang, Y.-J., Ed.; Lecture Notes in Computer Science; Springer International Publishing: Cham, Switzerland, 2015; Volume 9217, ISBN 9783319219776.

34. Baatz, M.; Schape, A. Object-Oriented and Multi-Scale Image Analysis in Semantic Networks. In Proceedings of the 2nd International Symposium on Operationalization of Remote Sensing, ITC, Enschede, The Netherlands, 16–20 August 1999.

35. Drăguţ, L.; Csillik, O.; Eisank, C.; Tiede, D. Automated parameterisation for multi-scale image segmentation on multiple layers. *ISPRS J. Photogramm. Remote Sens.* **2014**, *88*, 119–127. [CrossRef]

36. Lague, D.; Brodu, N.; Leroux, J. Accurate 3D comparison of complex topography with terrestrial laser scanner: Application to the Rangitikei canyon (N-Z). *ISPRS J. Photogramm. Remote Sens.* **2013**, *82*, 10–26. [CrossRef]

37. Charles, R.Q.; Su, H.; Kaichun, M.; Guibas, L.J. PointNet: Deep Learning on Point Sets for 3D Classification and Segmentation. In Proceedings of the 2017 IEEE Conference on Computer Vision and Pattern Recognition (CVPR), Honolulu, HI, USA, 21–26 July 2017; IEEE: Piscataway Township, NJ, USA, 2017; pp. 77–85.
38. Qi, C.R.; Yi, L.; Su, H.; Guibas, L.J. PointNet++: Deep Hierarchical Feature Learning on point sets in a metric space. *arXiv* **2017**, arXiv:1706.02413, 5105–5114.
39. Maturana, D.; Scherer, S. VoxNet: A 3D Convolutional Neural Network for Real-Time Object Recognition. In Proceedings of the 2015 IEEE/RSJ International Conference on Intelligent Robots and Systems (IROS), Hamburg, Germany, 28 September–2 October 2015; IEEE: Piscataway Township, NJ, USA, 2015; pp. 922–928.
40. Ben-Shabat, Y.; Lindenbaum, M.; Fischer, A. 3DmFV: Three-Dimensional Point Cloud Classification in Real-Time Using Convolutional Neural Networks. *IEEE Robot. Autom. Lett.* **2018**, *3*, 3145–3152. [CrossRef]
41. Hackel, T.; Savinov, N.; Ladicky, L.; Wegner, J.D.; Schindler, K.; Pollefeys, M. Semantic3D.Net: A New Large-Scale Point Cloud Classification Benchmark. In Proceedings of the ISPRS Annals of the Photogrammetry, Remote Sensing and Spatial Information Sciences, Melbourne, Australia, 26–27 October 2017; Volume 4, pp. 91–98.
42. Hu, S.M.; Cai, J.X.; Lai, Y.K. Semantic labeling and instance segmentation of 3d point clouds using patch context analysis and multiscale processing. *IEEE Trans. Vis. Comput. Graph.* **2018**, *14*, 2485–2498. [CrossRef]

remote sensing

Article

Assessing the Potential of Long, Multi-Temporal SAR Interferometry Time Series for Slope Instability Monitoring: Two Case Studies in Southern Italy

Fabio Bovenga [1,*], **Ilenia Argentiero** [1], **Alberto Refice** [1], **Raffaele Nutricato** [2], **Davide O. Nitti** [2], **Guido Pasquariello** [1] and **Giuseppe Spilotro** [1]

[1]   National Research Council of Italy, CNR-IREA, Via Amendola 122, 70126 Bari, Italy; argentiero.i@irea.cnr.it (I.A.); refice.a@irea.cnr.it (A.R.); pasquariello.g@irea.cnr.it (G.P.); giuseppe.spilotro@unibas.it (G.S.)
[2]   GAP S.R.L., 70125 Bari, Italy; raffaele.nutricato@gapsrl.eu (R.N.); davide.nitti@gapsrl.eu (D.O.N.)
*   Correspondence: bovenga.f@irea.cnr.it; Tel.: +39-080-592-9425

Citation: Bovenga, F.; Argentiero, I.; Refice, A.; Nutricato, R.; Nitti, D.O.; Pasquariello, G.; Spilotro, G. Assessing the Potential of Long, Multi-Temporal SAR Interferometry Time Series for Slope Instability Monitoring: Two Case Studies in Southern Italy. *Remote Sens.* **2022**, *14*, 1677. https://doi.org/10.3390/rs14071677

Academic Editor: Alex Hay-Man Ng

Received: 2 March 2022
Accepted: 23 March 2022
Published: 31 March 2022

Abstract: Multi-temporal SAR interferometry (MTInSAR), by providing both mean displacement maps and displacement time series over coherent objects on the Earth's surface, allows analyzing wide areas, identifying ground displacements, and studying the phenomenon evolution at a long time scale. This technique has also been proven to be very useful for detecting and monitoring slope instabilities. For this type of hazard, detection of velocity variations over short time intervals should be useful for early warning of damaging events. In this work, we present the results obtained by using both COSMO-SkyMed (CSK) and Sentinel-1 (S1) data for investigating the ground stability of two hilly villages located in the Southern Italian Apennines (Basilicata region), namely the towns of Montescaglioso and Pomarico. In these two municipalities, landslides occurred in the recent past (in Montescaglioso in 2013) and more recently (in Pomarico in 2019), causing damage to houses, commercial buildings, and infrastructures. SAR datasets acquired by CSK and S1 from both ascending and descending orbits were processed using the SPINUA MTInSAR algorithm. Mean velocity maps and displacement time series were analyzed, also by means of innovative ad hoc procedures, looking, in particular, for non-linear trends. Results evidenced the presence of nonlinear displacements in correspondence of some key infrastructures. In particular, the analysis of accelerations and decelerations of PS objects corresponding to structures affected by recent stabilization measures helps to shed new light in relation to known events that occurred in the area of interest.

Keywords: SAR interferometry; slope instability; ground stability monitoring; Sentinel-1; COSMO-SkyMed; time series analysis

1. Introduction

For more than two decades, multi-temporal SAR interferometry (MTInSAR) techniques have provided both mean displacement maps and displacement time series over coherent objects on the Earth's surface, allowing analysis of wide areas to identify ground displacements, and to study phenomena evolution at decadal time scales [1,2]. These techniques are among the most promising and reliable in the detection and monitoring of slope instabilities [3,4], and a long record of applications and case studies can be found in the literature, especially in Italy (e.g., [5] and references therein). Landslides are among the most difficult types of ground displacements to be investigated through remote sensing techniques, as their velocity rates may cover several orders of magnitude, from very low creeping to sudden catastrophic failures [6], thus pushing the detection performances to the limits. Further challenges related to the application of MTInSAR analysis for slope instability monitoring concern (i) the lack of a sufficient spatial density of coherent targets (man-made structures, rock outcrops, homogeneous distributed scatterers) due to vegetation or variable land cover; and (ii) the advanced time series analysis required for identifying complex kinematics (related, for instance, to early warning signals) [7].

Currently, several satellite data are available, from both past and current missions, providing stacks of interferometric SAR images at different wavelengths, spatial resolutions, and revisit times [8]. Recent satellite missions have improved the monitoring of unstable slopes. L-band missions such as ALOS-2/PALSAR-2 [9] and SAOCOM [10] provide interesting monitoring capabilities over vegetated areas, despite the low sensitivity to small displacements and the limited revisit time. The high-resolution X-Band COSMO-SkyMed (CSK) constellation [11], and the upcoming COSMO-SkyMed second generation [12], acquire data with a spatial resolution reaching metric values, and provide revisit times of up to a few days, leading to an increase in the density of the measurable targets, thus improving the monitoring of local scale events and the detection of non-linear displacement trends. The Sentinel-1 (S1) C-band mission from the European Space Agency (ESA) [13] provides a spatial resolution comparable to that of previous ESA missions, but a nominal revisit time reduced to 6 days. On December 2021, a failure on Sentinel-1B occurred, leading to a doubling in the revisit time. Nonetheless, by offering regular global-scale coverage, better temporal resolution, and freely available imagery, S1 improves the performance of MTInSAR for ground displacement investigations. In particular, the short revisit time allows a better time series analysis by improving the temporal sampling and the chances to identify pre-failure signals characterized by a high-rate and non-linear behavior. Moreover, it allows collecting large data stacks in a short time, thus improving the MTInSAR performance in emergency (post-event) scenarios. These characteristics are very promising for early warning of slope failure events and monitoring subsequent displacement trends, in the larger framework of Disaster Risk Reduction (DRR). However, when dealing with large data stacks, the distinction between the sensitivity and significance of the measurements is essential. Indeed, detection of velocity changes can be related to changes in the equilibrium of forces affecting terrain stability, provided that all variations due to other natural phenomena (e.g., thermal expansion and differences in soil moisture) have been excluded. Consequently, reliable monitoring and early warning require a detailed analysis of the displacement time series looking for signals actually related to ground instabilities.

In this work, we present the results obtained by analyzing displacement time series from both CSK and S1 for investigating the ground stability of hilly villages located in the Southern Italian Apennines (Basilicata region). In the area of interest, landslides are endemic. Many events occurred in the recent past, causing damage to houses, commercial buildings, and infrastructures, and several have been investigated through various remote sensing techniques [14–18]. We focus here on two test sites in the province of Matera, one on the outskirts of the town of Montescaglioso, where a landslide occurred in 2013, and one near Pomarico, affected by an event in 2019.

SAR datasets acquired by CSK and S1 from both ascending and descending orbits were processed by using the SPINUA MTInSAR algorithm [19], in order to exploit the potential of these two satellite missions to investigate ground displacements related to the slope instabilities in pre- and post-failure stages. Monitoring of the displacement rates is the main tool to detect the type and causes of a slope failure [20] and to plan further restoration measures, if possible, of the involved areas. Mean velocity maps and displacement time series were analyzed, looking, in particular, for non-linear trends that are possibly related to relevant ground instabilities and, due to the high spatial resolution, useful in terms of early warning. This analysis was also supported by automated procedures recently developed for identifying targets with peculiar, nonlinear motions. Specifically, this article illustrates an example of slope pre-failure monitoring on the Pomarico landslide, an example of slope post-failure monitoring on the Montescaglioso landslide, and a few examples of structures in both locations (such as buildings and roads) affected by instability related to different causes (soil swelling, for example).

2. Materials and Methods

2.1. Test Sites Description

The Montescaglioso and Pomarico villages are located in southern Italy, in the southern portion of the Basilicata region (Figure 1). This area is characterized by regressive sediments, generally coarsening upward, consisting of sand and conglomerate levels unconformably overlying the Pliocene—Middle Pleistocene Sub-Apennine Clay Formation, filling a deep tectonic trough [21].

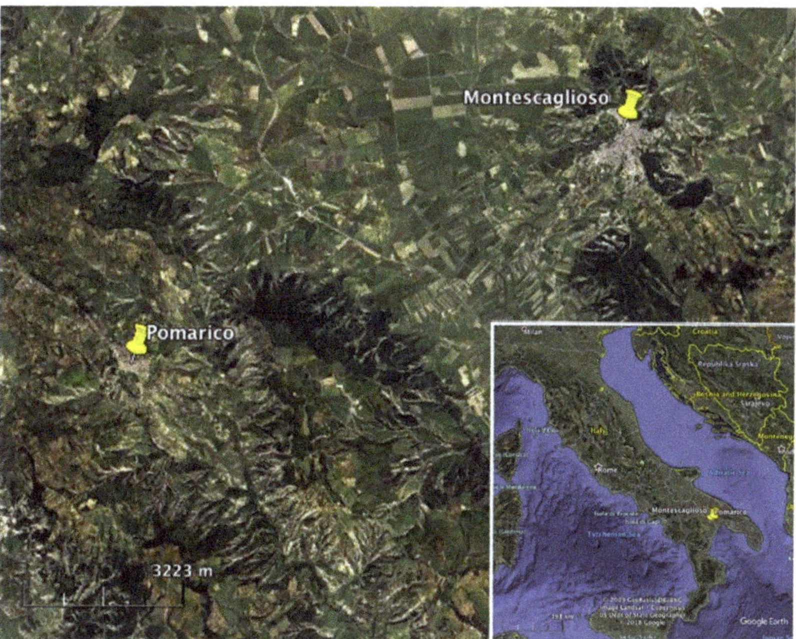

Figure 1. Location of Montescaglioso and Pomarico villages.

The tectonic history has influenced the geological setting of the area. In the presence of monoclinal strata, the hydraulic conditions worsened by the slope may coincide to cause sliding of large clods on the underlying and impermeable clays. This is the case of the Montescaglioso landslide, in addition to other landslides that occurred in neighboring villages, such as Aliano [22].

Instead, the presence of cemented overburden leads to progressive failure phenomena: first there are fissures of the rigid cover layer and subsequently the destructive evolution of the phenomenon. An example is the Pomarico landslide (2019) [18], in addition to those of Pisticci (1976) [23] and Senise (1986) [24,25], which also caused casualties.

Furthermore, the presence of ancient and quiescent landslides, as reported in national and regional databases [26–28], contributes to an anisotropy of the masses which, correlated to the strength and permeability, is relevant in the kinematics of landslides: landslide bodies tend to expand with a retrogressive behavior, often after intense rainfall periods [29,30] and evolve downstream in earthflows [31,32].

2.2. MTInSAR Datasets

Table 1 reports schematically the datasets used in the present work. S1 data are acquired in interferometric wide-swath (IW) mode, with a ground pixel size of about 5×20 m^2 (range $\times$ azimuth). CSK data are acquired in stripmap mode, with a ground pixel size of about 3×3 m^2. CSK scenes are acquired with an average time interval of 16 days, whereas the S1 acquisition schedule is more frequent, with an acquisition interval of 12 days, and reaching 6 days from 2017, when the second satellite of the constellation, Sentinel-1B, was launched. In fact, S1 series cover a shorter time interval than CSK, as they all start in October 2014, whereas CSK series start two years earlier, but with a considerably higher number of acquisitions, and almost double in the case of the Montescaglioso ascending dataset.

Table 1. Dataset characteristics.

Test Site	Sensor	Geometry	N. Images	Time Interval
	S1	Ascending	158	15/10/2014–15/05/2018
Pomarico	S1	Descending	141	21/10/2014–21/05/2018
	CSK	Ascending	96	28/10/2012–23/06/2018
	S1	Ascending	182	15/10/2014–29/12/2018
Montescaglioso	S1	Descending	159	21/10/2014–29/12/2018
	CSK	Ascending	96	28/10/2012–23/06/2018

All data series cover a time period that ends in 2018 since they are those available at the time when the MTInSAR processing was carried out. Although several more years would have been useful for assessing the ground instability of the case studies at a more recent time, the results derived by exploiting the available datasets are suitable for the time series analysis we present in the following, and perfectly suit the aim of this work.

2.3. MTInSAR Time Series Analysis

As previously mentioned, all data series were processed through the SPINUA MTInSAR processing suite [19]. As with most MTInSAR processing algorithms, the result of the processing consists of a large set of SAR pixels corresponding to coherent objects on the ground, characterized by their coordinates and by a merit figure such as the temporal coherence, and featuring the residual height and the mean line-of-sight (LOS) velocity, in addition to the complete time series of relative LOS displacements at each acquisition date. The temporal coherence is defined as $\gamma_t = \frac{1}{N_t}\left|\sum_{j=1}^{N_t} \exp i(\phi_j - \hat{\phi}_j)\right|$, where ϕ_j, with $j = 1, \ldots, N_t$, are the InSAR phases of the time series, and $\hat{\phi}_j$ are obtained from a model trend.

The model has to be postulated a priori in order to define the coherence figure. The most common choice, often adopted to minimize processing times, is to use a linear model, or e.g., a low-order polynomial. Clearly, such models are not suited for the type of nonlinearities involved in ground displacement phenomena such as those investigated here. Therefore, when no ad hoc, computation-intensive processing can be performed on limited datasets involving, e.g., libraries of more complex phase models, selecting targets to investigate time series trends is usually carried out by adopting low threshold values for the provided γ_t figure (which results in a large number of selected targets), and relying on the analysis of experts with knowledge of the area, or by integration of ancillary data taken on the ground or by other means.

In these cases, automated procedures may help identifying a few targets with peculiar, nonlinear motions that may signal, e.g., previously undetected phenomena, or reactivations/deactivation phases of known landslides. Most approaches adopt some parameterization of the time series trends, relying on statistical tests to assess the most suitable representation of each time series, e.g., through piecewise-linear or polynomial functions (e.g., [33,34]).

In particular, in [34] a procedure is proposed, based on the Fisher distribution, that is able to automatically detect nonlinearities such as piecewise-linear trends, which are typical of warning signals preceding the failure of natural and artificial structures. This nonlinear trend analysis (hereafter referred to as NLTA) processes the displacement time series derived by a generic MTInSAR algorithm and provides as output the coefficients of the polynomial model of degree $\bar{n}_p$ selected for describing the time series, and an index F_A

based on the Fisher statistics, which allows evaluating the quality of the approximation and the temporal coherence $\gamma_{t,\,OUT}$ computed by using the selected polynomial model. The polynomial degree is selected as:

$$\bar{n}_p = \min\left(\left\{n_p \middle| F(n_p) \leq F_\alpha \wedge F_A(n_p) \leq F_{A\alpha}\right\}\right) \tag{1}$$

where F is the standard Fisher test, and F_α and $F_{A\alpha}$ are threshold values that can be set according to both the number of samples (N) composing the time series and the significance level α of the statistical test [34]. In this work, the threshold values are set to $F_\alpha = F_{A\alpha} = 4$. The polynomial model selected by the NLTA procedure generally improves the fit with respect to the model provided by the MTInSAR algorithm that generated the time series, leading to higher values of the output coherence $\gamma_{t,\,OUT}$ with respect to the input coherence $\gamma_{t,\,IN}$. According to this, a subset of coherent targets showing interesting temporal behavior (PS_{NLTA}), can be derived by exploring the values of $\gamma_{t,\,OUT}$ and $\gamma_{t,\,IN}$ as, for instance, in the following:

$$PS_{NLTA} = \left\{\gamma_{t,\,IN} < \gamma_{t,IN}^{Th} \wedge \gamma_{t,\,OUT} \geq \gamma_{t,OUT}^{Th} \wedge (\gamma_{t,\,OUT} - \gamma_{t,\,IN}) \geq \Delta\gamma^{Th}\right\} \tag{2}$$

where $\gamma_{t,IN}^{Th}$, $\gamma_{t,OUT}^{Th}$, and $\Delta\gamma^{Th}$ are threshold values that for the present study are set respectively to 0.7, 0.8, and 0.1.

Furthermore, recently, an index based on the fuzzy entropy (FE) has been used [35] for measuring the degree of regularity of a given time series; the FE basically evaluates the gain in "information" by comparing segments of $m + 1$ samples with segments of m samples, according to some distance measure. The FE index, without postulating any a priori model, allows highlighting time series which show "interesting", i.e., locally smooth, trends, including strong non linearities, jumps related to phase unwrapping errors, and the so-called partially coherent scatterers. Also in this case, a subset of interesting coherent targets (PS_{FE}) can be selected through the FE index values according to a threshold value derived in [35] for time series of the type used in this work:

$$PS_{FE} = \{FE < 0.5\} \quad \text{with } m = 2 \tag{3}$$

Both NLTA and the FE index were adopted for analyzing the MTInSAR displacement time series described in Section 2.2. In particular, the MTInSAR targets were selected through a model-independent procedure based on FE [35] and classified according to the polynomial order selected through the NLTA procedure [34]. By identifying and correcting phase unwrapping errors through FE, and by refining the displacement model through NLTA, the proposed post-processing scheme allows decreasing the residual processing errors and thus also the noise affecting the final displacement estimation. This is proved by the final temporal coherence values, which are generally higher than those coming from the standard MTInSAR processing [34,35]. This estimation refinement leads to an increase in the accuracy of the derived displacement products. Moreover, this post-processing scheme allows a focus on a smaller set of coherent targets affected by nonlinear displacements, and potentially deserving further geophysical or geotechnical analysis. Finally, it is well suited for catching warning signals such as, for example, those preceding the failure of natural and artificial structures.

3. Results

In this section we present the displacement maps and time series obtained over Montescaglioso and Pomarico villages.

3.1. Montescaglioso Test Site

Figure 2 shows the mean LOS displacement rate maps derived by processing, through the SPINUA algorithm, the S1 ascending (Figure 2a), S1 descending (Figure 2b), and CSK ascending (Figure 2c) datasets over the Montescaglioso test site.

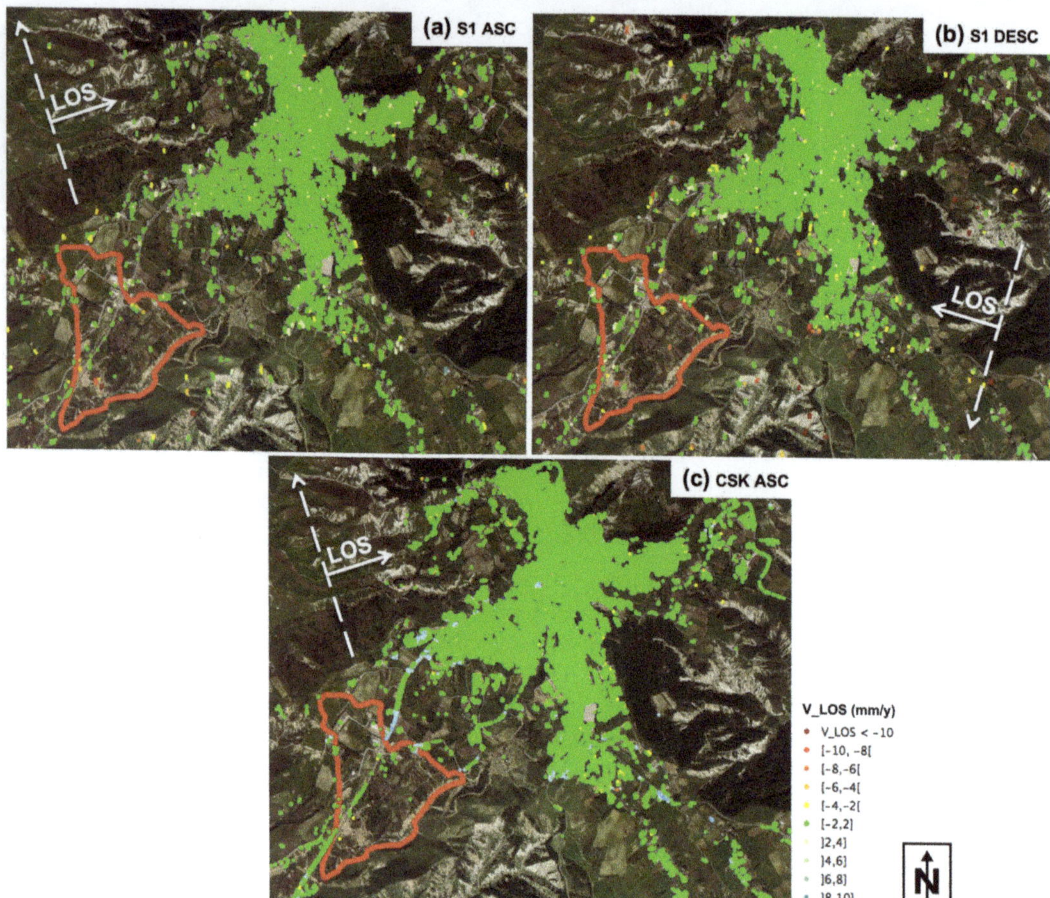

Figure 2. Mean LOS displacement rate maps over Montescaglioso village: (**a**) Sentinel-1 ascending dataset; (**b**) Sentinel-1 descending dataset; (**c**) COSMO-SkyMed ascending dataset. Dashed arrows represent orbit directions. The ground projection of the LOS vector is also shown. The red line delimits the known landslide body.

Differences between previously published MTInSAR results [36,37], covering a period up to 2013, and ours, reaching 2018, are minimal: all three maps show mostly stable points, i.e., green dots, with low displacement rates, ranging between −2 and +2 mm/y. This means that no major, large-scale processes are ongoing on the area. The target spatial distributions confirm that CSK is more effective over artificial structures (buildings, roads). S1 data are able to capture signals from targets located on the landslide body (indicated by the red contour on the maps). Both CSK and S1 provide interesting deformation time series with nonlinear trends, which are investigated in detail in the following.

A few points with high displacement rates, reaching in some cases ±10 mm/y (red and blue dots), can be discerned on the maps in Figure 2. Starting from these local displacements, we focused on four different situations: (i) post-landslide monitoring; (ii) monitoring of displacement evolution over a built-up area (a cemetery); (iii) monitoring of a structurally damaged building; (iv) monitoring of displacement evolution over a section of a connecting road.

3.1.1. Post-Landslide Monitoring

The SW hillslope of the Montescaglioso village is characterized by two main natural drainage networks and by a continuous cover layer of variable depth (sands, conglomerates and materials produced by human activities) overlying the impermeable Sub-Apennine Clay Formation in a monoclinal setting. The landslide, which occurred on 3 December 2013 after 56 h of continuous rain, covers an area of about 42 ha, caused severe failures along the road connecting Montescaglioso to the SP175 provincial road, and damaged some buildings for residential and productive use [17].

Despite the relatively wide area affected by the phenomenon and the presence of several man-made artifacts, the three datasets (S1 ascending and descending, CSK ascending) in Figure 2 show few targets over the landslide area. However, average displacement rates over these few points indicate a situation of general ground stability in the landslide area. Focusing on the S1 ascending displacement map (Figure 3a), it is possible to identify local displacements in correspondence of the Capoiazzo stream scarp (targets in the area delimited by a white polygon in Figure 3a) occurring with rates ranging between −4 and −10 mm/y.

The displacement time series in this area were further processed using both the *FE* and the NLTA, introduced in Section 2.3 and devoted to the automatic identification of nonlinear trends. Figure 3b,c show, respectively, the targets selected though the *FE* index according to (3), and those selected though the NLTA according to (2). The colors are representative of the *FE* values and of the polynomial degree selected by the NLTA, according to the color bars in the figures. In order to demonstrate the potential provided by these advanced approaches in support of the analysis of MTInSAR products, the time series corresponding to two targets, labeled P1 and P2, are shown in Figure 3. Target P1 was selected through the *FE* index and its displacement time series is sketched in yellow squares in Figure 3d. A discontinuity of about 28 mm is clearly visible, occurring in August 2018, which is due to an error in the phase unwrapping processing step. The red squares represent the displacement values obtained by subtracting the interferometric range of ambiguity ($\lambda/2 = 28$ mm) from the samples after the discontinuity, and thus provide the correctly unwrapped values. Hence, in this case, the *FE* index selection method was able to identify a target with a mostly smooth trend, but affected by a processing error, that, after proper correction, shows an interesting nonlinear continuous displacement trend, which can be now suitably fitted by a polynomial model, shown by the dotted black line in the figure. The NLTA was not able to select this target, since the discontinuity introduced by the phase unwrapping error does not allow the satisfactory modeling of the time series using a polynomial model. On the contrary, target P2, not affected by such an error, was properly selected by the NLTA, which models its displacement trend using a polynomial model of the third degree. Figure 3e shows the displacement values of target P2 as black dots and the polynomial curve as a red dotted line. Due to this more accurate model, the target temporal coherence increases, in the terms explained in Section 2.3. Both displacement times series of targets P1 (corrected for the phase unwrapping error) and P2 show a continuous trend with a marked change in rate in the first half of the 2018, which causes an increase in downward velocities.

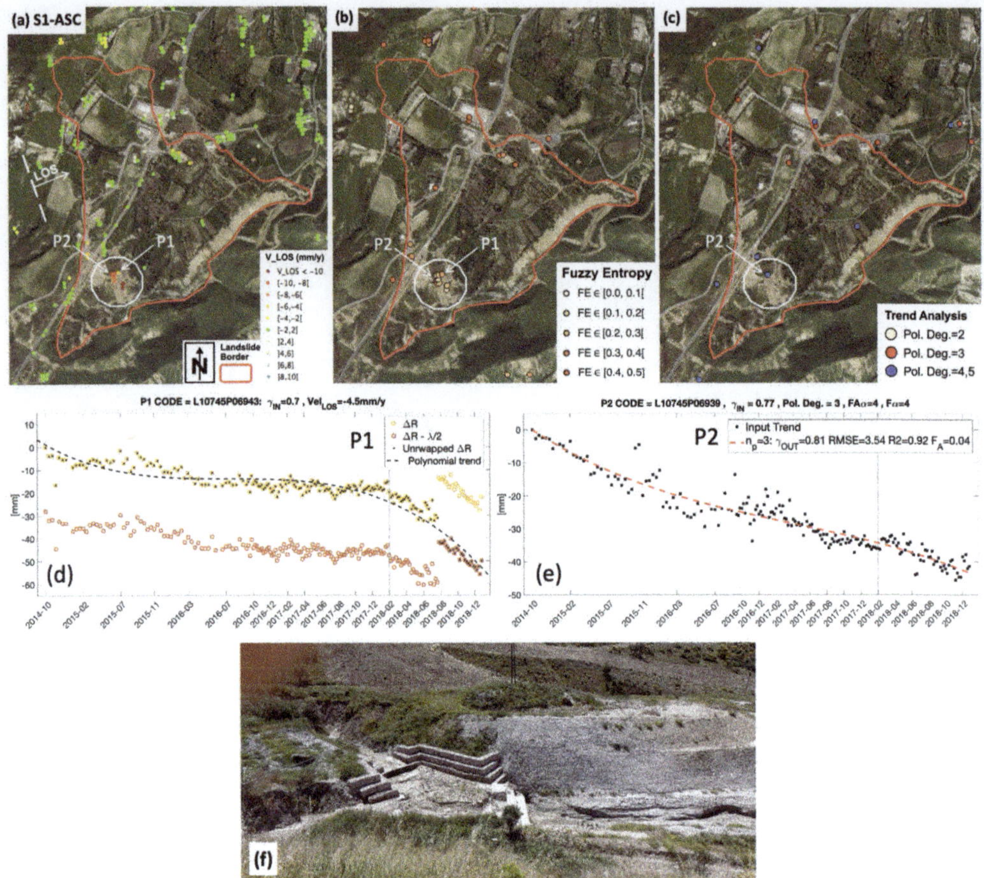

Figure 3. (**a**) Mean LOS displacement rate map over the Montescaglioso landslide derived by processing the Sentinel-1 ascending dataset; (**b**,**c**) targets selected respectively through the FE index according to (3), and through the NLTA procedure according to (2); (**d**) displacement time series corresponding to the target labeled P1 in panel (**b**): the yellow and red squares represent, respectively, the original displacement values and those obtained by subtracting the range of ambiguity ($\lambda/2 = 28$ mm) from the values after the discontinuity; (**e**) displacement time series corresponding to the target labeled P2 in panel (**c**): the black dots and the red dotted line represent, respectively, the displacement values and the polynomial curve derived through the NLTA; (**f**) ground picture of the area delimited by the white polygon in panels (**a**–**c**).

3.1.2. Cemetery Area

The MTInSAR datasets show SAR pixels over the Montescaglioso cemetery with high displacement rates (blue dots in Figure 4a) ranging between 6 and 10 mm/y. Local displacements were identified with MTInSAR datasets on the NW slope of the hill. Several displacement time series show a rate change at the end of 2013 with an increase in velocity, and a further one towards the end of 2014, with a decrease in velocity. As example of such behavior, in Figure 4b the displacement time series of the target labeled P1 in Figure 4a is reported.

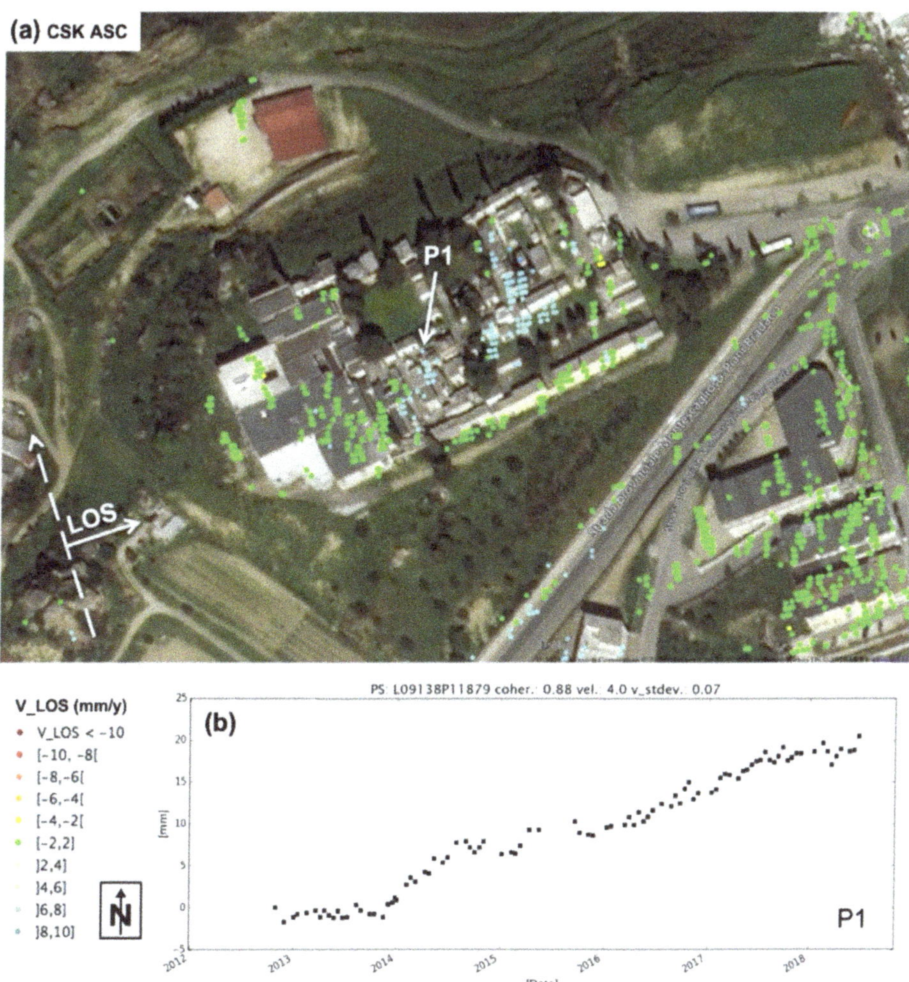

Figure 4. (**a**) Mean LOS displacement rate map over the cemetery in Montescaglioso village derived by processing the COSMO-SkyMed ascending dataset; (**b**) displacement time series corresponding to the target labeled P1 in panel (**a**).

3.1.3. Structurally Damaged Building

Figure 5a shows an enlargement of the mean LOS displacement map in Figure 2c, derived by the CSK ascending dataset, over an area containing a structurally damaged building, which was uninhabitable since 2013 (delimited by the dotted line in Figure 5a). The analyzed displacement time series shows a continuous and constant increase in displacements (with a mean velocity of 5.5 mm/y), as shown by temporal trends of P1 (Figure 5b).

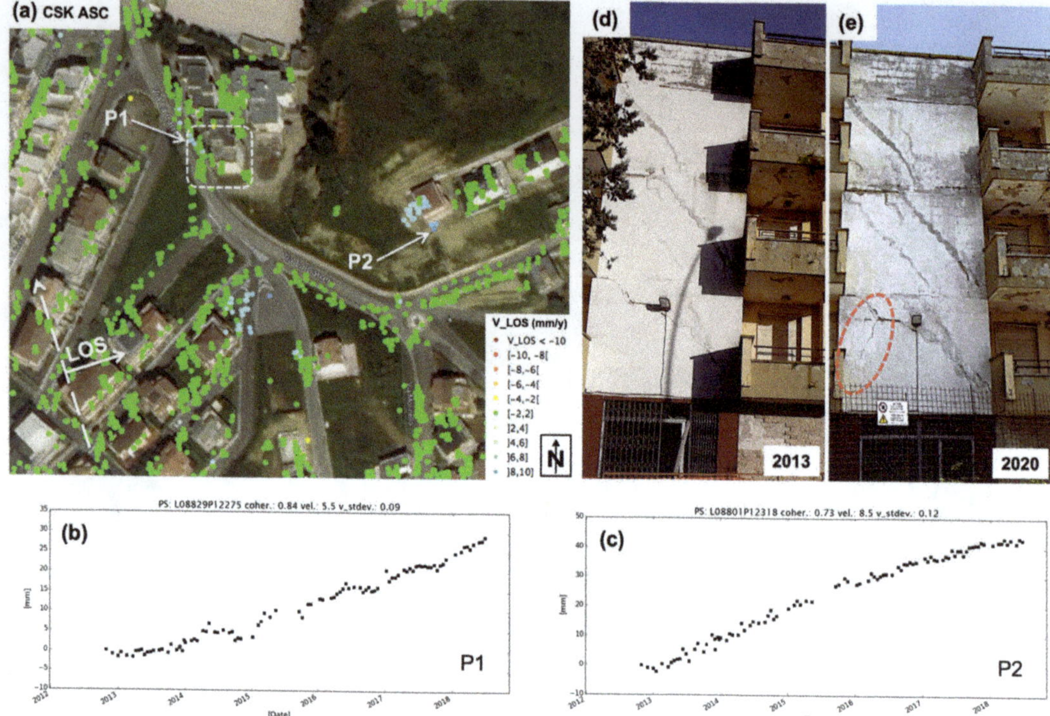

Figure 5. (**a**) Mean LOS displacement rate map over an area in Montescaglioso village derived by processing COSMO-SkyMed ascending dataset; (**b**,**c**) displacement time series corresponding to targets labeled, respectively, P1 and P2 in panel (**a**); (**d**,**e**) pictures of the building corresponding to target P1 taken in 2013 and 2020.

3.1.4. Connecting Road

Figure 6a shows another enlargement of the displacement rate in Figure 2c) over a road section linking the villages of Montescaglioso and Ginosa (the latter located in the Puglia region). The blue targets along the road show displacements with rates ranging between 6 and 10 mm/y, increasing along the LOS direction, which is consistent with a prevailing westward horizontal component of the ground movement. An example of such displacement trends is sketched in Figure 6b, corresponding to the target labeled P1 in Figure 6a. The point exhibits a rather steady upward trend of about 8.5 mm/y.

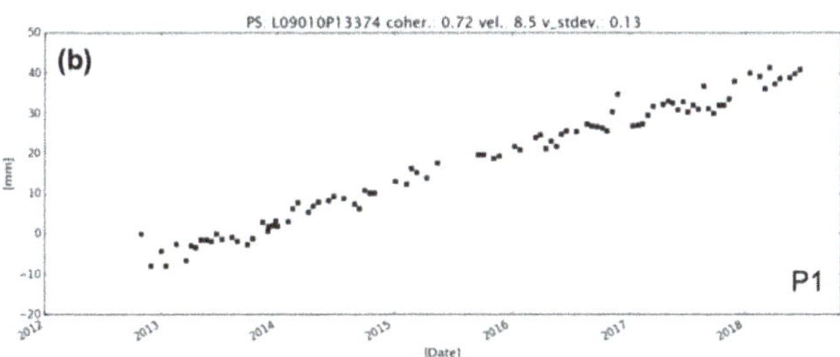

Figure 6. (**a**) Mean CSK ascending LOS displacement rate map over a portion of the road connecting the towns of Montescaglioso and Ginosa; (**b**) displacement time series corresponding to target P1 in panel (**a**).

3.2. Pomarico Test Site

Figure 7 shows the displacement maps along the LOS direction derived by MTInSAR processing of S1 ascending (Figure 7a), S1 descending (Figure 7b), and CSK ascending (Figure 7c) datasets over the Pomarico test site. The mean displacement maps are basically in agreement, highlighting a general situation of stable ground conditions with low displacement rates, ranging between −2 and 2 mm/y (green dots), and several local displacements with high rates, reaching in some cases −10 mm/y (yellow to red dots). Both CSK and S1 provide interesting displacement time series with nonlinear trends. However, as in the previous case, CSK appears to detect more stable targets over structures such as buildings and roads with respect to S1. We focused on two different scenarios: (i) pre-event landslide monitoring and (ii) monitoring of an area on the NE side of the village with high local displacements.

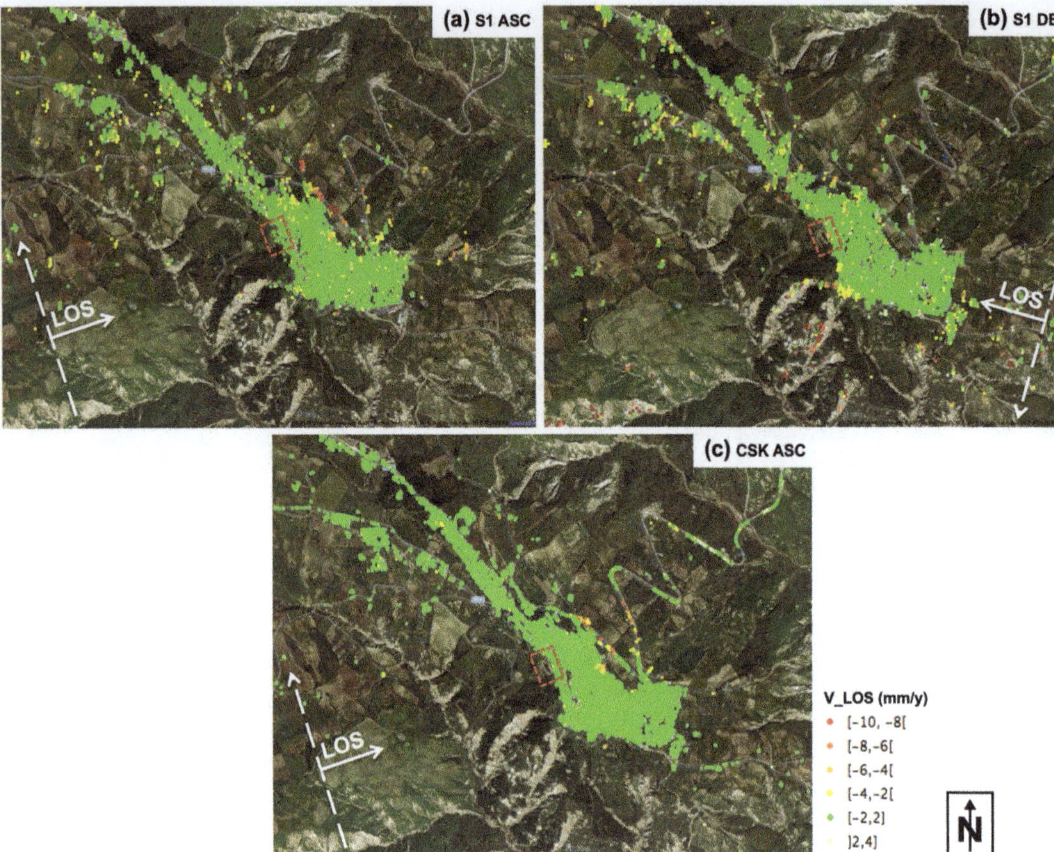

Figure 7. Mean LOS displacement rate maps over Pomarico village: (**a**) Sentinel-1 ascending dataset; (**b**) Sentinel-1 descending dataset; (**c**) COSMO-SkyMed ascending dataset.

3.2.1. Pre-Event Landslide Monitoring

The Pomarico landslide, delimited by the purple line on the enlarged maps in Figure 8a,b, occurred between 24 and 29 January 2019 after a long rainfall event; it is a retrogressive instability phenomenon that started with the fissures of the cemented cover layer and continued with a destructive evolution [18], which caused the collapse of a portion of the village (Figure 8c,d).

The mean displacement rate map sketched in Figure 8a shows moving targets located close to the landslide area. Figure 8b highlights a few targets selected through the NLTA procedure according to (2) and showing an approximating polynomial degree greater than 1 (nonlinear trend). Two targets labeled as P1 and P2 in the figure are selected as representative of the behavior of the two clusters of targets identified by the NLTA. Their displacement time series are represented by black dots in the plots in Figure 8e,f and Figure 8f, respectively. The dotted red lines in the same panels represent the polynomial curves selected by NLTA for modeling the displacement trends. In both cases, the temporal coherence increases with respect to the initial linear model trend, due to the improved kinematic models.

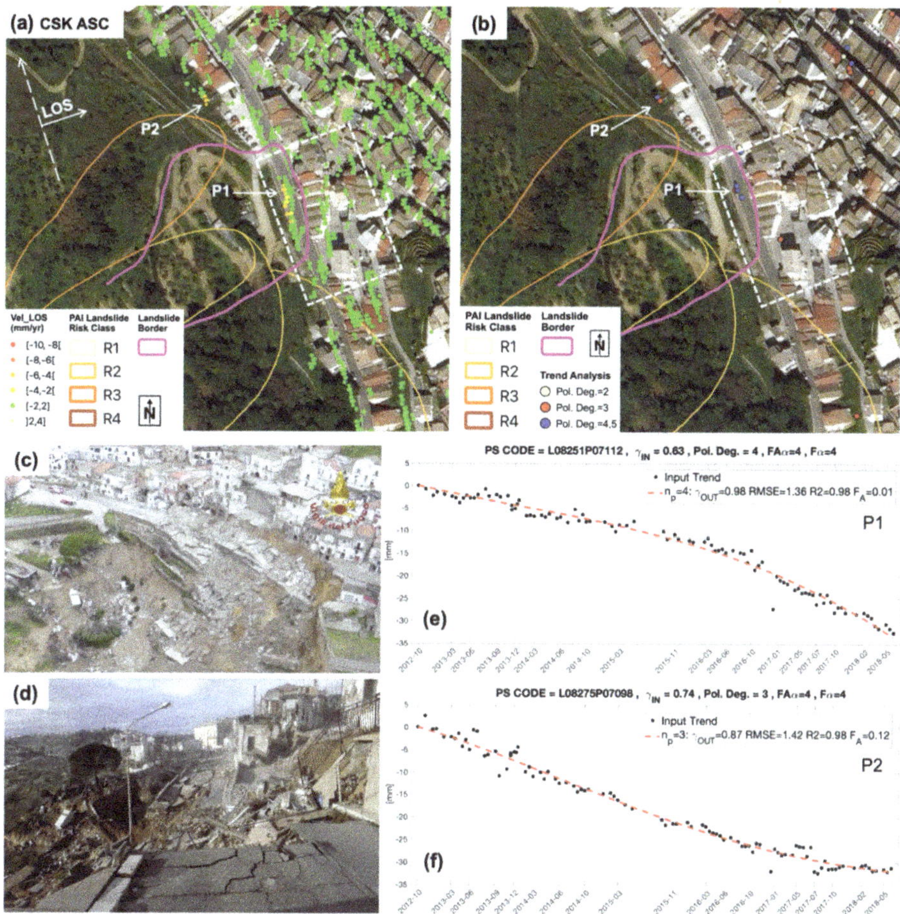

Figure 8. (a) Enlarged mean LOS displacement rate map over the Pomarico landslide derived by processing the COSMO-SkyMed ascending dataset; (b) map of targets selected though the NLTA according to (2) and showing a polynomial degree greater than 1; (c,d) pictures of the urban area affected by the landslide and bordered by the white rectangle in (a); (e,f) displacement time series corresponding to targets labeled, respectively, P1 and P2 in panel (a,b): black dots and the red dotted lines represent, respectively, the displacement values and the polynomial fit curves derived though the NLTA. The PAI (Piano di Assetto Idrogeologico) landslide risk map by the Basilicata section of the Southern Apennine District Basin Authority [27,28] is plotted with colored lines delimiting the risk classes areas, according to the colormap in the figure.

The selected P1 and P2 targets show displacement time series with opposite behaviors, which are representative of the two clusters of targets located, respectively, inside and outside the landslide: the former shows a linear trend with a rate of about −4.5 mm/y until the end of 2016, then it shows an increased downward linear rate of about −9.6 mm/y; the latter a linear trend with a rate higher than target P1, of about −6.5 mm until the beginning of 2017, then it shows a decrease in the absolute displacement rate, which reaches about −3.1 mm/y.

3.2.2. NE Side Area

Figure 9a shows an enlarged mean LOS displacement map of the CSK ascending dataset over the NE side of the Pomarico village. Targets exhibiting stronger local displacements are aligned on two sections of the road, the former upslope, close to the village, and the latter further downslope. Targets show mean displacement rates ranging between −2 and −10 mm/y (Figure 9a), and the corresponding time series show continuous and linear displacements starting from 2015 (Figure 9c,d).

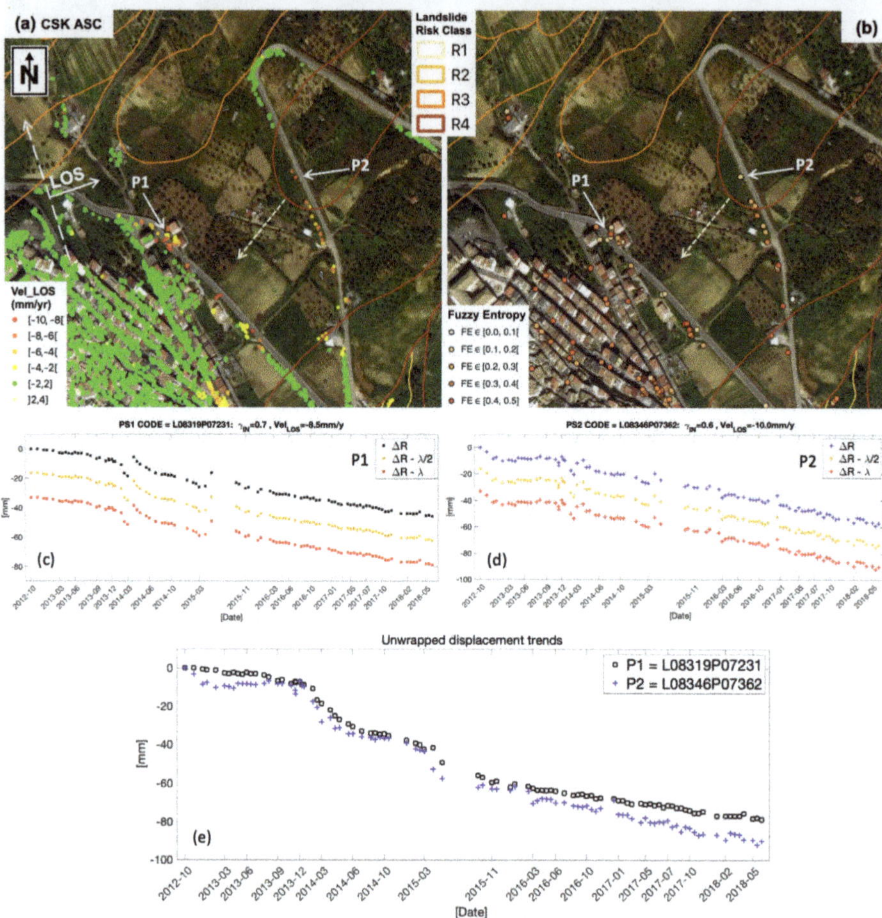

Figure 9. (a) Enlargement of the mean LOS displacement rate map derived by processing the COSMO-SkyMed ascending dataset over an area NE of the Pomarico village; (b) map of targets selected according to the *FE* index with a threshold of 0.5; (c) displacement time series (black dots) and its replicas obtained by subtracting $\lambda/2$ (yellow dots) and λ (red dots), corresponding to the target labeled P1 in panel (a); (d) displacement time series (blue crosses) and its replicas obtained by subtracting $\lambda/2$ (yellow crosses) and λ (red crosses), corresponding to the target labeled P2 in panel (a); (e) displacement time series corresponding to target P1 (black squares) and P2 (blue crosses) after correction of phase unwrapping errors. The PAI landslide risk map by the Basilicata section of the Southern Apennine District Basin Authority [27,28] is plotted with colored lines delimiting the risk classes areas according to the colormap in the figure.

Figure 9b shows the targets selected through the *FE* index according to (3). As in the example reported in Section 3.1.1, the *FE* index allows selecting targets with interesting nonlinear displacement trends, such as those sketched in Figure 9c,d, that refer, respectively, to the targets labeled P1 and P2 in Figure 9b. In both cases, like in the example in Section 3.1.1, the displacement time series are affected by discontinuities due to phase unwrapping errors, which can be corrected by exploring their replicas obtained by subtracting integer multiples of the range of ambiguity ($\lambda/2 = 16.5$ mm). For example, in Figure 9c, black dots represent the original displacement values of target P1, orange dots represent the displacements obtained by subtracting the range of ambiguity ($\lambda/2$), which accounts for the first discontinuity occurring around March 2014, and red dots represent the displacements obtained by subtracting twice the range of ambiguity (λ), which accounts for the second discontinuity occurring around April 2015. The displacement time series corrected by the phase unwrapping errors are sketched in Figure 9e in black squares for target P1 and blue crosses for target P2. Now the two targets, without discontinuities, show similar nonlinear trends of alternating periods of accelerating and decelerating displacements.

4. Discussion

In this section we discuss the reported examples of: (i) slope pre-failure monitoring on Pomarico landslide: (ii) slope post-failure monitoring on Montescaglioso landslide; and (iii) monitoring of structures (e.g., buildings and roads) affected by instability related to different causes.

4.1. Pre-Failure Monitoring

In the Pomarico village, as mentioned previously, between 25 and 29 January 2019 a complex landslide occurred: the roto-translational movement upslope, attenuated by a pile bulkhead along the roadside, evolved into a downstream earthflow (Figure 8a) [18]. The NLTA used for selecting targets (namely P1 and P2 in Figure 8b) evidenced nonlinear displacement trends related to the landslide event. The time series of the targets falling within the landslide area, in particular target P1 (Figure 8a,e), show continuous displacements. The area was in fact already classified as a medium-risk, R2 area (Figure 8a) in the regional landslide risk map (the regional Hydrogeological Management Plan, or PAI—Piano di Assetto Idrogeologico) by the Basilicata section of the Southern Apennine District Basin Authority [27,28]. Since the end of 2016, there has been an increase in downslope velocity. This may have created a contrast with the high-risk, R3 area (Figure 8a), causing it to decrease in velocity, as demonstrated by the time series of the target P2 (Figure 8f). These temporal trends describe the retrogressive behavior of the landslide, namely the expansion of the instability process beyond the edges of the medium-risk, R2 area (Figure 8a).

4.2. Post-Failure Monitoring

Several independent investigations have been carried out in recent years on the Montescaglioso landslide using different techniques, including LiDAR, SAR Interferometry and Pixel Offset [17,37–40], which present no major differences from our results. In particular, our analysis confirms the results of [17]: the landslide area shows a situation of substantial ground stability with only local displacements (Figure 3a). In this particular case, the *FE* index was able to recognize coherent targets affected by phase unwrapping errors. After performing proper corrections, these targets show interesting nonlinear continuous displacement trends that can be further analyzed using the NLTA. These displacements (targets in the area bordered with the white line in Figure 3a) are consistent with a prevailing vertical component, probably due to erosion phenomena of the Capoiazzo stream scarp. This behavior is confirmed by the displacement time series, which show a generally continuous trend. After a local survey, it was possible to identify the most likely cause of the displacement rate change in the first half of the 2018, shown in the displacement time series of P1 and P2 targets (Figure 3d,e): the beginning of major earthmoving works for the reconstruction of the Capoiazzo streambed (Figure 3f).

4.3. Displacement Evolution Monitoring

The cemetery area in the Montescaglioso village is located on the top of a hill, a residual strip of marine terraces, which consist of conglomerates and sands. Downslope of the area, monoliths of inclined cemented conglomerates highlight the recent instability of the terrace edges. Local displacements identified with MTInSAR datasets over this test site concern the NW slope of the hill. The direction of displacements is consistent with a prevailing west horizontal component, which is also confirmed by local surveys (Figure 10b–d).

Figure 10. (a) Enlargement of the mean LOS CSK PS displacement map over the Montescaglioso cemetery area (same as Figure 4a); (b–d) ground pictures of cracks consistent with displacements detected during a local survey, (b) on the NE part of the cemetery area, (c) on the road that runs alongside the cemetery and goes towards the valley, (d) on the wall of a chapel inside the cemetery; (e) satellite images (from Google Earth) showing the evolution from 2010 to 2019 of the N slope.

The cause of these displacements is to be found downslope of the NW slope, where satellite images (Figure 10e) show a continuous erosion of the slope toe. Moreover, Figure 10e is valuable for interpreting the temporal trend of the P1 targets (Figure 4b): (i) at the end of 2013 there was an increase in displacement trends, probably linked to the same heavy rainfall that triggered the Montescaglioso landslide; (ii) towards the end of 2014, a decrease in velocity occurred, attributed to some earthworks shown in Figure 10e (2015); (iii) the erosion subsequently continued, as the trend and Figure 10e (2019) show. These phenomena, which were also confirmed by local surveys, may evolve in the future, causing a landslide.

The NE side of Pomarico village is an area affected by various instability phenomena, mostly retrogressive roto-translational phenomena due to the presence of cemented cover layers [29,41]. The displacements identified by the CSK ascending dataset in this test site (Figure 9a) were also verified by local surveys. Along the road, which runs alongside the village, there are cracks and damaged structures (Figure 11) indicating continuous displacements. Again, in this case the *FE* index was able to recognize targets affected by phase unwrapping errors, that were corrected in order to derive reliable displacement time series. The analysis of these targets, characterized by a direction with a horizontal and downwards component (Figure 9a) and by similar trends of displacement time series (Figure 9e), suggests a retrogressive behavior of the instability process identified by the very-high-risk, R4 area. This hypothesis is confirmed by the cracks on the retaining walls and on the roads (Figure 11b,c) along the white dotted arrow in Figure 9a, detected during the local survey. This instability process, which is slowly expanding upslope, may cause further damage in the future and eventually evolve into a landslide.

Figure 11. (**a**) Enlargement of the mean LOS S1 ascending PS displacement map of the Pomarico NE side area (same area as Figure 9a); (**b–e**) ground pictures of (**b**) cracks on the retaining wall and on the road upstream, (**c**) cracks on the retaining wall and on the road downstream, (**d**) cracks on the road, (**e**) damaged sidewalk.

4.4. Structure Instabilities

The analysis of the MTInSAR time series was also carried out on two structures subject to already known instability phenomena, located in the Montescaglioso village.

In the case of the structurally damaged buildings (Figure 5), the cause of the displacements is due to the swelling phenomenon of the sub-Apennine clays [42], which outcrop in this area. The slope of the hill was reshaped and excavated for the building construction. The lightening of the underlying clay, also exposed to rainwater, has caused the swelling phenomenon with structural consequences on the building, such as differential settlements and consequent cracking (Figure 5d). Swelling, which is a very slow phenomenon that completely exhausts over many years, continued to cause small and constant displacements, as shown by the evolution of the crack pattern (Figure 5e) and displacement time series of the target labeled P1 in Figure 5b. Furthermore, the phenomenon affects the whole area around the hill, where the time series show trends consistent with an uplift, as shown by the trend of the target labeled P2 (Figure 5c). This analysis highlights the high sensitivity of the methodology, already observed in other sites subject to the swelling phenomenon, although in profoundly different contexts [43].

In the case of the road section in Figure 6, the analysis of the MTInSAR time series has detected continuous surface displacements, which have caused landslips (Figure 12a). These require continuous maintenance interventions (Figure 12b), as the potential evolution of the instability phenomenon represents both a danger for users and a problem for land management. Certainly, the study of MTInSAR time series is valuable for monitoring the evolution of instability phenomena along the road corridors. Furthermore, the combination of these analyses with vulnerability maps, such as those presented in [44], may provide support to regional planning by identifying the priorities for maintenance interventions. This approach becomes essential in areas characterized by distant urban centers and few connecting roads in a hilly environment, which is common in many regions.

Figure 12. Pictures of the road section in Montescaglioso village corresponding to the area where target P1 in Figure 6 is located.

4.5. Final Comments

At present, MTInSAR products may consist of millions of targets covering a large area, each associated with time series covering long time periods (potentially more than 20 years). In order to fully exploit such products, methods are needed for the automatic identification of targets with nonlinear displacement time series that may be relevant for ground phenomena such as reactivation/deactivation phases of known landslides. We adopted two approaches recently introduced in the literature, one based on the fuzzy entropy indicator, the other performing nonlinear trend analysis based on the Fisher statistics. The *FE* index was able to recognize coherent targets affected by phase unwrapping errors, which should be corrected to provide reliable displacement time series to be further analyzed. The NLTA was used for classifying targets according to the optimal degree of a polynomial function describing the displacement trend. This allowed selecting targets showing nonlinear displacement trends related to the several ground and structure instabilities as pre-failure, post-failure, and evolution signals. Thus, both approaches were very effective in supporting the analysis of ground displacements provided by MTInSAR, since they helped focusing on a smaller set of coherent targets identifying areas or structures on the ground that deserved further detailed geotechnical investigations. Moreover, the joint exploitation of MTInSAR datasets acquired at different wavelengths, resolutions, and revisit times provided valuable insights, with CSK more effective over man-made structures, and S1 over outcrops.

The analysis of pre-failure signals related to the Pomarico landslide showed that CSK has been able to capture better the building deformations preceding the landslide and the collapse. This allowed the understanding of the phenomenon evolution, highlighting a change in velocities that occurred two years before the collapse. This variation probably influenced the dynamics of the landslide leading to the collapse of an area considered to be at a medium-risk level by the regional landslide risk map. MTInSAR datasets also provided valuable results in the monitoring of the displacement evolution. CSK has been able to capture the building and structure deformations both in the Montescaglioso cemetery area and in the NE area of the Pomarico village. Displacement trends of selected targets have been confirmed in both test sites by on-site surveys. The analysis of the signals in the Montescaglioso cemetery area showed the displacement evolution of a site not included in the regional landslide risk map but of interest for local authorities and already affected by restoration works in recent years. The analysis of the NE side of the Pomarico village confirmed the very-high-risk R4 level of the area and showed the retrogressive evolution of the phenomenon, expanding beyond the borders currently established by the regional landslide risk map. These results clearly confirm the valuable use of MTInSAR products as a tool that is additional to the established techniques for studying the dynamics of instability phenomena and their evolution. The analysis of MTInSAR-based displacement time series, possibly performed though ad hoc automated procedures, can provide very useful information for long-term monitoring, management, and risk assessment at the regional level, when combined with planning tools such as the PAI.

The better sensitivity of CSK to capture building and structure deformations was confirmed by the analysis of structures with instability signals, i.e., the building and the road connection in the Montescaglioso village. In the first case, the signal analysis described a very complex phenomenon, due to clay swelling, which led to the evacuation of the building. In the second case, the analysis detected continuous surface displacements that require continuous maintenance interventions. Both cases represent examples of how the analysis of MTInSAR products can support the instability monitoring even at a local level. In particular, the first case concerns the application to slow phenomena extended over time, and shows the potential of the MTInSAR displacement time series analysis for investigating the behavior of structures subjected to yielding (e.g., buildings, bridges, viaducts). It can be a useful tool in planning maintenance activities, for instance, by supporting local authorities in identifying the structures that require timely technical inspections.

The analysis of post-failure signals related to the Montescaglioso landslide showed that S1 provided useful indications within the Montescaglioso landslide body. The selected trends confirmed the stability of the landslide area with some local displacements due to restoration works. In this case, the value of the MTInSAR displacement time series analysis emerged in the assessment phase of post-landslide stability, resulting in a useful support tool in the planning of safety measures in landslide areas.

5. Conclusions

This work was devoted to the exploration of long MTInSAR displacement time series derived from both CSK and S1 for investigating ground instabilities related, in particular, to landslide events. In fact, in order to fully exploit the information content of MTInSAR products covering long time periods, it is crucial to analyze not only the mean rate of displacements, but also their full time series, looking for signals related to the ground phenomena of interest that can be also highly non-linear. Examples are periodic signals related to structural thermal oscillations or seasonal ground subsidence, in addition to high-rate and non-linear trends that characterize the warning signals related to landslides or the pre-failure of artificial infrastructures. However, this detailed analysis is often hindered by the large number of coherent targets (up to millions) required to be inspected by expert users to recognize different signal components and also possible artifacts affecting the MTInSAR products, such as, for instance, those related to phase unwrapping errors.

In the present work the displacement time series analysis was supported by automated procedures recently developed for identifying targets with peculiar, nonlinear signals, one based on fuzzy entropy, and the other performing nonlinear trend analysis based on the Fisher statistics. This allowed the focus on a smaller set of coherent targets showing nonlinear displacements and the identification of trends affected by residual unwrapping errors.

We investigated the ground stability of two hilly villages located in the Southern Italian Apennines, namely the towns of Montescaglioso and Pomarico, where landslides occurred in the recent past, causing damage to houses, commercial buildings, and infrastructures. Specifically, we presented an example of slope pre-failure monitoring on the Pomarico landslide, an example of slope post-failure monitoring on the Montescaglioso landslide, and a few examples of buildings and roads affected by instability related to different causes (soil swelling, for example). The analysis of MTInSAR displacement time series showed that the joint exploitation of different MTInSAR datasets provides valuable information about the ground instabilities, with CSK more effective over man-made structures, and S1 over outcrops.

These results once more confirm the potential role of MTInSAR technology in supporting both regional and local civil protection organizations in all phases of risk management: identification, analysis and monitoring, assessment, and mitigation. To achieve this aim, it is very important that MTInSAR datasets are regularly acquired over long time periods with different wavelengths and acquisition modes, and to perform detailed analysis of displacement time series through reliable and automated procedures. Finally, despite the potential of both FE and NLTA in selecting reliable targets and characterizing their temporal behavior, an operational service devoted to slope instability early warning still needs more development. In particular, concerning the time series analysis, the NLTA should be followed by specific processing steps aimed, for instance, at deriving the time corresponding to a breakpoint [33], at forecasting future displacement behavior [45], and at predicting failure time for landslides [46]. The development of these processing steps is also the subject of ongoing research by exploring machine learning techniques.

Author Contributions: Conceptualization, F.B. and I.A.; methodology, F.B., A.R. and G.P.; investigation, I.A. and G.S.; formal analysis, F.B.; resources and software, R.N. and D.O.N.; validation, F.B., I.A. and G.S.; writing—original draft preparation, F.B. and I.A.; writing—review and editing, A.R., R.N., D.O.N. and G.S. All authors have read and agreed to the published version of the manuscript.

Funding: This work was supported by the Italian Ministry of Education, University and Research, D.D. 2261 del 6.9.2018, Programma Operativo Nazionale Ricerca e Innovazione (PON R&I) 2014–2020 under Project OT4CLIMA (ID ARS01_00405).

Acknowledgments: The authors thank GAP srl and Planetek Italia srl for providing MTInSAR time series through the Rheticus® platform. The Sentinel-1 data were provided through the Copernicus Program of the European Union. Project carried out using CSK® Products, © of the Italian Space Agency (ASI), delivered under a license to use by ASI. Finally, the authors thank M. Mottola for her support.

Conflicts of Interest: The authors declare no conflict of interest.

References

1. Crosetto, M.; Monserrat, O.; Cuevas-González, M.; Devanthéry, N.; Crippa, B. Persistent Scatterer Interferometry: A review. *ISPRS J. Photogramm. Remote Sens.* **2015**, *115*, 78–89. [CrossRef]
2. Singh Virk, A.; Singh, A.; Mittal, S.K. Advanced MT-InSAR Landslide Monitoring: Methods and Trends. *J. Remote Sens. GIS* **2018**, *7*, 1–6. [CrossRef]
3. Zhao, C.; Lu, Z. Remote Sensing of Landslides—A Review. *Remote Sens.* **2018**, *10*, 279. [CrossRef]
4. Wasowski, J.; Bovenga, F. Remote Sensing of Landslide Motion with Emphasis on Satellite Multitemporal Interferometry Applications: An Overview. In *Landslide Hazards, Risks, and Disasters*; Elsevier Inc.: Amsterdam, The Netherlands, 2015; ISBN 9780123964755.

5. Solari, L.; Del Soldato, M.; Raspini, F.; Barra, A.; Bianchini, S.; Confuorto, P.; Casagli, N.; Crosetto, M. Review of Satellite Interferometry for Landslide Detection in Italy. *Remote Sens.* **2020**, *12*, 1351. [CrossRef]

6. Cruden, D.M.; Varnes, D.J. Landslide types and processes. *Spec. Rep. Natl. Res. Counc. Transp. Res. Board* **1996**, *247*, 36–75.

7. Wasowski, J.; Bovenga, F. Investigating landslides and unstable slopes with satellite Multi Temporal Interferometry: Current issues and future perspectives. *Eng. Geol.* **2014**, *174*, 103–138. [CrossRef]

8. Bovenga, F.; Belmonte, A.; Refice, A.; Pasquariello, G.; Nutricato, R.; Nitti, D.O.; Chiaradia, M.T. Performance analysis of satellite missions for multi-temporal SAR interferometry. *Sensors* **2018**, *18*, 1359. [CrossRef]

9. Raucoules, D.; Tomaro, F.; Foumelis, M.; Negulescu, C.; de Michele, M.; Aunay, B. Landslide observation from ALOS-2/PALSAR-2 Data (image correlation techniques and sar interferometry). Application to salazie circle landslides (La Reunion Island). In Proceedings of the IGARSS 2018–2018 IEEE International Geoscience and Remote Sensing Symposium, Valencia, Spain, 22–27 July 2018; IEEE: Piscataway, NJ, USA, 2018; pp. 506–509.

10. Roa, Y.; Rosell, P.; Solarte, A.; Euillades, L.; Carballo, F.; García, S.; Euillades, P. First assessment of the interferometric capabilities of SAOCOM-1A: New results over the Domuyo Volcano, Neuquén Argentina. *J. S. Am. Earth Sci.* **2021**, *106*, 102882. [CrossRef]

11. Covello, F.; Battazza, F.; Coletta, A.; Lopinto, E.; Fiorentino, C.; Pietranera, L.; Valentini, G.; Zoffoli, S. COSMO-SkyMed an existing opportunity for observing the Earth. *J. Geodyn.* **2010**, *49*, 171–180. [CrossRef]

12. Lorusso, R.; Fasano, L.; Dini, L.; Facchinetti, C.; Varacalli, G.N. "COSMO-SkyMed di Seconda Generazione"—Civilian product specifications. In Proceedings of the Proceedings of the International Astronautical Congress, IAC, Bremen, Germany, 1–5 October 2018; International Astronautical Federation (IAF): Paris, France, 2018.

13. Torres, R.; Snoeij, P.; Geudtner, D.; Bibby, D.; Davidson, M.; Attema, E.; Potin, P.; Rommen, B.; Floury, N.; Brown, M.; et al. GMES Sentinel-1 mission. *Remote Sens. Environ.* **2012**, *120*, 9–24. [CrossRef]

14. Carlà, T.; Raspini, F.; Intrieri, E.; Casagli, N. A simple method to help determine landslide susceptibility from spaceborne InSAR data: The Montescaglioso case study. *Environ. Earth Sci.* **2016**, *75*, 1492. [CrossRef]

15. Bentivenga, M.; Giano, S.I.; Murgante, B.; Nolè, G.; Palladino, G.; Prosser, G.; Saganeiti, L.; Tucci, B. Application of field surveys and multitemporal in-SAR interferometry analysis in the recognition of deep-seated gravitational slope deformation of an urban area of Southern Italy. *Geomat. Nat. Hazards Risk* **2019**, *10*, 1327–1345. [CrossRef]

16. Bentivenga, M.; Bellanova, J.; Calamita, G.; Capece, A.; Cavalcante, F.; Gueguen, E.; Guglielmi, P.; Murgante, B.; Palladino, G.; Perrone, A.; et al. Geomorphological and geophysical surveys with InSAR analysis applied to the Picerno earth flow (southern Apennines, Italy). *Landslides* **2020**, *18*, 471–483. [CrossRef]

17. Pellicani, R.; Argentiero, I.; Manzari, P.; Spilotro, G.; Marzo, C.; Ermini, R.; Apollonio, C. UAV and Airborne LiDAR Data for Interpreting Kinematic Evolution of Landslide Movements: The Case Study of the Montescaglioso Landslide (Southern Italy). *Geosciences* **2019**, *9*, 248. [CrossRef]

18. Doglioni, A.; Casagli, N.; Nocentini, M.; Sdao, F.; Simeone, V. The landslide of Pomarico, South Italy, occurred on January 29th 2019. *Landslides* **2020**, *17*, 2137–2143. [CrossRef]

19. Bovenga, F.; Nutricato, R.; Refice, A.; Guerriero, L.; Chiaradia, M.T. SPINUA: A flexible processing chain for ERS/ENVISAT long term interferometry. In Proceedings of the 2004 Envisat & ERS Symposium, Salzburg, Austria, 6–10 September 2004; ESA Special Publication: Salzburg, Austria, 2005; pp. 473–478.

20. Spilotro, G.; Coviello, L.; Trizzino, R. Post failure behaviour of landslide bodies. In *Landslides in Research, Theory and Practice, Proceedings of the 8th International Symposium on Landslides, Cardiff, UK, 26–30 June 2000*; Bromhed, E., Dixon, E.N., Ibsen, M.L., Eds.; Thomas Telford: London, UK, 2000; Volume 3, pp. 1379–1386.

21. Caputo, R.; Bianca, M.; D'Onofrio, R. Ionian marine terraces of southern Italy: Insights into the Quaternary tectonic evolution of the area. *Tectonics* **2010**, *29*, TC4005. [CrossRef]

22. Cherubini, C.; Cotecchia, V.; Guerricchio, A.; Mastromattei, R. The stability conditions of the town of Aliano (Southern Italy). In Proceedings of the 7th International IAEG Congress, Lisbon, Portugal, 5–9 September 1994; Oliveira, R., Rodrigues, L.F., Coelho, G., Cunha, A.P., Eds.; A.A. Balkema: Rotterdam, The Netherlands, 1994; pp. 2145–2153.

23. Bozzano, F.; Floris, M.; Polemio, M. An Interpretation of Slope Dynamics in Pisticci (Southern Italy). In *Landslides in Research, Theory and Practice, Proceedings of the 8th International Symposium on Landslides, Cardiff, UK, 26–30 June 2000*; Bromhed, E., Dixon, E.N., Ibsen, M.L., Eds.; Thomas Telford: London, UK, 2000; Volume 1, pp. 171–176.

24. Troncone, A.; Conte, E.; Donato, A. Two and three-dimensional numerical analysis of the progressive failure that occurred in an excavation-induced landslide. *Eng. Geol.* **2014**, *183*, 265–275. [CrossRef]

25. Troncone, A.; Conte, E.; Donato, A. Three-dimensional Finite Element Analysis of the Senise Landslide. *Procedia Eng.* **2016**, *158*, 212–217. [CrossRef]

26. ISPRA—Istituto Superiore per la Protezione e la Ricerca Ambientale IFFI Project—English. Available online: https://www.isprambiente.gov.it/en/projects/soil-and-territory/iffi-project (accessed on 28 September 2021).

27. Autorità di Bacino Distrettuale dell'Appennino Meridionale PAI—Piano Stralcio per la Difesa dal Rischio Idrogeologico. Available online: http://rsdi.regione.basilicata.it/viewGis/?project=45774E9D-93DF-6578-E022-46605663079B (accessed on 28 September 2021).

28. Autorità di Bacino Distrettuale dell'Appennino Meridionale PAI—Aggiornamento 2019. Available online: http://www.adb.basilicata.it/adb/pStralcio/piano2019_adoz.asp (accessed on 28 September 2021).

29. Floris, M.; Bozzano, F. Evaluation of landslide reactivation: A modified rainfall threshold model based on historical records of rainfall and landslides. *Geomorphology* **2008**, *94*, 40–57. [CrossRef]

30. Pellicani, R.; Spilotro, G.; Ermini, R.; Sdao, F. The large montescaglioso landslide of December 2013 after prolonged and severe seasonal climate conditions. In *Landslides and Engineered Slopes. Experience, Theory and Practice, Proceedings of the 12th International Symposium on Landslides, Napoli, Italy, 12–19 June 2016*; Aversa, S., Cascini, L., Picarelli, L., Scavia, C., Eds.; CRC Press: Croydon, UK, 2016; Volume 3, pp. 1591–1597.

31. Troncone, A. Numerical analysis of a landslide in soils with strain-softening behaviour. *Géotechnique* **2005**, *55*, 585–596. [CrossRef]

32. Spilotro, G.; Fidelibus, M.D.; Pellicani, R.; Argentiero, I.; Parisi, A. La stabilità degli abitati su placche rigide su argille plio-pleistoceniche. Esperienze dai centri storici della Basilicata e della Puglia. *Geol. Dell'Ambiente* **2017**, *1*, 47–53.

33. Berti, M.; Corsini, A.; Franceschini, S.; Iannacone, J.P. Automated classification of Persistent Scatterers Interferometry time series. *Nat. Hazards Earth Syst. Sci.* **2013**, *13*, 1945–1958. [CrossRef]

34. Bovenga, F.; Pasquariello, G.; Refice, A. Statistically-based trend analysis of MTInSAR displacement time series. *Remote Sens.* **2021**, *13*, 2302. [CrossRef]

35. Refice, A.; Pasquariello, G.; Bovenga, F. Model-Free Characterization of SAR MTI Time Series. *IEEE Geosci. Remote Sens. Lett.* **2022**, *19*, 4004405. [CrossRef]

36. Bozzano, F.; Caporossi, P.; Esposito, C.; Martino, S.; Mazzanti, P.; Moretto, S.; Mugnozza, G.S.; Rizzo, A.M. Mechanism of the Montescaglioso Landslide (Southern Italy) Inferred by Geological Survey and Remote Sensing. In *Proceedings of the Advancing Culture of Living with Landslides. WLF 2017*; Mikos, M., Tiwari, B., Yin, Y., Sassa, K., Eds.; Springer: Cham, Switzerland, 2017; pp. 97–106.

37. Raspini, F.; Ciampalini, A.; Del Conte, S.; Lombardi, L.; Nocentini, M.; Gigli, G.; Ferretti, A.; Casagli, N. Exploitation of Amplitude and Phase of Satellite SAR Images for Landslide Mapping: The Case of Montescaglioso (South Italy). *Remote Sens.* **2015**, *7*, 14576–14596. [CrossRef]

38. Manconi, A.; Casu, F.; Ardizzone, F.; Bonano, M.; Cardinali, M.; De Luca, C.; Gueguen, E.; Marchesini, I.; Parise, M.; Vennari, C.; et al. Brief communication: Rapid mapping of landslide events: The 3 December 2013 Montescaglioso landslide, Italy. *Nat. Hazards Earth Syst. Sci.* **2014**, *14*, 1835–1841. [CrossRef]

39. Caporossi, P.; Mazzanti, P.; Bozzano, F. Digital Image Correlation (DIC) analysis of the 3 December 2013 Montescaglioso landslide (Basilicata, Southern Italy): Results from a multi-dataset investigation. *ISPRS Int. J. Geo-Inf.* **2018**, *7*, 372. [CrossRef]

40. Elefante, S.; Manconi, A.; Bonano, M.; De Luca, C.; Casu, F. Three-dimensional ground displacements retrieved from SAR data in a landslide emergency scenario. In Proceedings of the 2014 IEEE Geoscience and Remote Sensing Symposium, Quebec City, QC, Canada, 13–18 July 2014; pp. 2400–2403. [CrossRef]

41. Bozzano, F.; Cherubini, C.; Floris, M.; Lupo, M.; Paccapelo, F. Landslide phenomena in the area of Pomarico (Basilicata-Italy): Methods for modelling and monitoring. *Phys. Chem. Earth* **2002**, *27*, 1601–1607. [CrossRef]

42. Fidelibus, M.; Argentiero, I.; Canora, F.; Pellicani, R.; Spilotro, G.; Vacca, G. Squeezed Interstitial Water and Soil Properties in Pleistocene Blue Clays under Different Natural Environments. *Geosciences* **2018**, *8*, 89. [CrossRef]

43. Refice, A.; Pasquariello, G.; Bovenga, F.; Festa, V.; Acquafredda, P.; Spilotro, G. Investigating uplift in Lesina Marina (Southern Italy) with the aid of persistent scatterer SAR interferometry and in situ measurements. *Environ. Earth Sci.* **2016**, *75*, 243. [CrossRef]

44. Pellicani, R.; Argentiero, I.; Spilotro, G. GIS-based predictive models for regional-scale landslide susceptibility assessment and risk mapping along road corridors. *Geomat. Nat. Hazards Risk* **2017**, *8*, 1012–1033. [CrossRef]

45. Hill, P.; Biggs, J.; Ponce-López, V.; Bull, D. Time-Series Prediction Approaches to Forecasting Deformation in Sentinel-1 InSAR Data. *J. Geophys. Res. Solid Earth* **2021**, *126*, e2020JB020176. [CrossRef]

46. Carlà, T.; Intrieri, E.; Raspini, F.; Bardi, F.; Farina, P.; Ferretti, A.; Colombo, D.; Novali, F.; Casagli, N. Perspectives on the prediction of catastrophic slope failures from satellite InSAR. *Sci. Rep.* **2019**, *9*, 14137. [CrossRef] [PubMed]

 *remote sensing*

Article

Threshold Definition for Monitoring Gapa Landslide under Large Variations in Reservoir Level Using GNSS

Shuangshuang Wu [1,2], Xinli Hu [1,*], Wenbo Zheng [3], Matteo Berti [2], Zhitian Qiao [2] and Wei Shen [2]

[1] Faculty of Engineering, China University of Geosciences, Wuhan 430074, China; wshuang@cug.edu.cn
[2] Department of Biological, Geological and Environmental Sciences, University of Bologna, 40126 Bologna, Italy; matteo.berti@unibo.it (M.B.); zhitian.qiao2@unibo.it (Z.Q.); wei.shen2@unibo.it (W.S.)
[3] School of Engineering, University of Northern British Columbia, Prince George, BC V2N 4Z9, Canada; Wenbo.Zheng@unbc.ca
* Correspondence: huxinli@cug.edu.cn; Tel.: +86-13907152610

Abstract: The triggering threshold is one of the most important parameters for landslide early warning systems (EWSs) at the slope scale. In the present work, a velocity threshold is recommended for an early warning system of the Gapa landslide in Southwest China, which was reactivated by the impoundment of a large reservoir behind Jinping's first dam. Based on GNSS monitoring data over the last five years, the velocity threshold is defined by a novel method, which is implemented by the forward and reverse double moving average of time series. As the landslide deformation is strongly related to the fluctuations in reservoir water levels, a crucial water level is also defined to reduce false warnings from the velocity threshold alone. In recognition of the importance of geological evolution, the evolution process of the Gapa landslide from topping to sliding is described in this study to help to understand its behavior and predict its potential trends. Moreover, based on the improved Saito's three-stage deformation model, the warning level is set as "attention level", because the current deformation stage of the landslide is considered to be between the initial and constant stages. At present, the early warning system mainly consists of six surface displacement monitoring sites and one water level observation site. If the daily recorded velocity in each monitoring site exceeds 4 mm/d and, meanwhile, the water level is below 1820 m above sea level (asl), a warning of likely landslide deformation accelerations will be released by relevant monitoring sites. The thresholds are always discretely exceeded on about 3% of annual monitoring days, and they are most frequently exceeded in June (especially in mid-June). The thresholds provide an efficient and effective way for judging accelerations of this landslide and are verified by the current application. The work presented provides critical insights into the development of early warning systems for reservoir-induced large-scale landslides.

Keywords: threshold; landslide; early warning system; velocity; water level; GNSS

Citation: Wu, S.; Hu, X.; Zheng, W.; Berti, M.; Qiao, Z.; Shen, W. Threshold Definition for Monitoring Gapa Landslide under Large Variations in Reservoir Level Using GNSS. *Remote Sens.* **2021**, *13*, 4977. https://doi.org/10.3390/rs13244977

Academic Editors: Daniele Giordan, Guido Luzi, Oriol Monserrat and Niccolò Dematteis

Received: 29 September 2021
Accepted: 29 November 2021
Published: 8 December 2021

Publisher's Note: MDPI stays neutral with regard to jurisdictional claims in published maps and institutional affiliations.

1. Introduction

Landslides pose great threats to life and property. Many large-scale landslides have now been reactivated by the impoundment of large reservoirs behind high dams in southwestern China. The heights of these reservoir impoundments reach hundreds of meters. For example, the impoundment of Jinping's first reservoir is 230 m in height, and the fluctuating water level is 267% greater than that of the well-known Three Gorges Reservoir Area (TGRA) [1]. The framework of landslide monitoring has been established to provide a scientific basis for geohazard forecasting and warning in the TGRA [2]. However, many high dams with large reservoirs are now being constructed to the west of the TGRA, where the in-situ monitoring technologies and early warning strategies have not been applied and developed immediately. Consequently, large-scale landslides can be reactivated during reservoir construction and subsequent operation and may evolve into destructive failures [3]. Thus, designing and implementing monitoring and early warning systems (EWSs) are urgent for the safety of dam operations as well as that of residents and property on both river banks.

EWSs can be used for disaster mitigation based on landslide monitoring and prediction. Researchers have been encouraged to establish EWSs for various landslides around the world [4], because they are cost-effective and efficient. Concerning the mechanisms of landslides, scenario analysis, the evolution process, and long-term in situ monitoring laid an important foundation for the further development of EWSs. Researchers design and implement effective EWSs at regional and slope scales, respectively. They are established based on geological knowledge, field investigation, monitoring of geohazards with threats, and the choice of monitoring parameters [4–7]. For instance, an advancing regional warning model for rainfall-induced landslides, based on rainfall thresholds, has been proposed in a landslide-prone area in the Campania region, Italy [8]. An EWS has also been designed for the La Saxe rockslide with predefined displacement and/or velocity thresholds to offer the near real-time of failure [9].

There have been substantial advances in EWSs thanks to the development of modern monitoring technology. For instance, a global navigation satellite system (GNSS) has been installed on a deep-seated landslide in Slovenia to establish an EWS which has the advantages of low cost, open-source processing software, and automatic data collection over the Internet. The displacement data are correlated with rainfall data to reveal how different parts of the landslide react to precipitation and further develop the EWS [10]. Other commonly employed monitoring instruments include the ground-based interferometric synthetic aperture radar (GBInSAR), ShapeAccelArray (SAA), unmanned aerial vehicle (UAV), crack extensometers, and piezometers [11–13].

The importance of threshold definition in EWSs for landslides has been widely recognized [14]. Rainfall is the main trigger of landslides, and for this reason rainfall thresholds are the most well-known threshold definition methods in landslide and debris flows [15–17]. In contrast, water level thresholds are also used for reservoir-induced landslides, because these landslides' behaviors are strongly related to reservoir impoundment and fluctuations [18–20]. In this case, the two main landslide types are seepage-induced and buoyancy-induced, which are related to landslide geometry and materials [2]. Of course both rainfall and water levels influence the infiltration and pore water pressure within slopes, thus affecting the stability of landslides. Therefore, more parameters (e.g., soil moisture and underground water table) that directly affect the landslide body and sliding surface are needed to predict failure or establish EWSs [21–25]. In addition to various water-related thresholds, another effective threshold for EWSs at the slope scale is kinematics parameters, which can objectively demonstrate the behaviors of geohazards. Basically, measured displacement and its derivatives are most widely mined for threshold definition with the aid of useful on-site monitoring data [26–30].

A successful EWS at the slope scale encompasses proper early warning indicators, primarily using monitored displacement and its velocity and, sometimes, environmental quantities, such as critical water level and rainfall threshold [19,31,32]. The EWS model based on the three-stage creep theory of rock and soil materials is widely used in landslides. The model was first proposed by Saito in the 1960s and further developed with the popularity of the concepts of inverse velocity and displacement increment [26,33–35]. It was successfully applied on landslide forecasting with the improved accuracy of displacement increment and velocity. Systems can release alerts after observations reach default thresholds in order to save lives and properties and prevent environmental damage [36–38], but the prerequisite is that the landslide must have finished the constant deformation stage and entered the acceleration stage in Saito's model, because the velocity in the constant deformation phase needs to be predetermined for the establishment of an EWS.

On the other hand, the cumulative displacement of landslides under periodic reservoir water fluctuations always shows a constant period and some periodic (yearly or biennially) abrupt accelerations [39]. This cyclic displacement feature increases the difficulty of judging the landslide evolution stage. With this background, this study was carried out in a riverbank landslide located in the Yalong River, one of the tributaries of the upper Yangtze River, China. The Gapa landslide, with a volume of 13 million cubic meters, was chosen as a typical case by its type, scale, and large deformation. The warning level of the EWS is determined by five-year continuous monitoring with the aid of an improved Saito's model. The displacement data are thoroughly analyzed using the running average velocity to define a threshold. Both the velocity threshold and water level threshold are recommended for the Gapa landslide's EWS.

2. General Setting

2.1. Reservoir and Landslide Features in the Study Area

The western part of the upper Yangtze River is a landslide-prone area, especially the reservoir area. Many landslides were triggered due to large periodic water level changes of reservoirs in this area, seriously affecting the safety of bank residents and hydropower facilities. Seventy-seven landslides have been found in nine reservoirs in southwestern China along the Yalong River, Jinsha River, Dadu River, and Min River (Figure 1). These landslides developed behind hydropower dams, and 97% of these previously stable landslides have been triggered or reactivated by these reservoir impoundments. Thirty-six landslides are situated on the banks of the Yalong River. Among the 77 landslides, 42 landslides have volumes of more than 10 Mm^3. Southwestern China is also a bedrock landslide-prone area, due to its geomorphological and lithological features. Twenty-four landslides are formed by topping or bending, and 34 landslides are Triassic, in terms of the stratigraphic age of the sliding surface. These sliding-prone lithologies of the Triassic strata are sandstone, metamorphic sandstone interspersed with shale, slate, and coal layers.

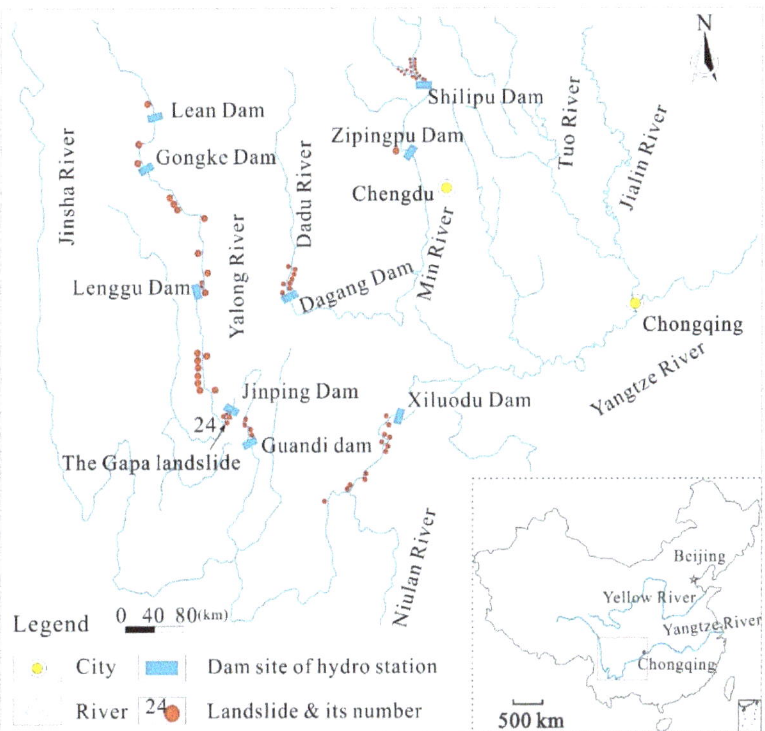

Figure 1. Riverbank landslide distribution in reservoir areas, Southwest China.

2.2. The Brief of the Gapa Landslide

The Gapa landslide is representative of the area, with almost all the main landslide features described above. The Gapa landslide is denoted as landslide no. 24 in Figure 1, behind the first Jinping hydropower dam.

Figure 2 presents the topographic map and typical cross-section of the Gapa landslide. The landslide is approximately 980 m long and, on average, 360 m wide from the front view, with a height of 470 m from the highest detected head scarp to its projected toe. The depth of the sliding surface is estimated to be 60 m. Accordingly, the estimated landslide volume is 13 Mm^3. The landslide consists of soil and blocky crushed rocks, mostly silty slate, mudstone, and conglomerate. It was formed by gravity in a pattern of slope bedding failures, and the bedding zone still exists today in bedrock at a certain depth below the sliding surface.

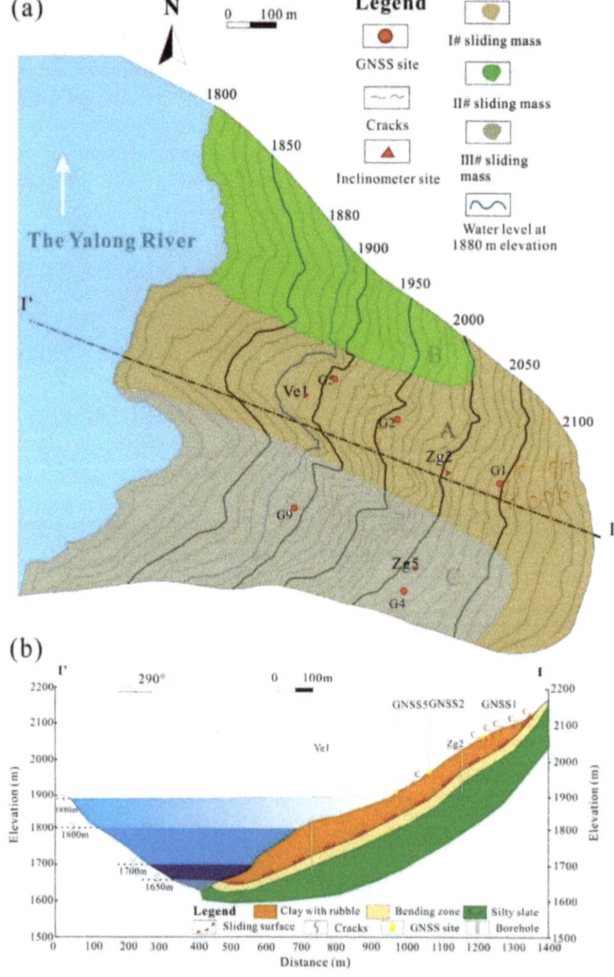

Figure 2. Engineering geological plan (**a**) and typical cross-section (**b**) of the Gapa landslide. (**a**) The landslide surface below 1800 m asl has been permanently submerged by the reservoir lake; (**b**) the reservoir impounded to 1880 m asl in three stages (1650–1700 m asl in November 2012, 1700–1800 m asl in July 2013, and 1800–1880 m asl in September 2014).

The first Jinping reservoir initially impounded in November 2012. The original lake level was 1650–1655 m asl, and it rapidly rose to 1700 m asl and remained at this lake level for a few months. Then, the reservoir was filled to 1800 m asl in August 2013. The reservoir impoundment in this area greatly affected the hydrogeological condition of these riverside slopes and landslides. After that the first impundment, the lake level cyclically fluctuated between 1800-m and 1840-m elevations in 2014 and between 1800-m and 1880-m elevations in 2015. The reservoir submerged half of the landslide at the lake level of 1880 m asl, as per Figure 2. These fluactuations caused continuous infiltration and discharge of reservoir water inside the slope. Local collapses occurred in the front, and cracks continuously developed and enlarged in the middle and upper parts of the landslide in 2014 and 2015. The reservoir operation recognized the need to increase our understanding of the evolutionary mechanisms and movement triggers of the Gapa landslide, as well as to implement a monitoring and early warning system for hazard control.

3. Evolution Process and Future Trend

3.1. Slope Behavior before the Impoundment

3.1.1. Historical Failure

The materials at the Gapa landslide's sliding surface were collected to determine the formation age. The result of soil thermoluminescence dating indicates that the landslide was formed between 14,600 and 23,500 years ago, which is during the late Pleistocene and early Holocene. Below the sliding surface, the bedrock can be divided into three sublayers, based on the bending degree, namely, strong bending, weak bending, and normal zones (Figure 3). The different bending degrees were illustrated by drill cores, which are neatly arranged with different fracture dips and block thicknesses, providing the bending evidence.

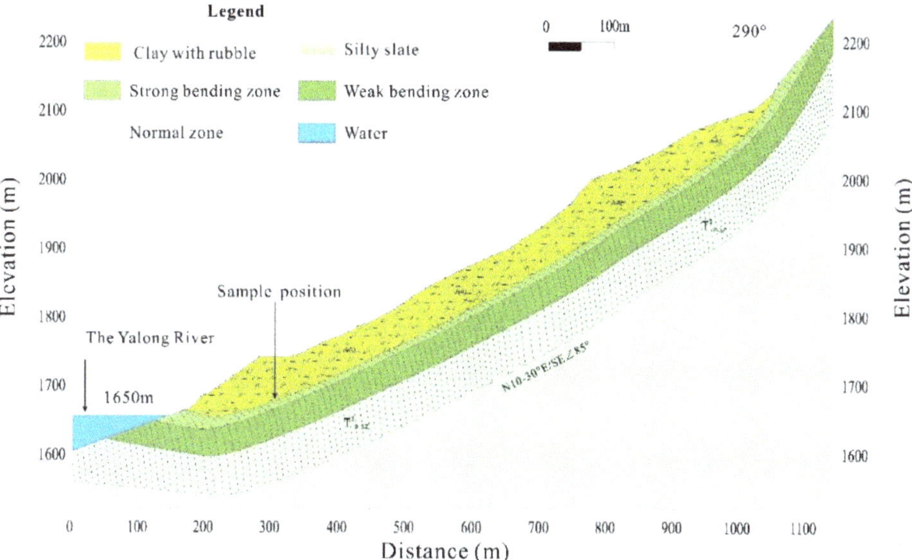

Figure 3. Cross-section of the Gapa landslide formed by bending pattern; the slope geometry refers to Zone A of the landslide before the reservoir impoundment.

3.1.2. Slow Movement before Impoundment

After the topping failure, it is hard to confirm the movement or evolution of the landslide without records, but satellite-based images and investigation provide insights into the landslide deformation in the 21st century. Figure 4 shows the images of the landslide in 2005 and 2011 from Google Earth. The difference in geomorphology between 2005 and 2011 is not significant. Before the reservoir impoundment (in November 2012), GPS stations were installed to monitor the deformation between 2007 and 2012. The GP2 station is near the location of the GNSS2 shown in Figure 5a; both are situated in the middle of the main cross-section, as shown in Figure 2. Figure 5b presents the cumulative displacement, with an average velocity of 14 mm/year during 2007–2012. According to the landslide velocity classification proposed by Cruden and Varnes (1996) [40], the Gapa landslide velocity before impoundment was below 15 mm/year and is considered to be very low. Without considering unknown earthquake occurrences or other influencing factors in history, it can be deduced that the Gapa landslide was formatted by rock slope topping and then deformed very slowly in the 21st century until the reservoir impoundment.

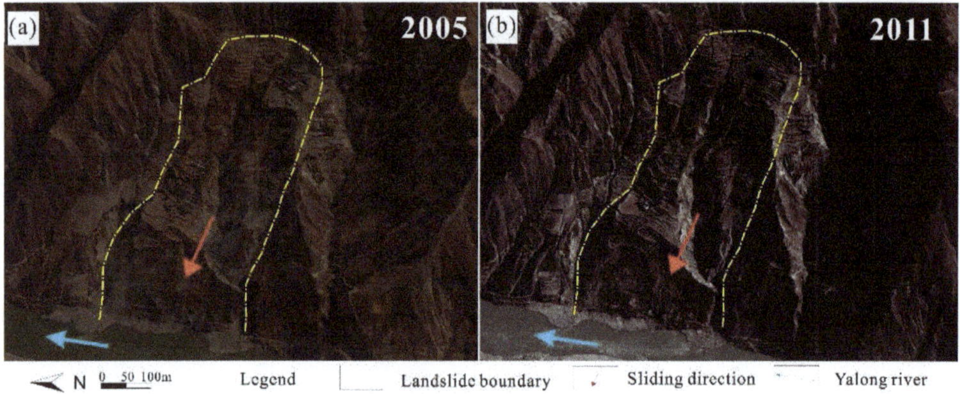

Figure 4. Satellite images of the Gapa landslide before reservoir impoundment (**a**) in 2005 and (**b**) in 2011 (Source: Google Earth).

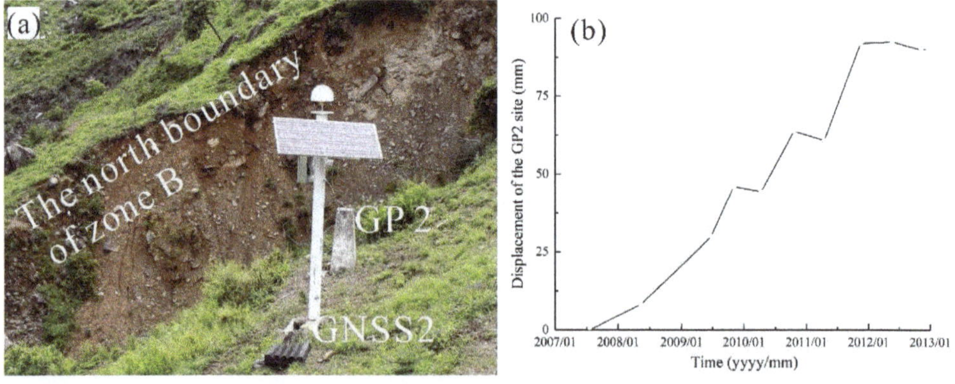

Figure 5. (**a**) GNSS2 unit and GP2 unit and their relative positions; (**b**) the cumulative displacement curve of GP2 between 2007 and 2012.

3.2. Recent Deformation Subjected to Impoundment

After the reservoir impoundment was completed, the Gapa landslide was expected to reactivate, due to the influence of reservoir water. Surface cracks developed on the ground highlighting the boundary of the landslide. Collapses occurred nearly above the highest water level. Consequently, field investigations were carried out to assess the landslide activity, and monitoring instruments were installed partially in 2016 and 2018 to set up a deformation monitoring system. The layout of the landslide monitoring is also shown in Figure 6. The monitoring plan includes three subsurface monitoring sites, six surface monitoring points, and an unmanned aerial vehicle (UAV).

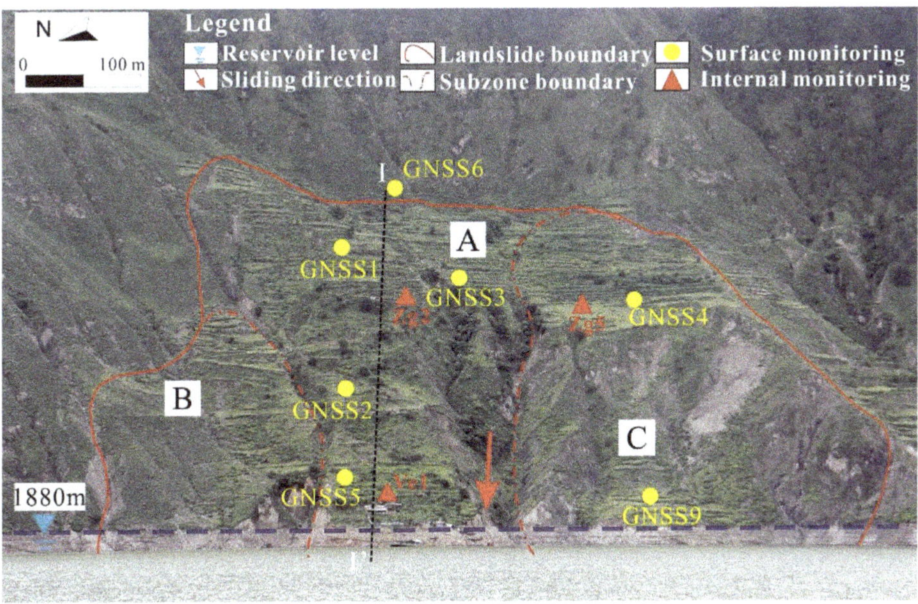

Figure 6. Overview and the monitoring layout of the current Gapa landslide.

The subsurface displacement was monthly monitored by inclinometers. Because the large deformation is beyond the allowable value of the instrument, none of the inclinometers can constantly access data, and they have been disused. However, the depth of the sliding surface was revealed by the available inclinometer measurements.

The crack distribution, using photos from the UAV with a camera, is illustrated in Figure 7. From the front view, the main boundary and zoning are highlighted with the red line, with a length of nearly 1 km and an average width of 2 m. In addition, two parallel tension cracks are on the upper part of the landslide at 2100 m and 2060 m asl, running through Zone B to Zones A and C. A transverse crack extends from Zone B to Zone A at 1900 m asl which does not pass to Zone C, due to the right-side gully and cliffs. Meanwhile, the UAV image shows the detailed cracks, which are rendered by solid yellow lines (Figure 7). In addition to those presented in the image, more than 30 visible cracks in total were found during the survey in 2018 with a certain distribution law. Most of the cracks continue to develop according to the later investigations in 2019 and 2020.

According to Kilburn and Petley (2003) and Xu et al. (2008), cracks on the ground surface are likely to gradually form a complete crack system with the increase of landslide deformation in space [41]. Unless the influences of external factors diminish, the cracking can grow at an accelerating rate until they coalesce into a major plane of failure; additionally, the cracking is enhanced by circulating water for deep-seated movements [42].

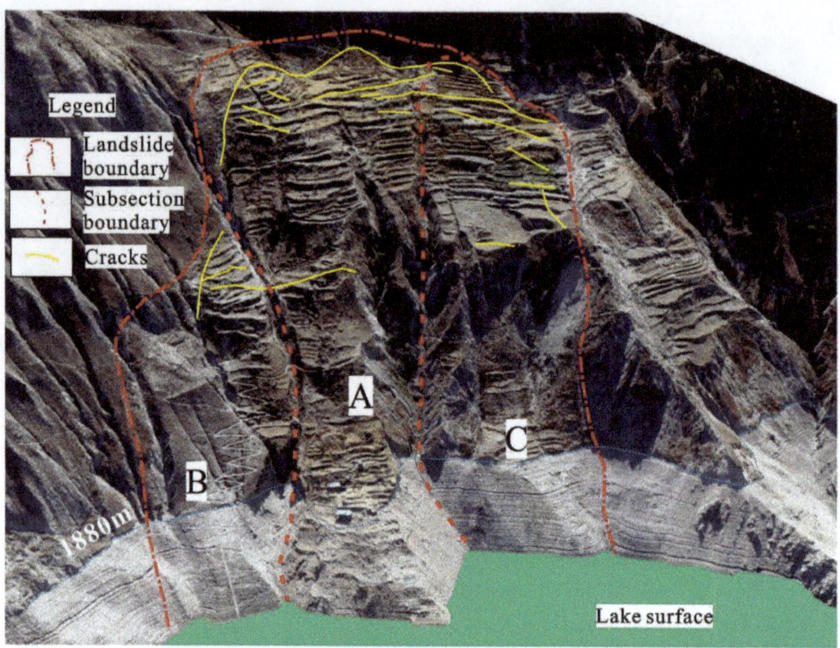

Figure 7. Main cracks of the overall Gapa landslide; this UAV image was captured in March 2019.

The surface displacements are daily recorded and obtained by a global navigation satellite system (GNSS). The GNSS processing is performed automatically in a server. Each unit sends raw GNSS data to the base station via a 3G signal. The base station is 11 km away from the landslide site, located in the dam safety center. GNSS data were processed in baseline mode by a batch least squares adjustment approach [10]. The observation was weighted with the inverse of the sine of the satellite elevation; its rate is 30 s, and the interval of the estimated coordinates is 24 h. However, the estimated daily displacement time series are formatted by the Yalong River Hydropower Development Company, which means that the company has direct access to the data, instead of every end-user.

The GNSS data between January 2016 and September 2020 were provided by the company for this study (over five years). The monitored main displacement is horizontal along the east-west, out of more than 60% of the combined displacement [43]. To check their validity, we reset one reference point on a supposedly stable terrain outside the landslide boundary (GNSS6, hereafter G6). The displacement data from the G6 site span from −8.5959 to 8.3742 mm between 2018 and 2020, converging on ±1 cm.

The surface horizontal displacement monitoring results are presented in Figure 8. During the first fluctuation cycle of the reservoir lake, the landslide deformation was unexpected and, therefore, was not monitored. The fluctuation cycle here includes different periods of the reservoir drawdown (Jan. to Jun.), filling (Jul. and Aug.), and maintaining at the highest level (Sep. to Dec.) in each year. From the second to the sixth fluctuation cycle, the movement of the landslide was daily recorded. The cumulative displacement increased during by the reservoir water fluctuations between 2016 and 2020. Some accelerated deformations are highlighted on the displacement curves during each fluctuation except the sixth fluctuation, whose lowest water level (1812 m elevation asl) is more than 10 m higher than usual, due to excessive rainfall. The acceleration always happened during lower water levels. Since the phase under a lower water level is short, the acceleration disappears after the water level increases.

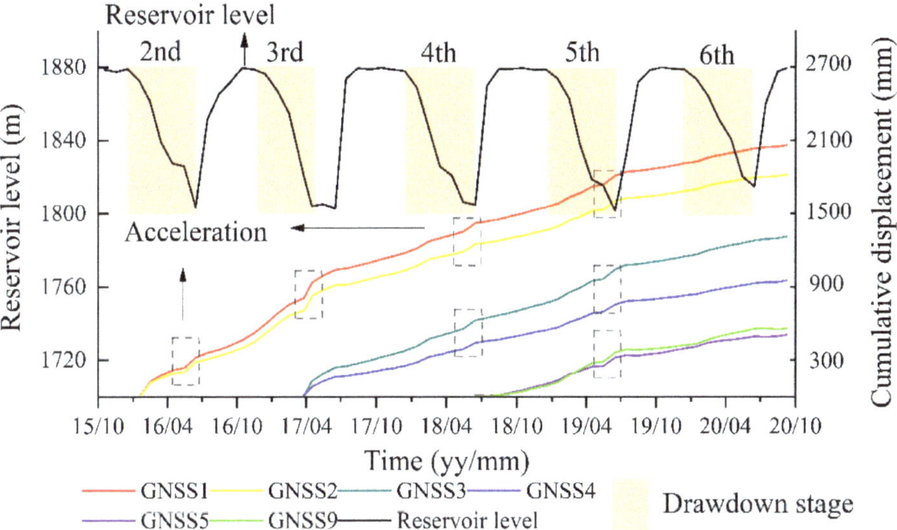

Figure 8. The surface displacement of GNSS units and water level from the 2nd to 6th fluctuation of the reservoir. Obvious accelerations indicated by the dotted box occur when the reservoir water drops to the lowest level.

Wu et al. (2021, 2022) and Hu et al. (2021) provided additional details about the impacts of reservoir water and its fluctuations on yearly periodic accelerations of the Gapa landslide [43–45]. The accelerations during the reservoir drawdown are mainly induced by: (1) the release of the buttressing effect of the reservoir lake; and (2) the outward seepage force inside the landslide before the dissipation of pore water pressure. Moreover, reservoir water dropping to the lowest level occurred within the local rainy season, which may also enlarge the acceleration behaviors [45]. More detailed displacement features are illustrated in Section 4.

3.3. Projected Displacement Trend

To investigate the evolution of the Gapa landslide, the displacement vs. time curve is plotted in order to explain the observed evolution stages (Figure 9). Five stages, including two possible displacement trends, are shown in the landslide lifespan curve: (1) topping failure (AB), which occurred during the late Pleistocene and early Holocene; (2) very slow motion (BC) until the reservoir impoundment in 2012; (3) new deformation (CD) caused by reservoir impoundment and fluctuations between 2013 to 2020. Notably, the interannual deformation rates in stages 2 and 3 have been compared by the typical monitoring sites GP2/GNSS2 (Figure 9). The velocities during stage 3 are dozens of times higher than during stage 2, indicating the reactivation of the Gapa landslide, but the velocity yearly or biennially decreases, instead of increasing, during the second and sixth fluctuations, which seems to be consistent with the features of the initial deformation phase based on the three-stage creep model (Satio 1969 [33]). In other words, if taking point C as the starting point of the old Gapa landslide reactivation, the displacement behavior of the reactivated Gapa landslide is unlikely to be the same as the Qianjiangping landslide or the Vajont landslide, which immediately collapsed after impoundment or the first reservoir water drawdown [46,47]. According to the Saito's model, the landslide has exhibited the initial stage. Based on this conceptual framework, predictions can be made about landslide deformation under the next long-term periodic fluctuations.

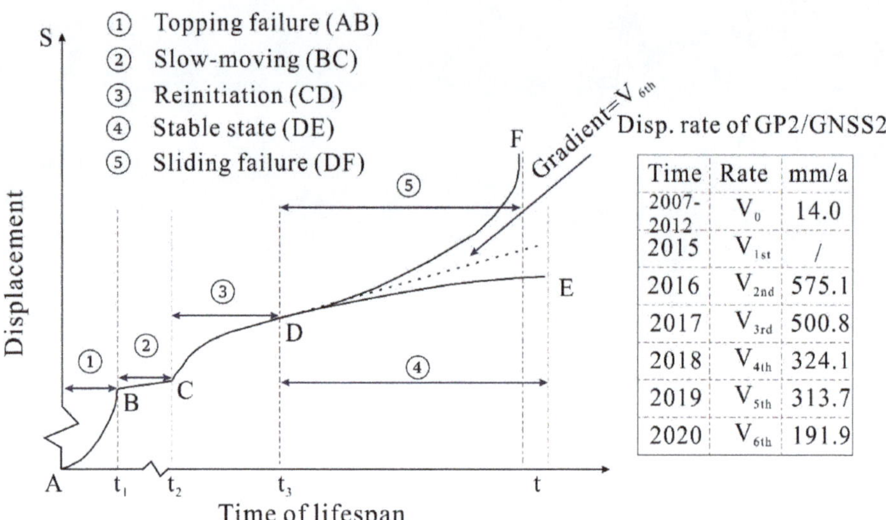

Time	Rate	mm/a
2007–2012	V_0	14.0
2015	V_{1st}	/
2016	V_{2nd}	575.1
2017	V_{3rd}	500.8
2018	V_{4th}	324.1
2019	V_{5th}	313.7
2020	V_{6th}	191.9

Figure 9. Displacement vs. lifespan curve of the Gapa landslide. In this figure, t_3 is the present time; therefore, stages AB, BC, and CD are supported by data, while stages DE or DF are inferred. The table on the right summarizes the yearly velocity of the GP2/GNSS2 unit between 2007 and 2020, indicating movement spans from constant rate, acceleration, and deceleration and corresponding to some period in stages BC and CD.

Thus, in addition to these three known stages, two likely stages—(4) stable state (DE) and (5) sliding failure (DF)—are suggested. Stage 4 means that the landslide can gradually adapt to the reservoir water influences after experiencing perennial deformations and finally recovers to the slow-moving state or reaches the stable state. Conversely, the reservoir water may lead to cumulative strain effects, reducing the mechanical strength within the landslide body. In this case, the Gapa landslide could experience a constant deformation stage and then acceleration deformation, as stage 5 illustrates.

At present, considering that the monitoring period is relatively short and thus has limited representativeness, the behavior of the landslide could change substantially in the following years. However, other factors, such as unexpected heavy rain, could jointly affect the landslide in the future. Based upon the above uncertainties, an EWS with a degree of security is strongly recommended, as well as effective thresholds.

4. Threshold Definition

4.1. Warning Model and Warning Level Determination

The modified creep model described above was used to develop the Gapa landslide early warning systems (EWSs) [36,37,48]. As shown in Figure 10, deformation before failure from the initial to acceleration stages can properly correspond to a four-class warning in the order of "attention level", "caution level", "vigilance level", and "alarm level". This EWS model can be established using in situ displacement monitoring. The data of displacement-time series can be presented in this way to judge the deformation stage of the monitored object and help to set the warning level, but the starting point of the monitoring time is often not at the onset of deformation [7]. Thus, this model cannot be realized without an understanding of the deformation behaviors and history, as well as of the evolution process, of the landslide unless the landslide has definitively entered the accelerated deformation stage.

Based on existing literature, the movement of the Gapa landslide is based on the motion of the deep-seated sliding surface [43]. Therefore, the EWS for the Gapa landslide under the periodic reservoir level fluctuations has now been preliminarily developed based on: (1) the knowledge of the landslide evolution process from topping to sliding; and (2) daily surface displacement data using GNSS monitoring over the last five years, as the motion at depths is close to that of the ground.

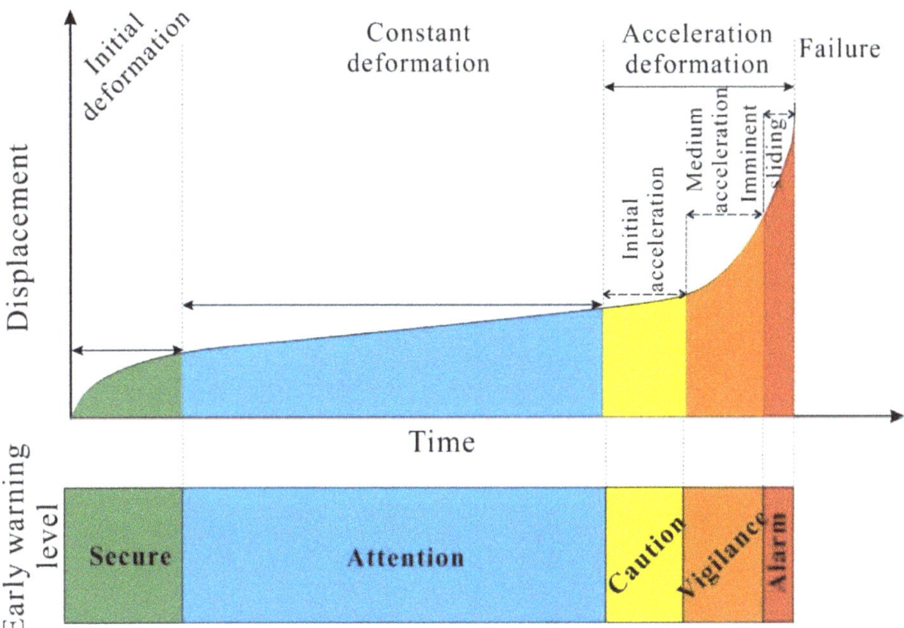

Figure 10. The warning model with four warning levels using the improved Saito's model (modified from [36]).

As stated above, the yearly rate of the landslide during the second and sixth fluctuations decreased by 40% (Figure 9), indicating a trend that the landslide movement now is experiencing the initial deformation stage and may enter the consistent deformation stage. However, it is hard to determine the ranges of deformation rates during the next couple of fluctuations or distinct futures, as this type of movement exhibited rapid acceleration during every reservoir drawdown period (Figure 8). Meanwhile, according to the literature, some reservoir landslides in Southwest China with a longer monitoring period have the adaptability or resistance to undergo the same scenario after they have deformed under the same and periodic fluctuations [49–51]. It is more likely that the deformation rate during the next couple of fluctuations will not exceed the velocity that occurred during the sixth fluctuation under the same single scenario, despite the uncertainties mentioned above. Therefore, it can be concluded that the level of warning is now properly releasing the attention level in terms of safety, as illustrated in Figure 10. The objectives of monitoring at the attention level are to determine whether acceleration is occurring or not, whether the acceleration phase is continuous, and whether it leads to tertiary deformation. Thresholds should, therefore, be defined at this level to issue warnings that acceleration has occurred or is likely to continue.

4.2. Velocity Threshold Based on Moving Average

The threshold must be conservative and minimize unnecessary warnings [52]. For this purpose, the displacement data are further analyzed to define the optimal threshold levels. Note that in this EWS model, the velocity, instead of the accumulative displacement, is applied to design the threshold, because the accumulative displacement can be easily influenced by the landslide scale and failure mode. Velocity values can be computed by original records of displacement according to the following equation:

$$v_j = \frac{S_j - S_{j-1}}{t_j - t_{j-1}} \tag{1}$$

where v is the displacement rate recorded between j and $j-1$ readings, S is the displacement recorded at reading j, and t_j is the date, expressed as a number corresponding to reading j. Note that the absence of daily displacement data is supplemented by linear interpolation.

The velocity vs. time series is illustrated in Figure 11, taking GNSS2 (G2) as an example. As gray scatters are shown in the figure, two features can be found: (1) the velocity points are chaotic, which hides the order or rule of the changes in velocity and time; and (2) many velocity points are at noise level and some below the zero line. The second feature is frequently observed in field monitoring data but without real meaning. Obviously, no thresholds can be directly picked by the gray scatters, but the two defects can be overcome by introducing the moving average (MA) method to deal with the original data, which is a useful method and commonly applied to data smoothing and decomposing in landslide fields [11,45,53,54]. The moving average displacement velocity can be calculated by the original time series and can be expressed as:

$$v'_t = \frac{v_t + v_{t-1} + \ldots + v_{(t-n+1)}}{n}, (t \geq n) \tag{2}$$

$$v''_t = \frac{v'_t + v'_{t-1} + \ldots + v'_{(t-n+1)}}{n}, (t \geq 2n - 1) \tag{3}$$

where v_t is the recorded data of the original time series at time t, v'_t is the value of the time series at time t by first moving average, v''_t is the value of the time series at time t by second moving average, and n is the moving interval for generating each data point, also known as moving window. Note that the moving window is the most important parameter in determining the MA trend, instead of the setting of the onset or the end of the time series. For example, n can be set between days and years to represent the short- and long-term trends of the velocity.

The seven-day running average is used once (Equation (2)) and twice (Equation (3)) to plot G2's velocity over short-term durations. Shown as black and orange scatters, most of the scatters distribute on the positive zone above the -1.3 and -0.2 mm/d delimited lines, respectively. Meanwhile, the change trends in velocity are obvious, with some accelerations and decelerations, which is beneficial for finding the most common rates and maximum rates.

Based on MA, short- and long-term G2 velocity trends are summarized in time series, as shown in Figure 12. The running values are chosen as 7, 30, 120, and 360 days to represent weekly, monthly, seasonally, and yearly velocity trends of G2, respectively. As the running value increases, the velocity curve smooths and the maximum rate decreases. Meanwhile, the maximum rates both in the short and long terms reduce as the fluctuation time increases over the monitoring period. For instance, the weekly maximum rate in 2016 and 2017 is 8.2 mm/d, then declines to 3.5 mm/d in 2018 and 2019, and further decays to 2 mm/d in 2020 (gray line). This finding can be drawn from the monthly maximum rate as well (blue line). From 2016 to 2020, while the number of fluctuations in the reservoir increases, the induced landslide deformation rate gradually decreases. The MA results are consistent with the previous observation (Figures 8 and 9), which is important for determining the threshold.

Meanwhile, the accessed data should be processed in a timely manner in an EWS, notably the weight of new data. An EWS cannot rely on the long-term MA, otherwise the time effectiveness of the EWS will be missing. Thus, the weekly running, which presents the short-term velocity trend, seems to be more suitable for finding a threshold than others.

Meanwhile, in order to make the seven-day running velocity more available, double running is necessary, because the double moving average (DMA) successfully captures variations in velocity and, simultaneously, reduced scatter [11], as shown in Figure 11. The DMA can also eliminate the time lag effect on the original time series compared with the single MA [45].

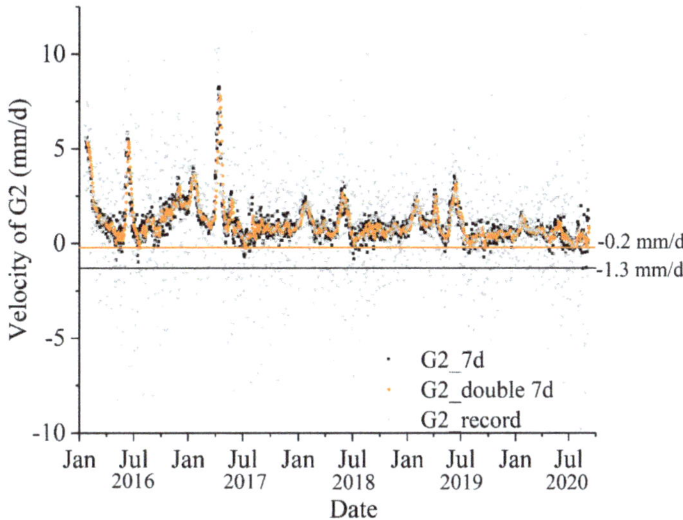

Figure 11. Original and MA/DMA velocity series of G2 unit during 2016~2020; −1.3 mm/d is the minimum MA velocity, and −0.2 mm/d is the minimum DMA velocity; the running window is 7 days.

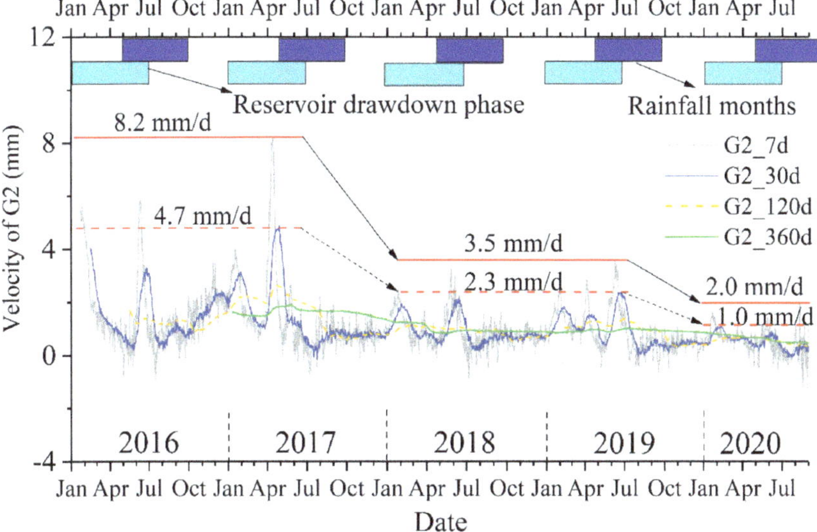

Figure 12. MA velocity series of G2 unit in short− and long−term running windows during 2016~2020. Seven (30) days are the short-term running window, and the yearly maximum MA velocities are 8.2 (4.7) mm/d in 2016 and 2017, 3.5 (2.3) mm/d in 2018 and 2019, and 2.0 (1.0) mm/d in 2020. In addition, 120 and 360 days are the long-term running windows, and maximum MA velocity is close to the mean in the velocity series.

The 8.2 mm/d velocity is the maximum weekly average velocity known to have occurred, and 2 mm/d is the maximum weekly average during the most recent fluctuation. Thus, all the velocity scatters are classified into velocity groups ranging from less than −8 mm/d to more than 8 mm/d to find the regularities of velocity distribution. The velocity groups between −8 and 8 mm/d are (−8, −2], (−2, 0], (0, 2], and (2, 8]. The statistics results of velocity in records, seven-day average, 30-day average, double seven-day average, double five-day average, and double three-day average are compared (Figure 13). The velocity group (0, 2] has the maximum number of frequencies in all situations, with percentages of 30.1%, 80.6%, 87.6%, 85%, 87.7%, and 86.4%, respectively. More than 80% of velocities in the running average are limited below 2 mm/d. Therefore, not only in the sixth fluctuation, the running average velocities are below 2 mm/d, but the majority of velocities are less than 2 mm/d. For the analysis result of G2, it seems that the running average velocity of 2 mm/d is an effective value for acceleration warning, but note that the velocity of more than 2 mm/d in original records has a percentage of 66.8%. It is not reasonable to directly use the threshold (2 mm/d) for warning in monitoring, as it brings too many false warnings.

In addition, in the cases of five-day and three-day double moving average velocities (Figure 13e,f), the percentages of measurement points below 0 mm/d are 4.9% and 10.8%, which are 2.1% and 8.0% higher than the seven-day double moving average, respectively. For maximizing the use of information from all data points in the double running average, the great majority of the points after the running average should be greater than zero to obtain a specific meaning. Based on this purpose, the smallest window for the short-term moving average is seven days in this study.

For the threshold values of the whole monitoring system, we pre-checked the Pearson correlation coefficient (PCC) of the cumulative displacement and velocity between each remaining GNSS unit and G2 unit. The results show that the displacement coefficients are more than 0.95 and the velocity coefficients can also exceed 0.8, both suggesting that the other GNSS units have a strong similarity with the G2 unit in variations of displacement and velocity. Therefore, the threshold value of 2 mm/d was tested for the G1, G4, G5, and G9 units. The triggering factor of the deformation velocity is the fluctuations of reservoir levels. Therefore, the time series, including velocity, water level, and rate of change in water level, are presented, featured in Figure 14. Taking the double seven-day running average results of G1 as an example (Figure 14a), running average velocity scatters are distributed in three areas. Most of them are less than 2 mm/d. Those above the 2 mm/d line concentrate on the areas right of the 1819 m line and left of the 1860 m line, respectively, which represents displacement acceleration. Taking the rate of water level change factor into account, accelerated behaviors occurred at the beginning of the reservoir drawdown (left of 1860 m line) and during the lower water level (right of 1819 m line); however, the rate of water level change correlates more poorly with the accelerated deformation than the effect of water level, as many colored points with a bigger filling rate or dropping rate are below the 2 mm/d line. The water level controls the accelerated deformation, instead of the water level change rate, which is similar to the Muyubao landslide in the TGRA [19].

Note that we have compared the results using the seven-day and 10-day running averages on the G2 monitoring site (Figure 14b,c). They have a similar three-area feature as described in G1, and the critical velocity of the two is the same. Thus, using a 10-day running window for MA would not obtain a lesser delimited velocity than the seven-day interval, but it requires more data and more monitoring time in practical application, which is contrary to the timeliness capability of EWS. Thus the seven-day is determined as the unique short-term running window for MA.

Besides the G1 and G2 units, similar findings can also be obtained by the rest of the GNSS points, but the difference is the onset of the monitoring time of these GNSS sites, which affects the highest running average velocity. G1 and G2 started monitoring in January 2016; G3 and G4 were recorded from March 2017; and G5 and G9 were installed in June 2018. The delimitation for velocities of the G5 site by the 2 mm/d line is higher than reality (Figure 14d). However, if the G5 site had been monitored earlier, 2 mm/d would be proper. As the monitoring starts later, the acceleration area on the left of the 1860 m line gradually disappears, but the acceleration areas on the right of 1817–1819 m lines have been presented without exceptions. Below or above a critical water level is a criterion condition to judge the initiation or termination motions of many reservoir-induced landslides [55]. For all monitoring units, it can be concluded that the boundary of the acceleration area is delimited below the 1820 m water level. Thus, we selected a higher integer with a value of 1820 m as the critical level to unify the water level thresholds.

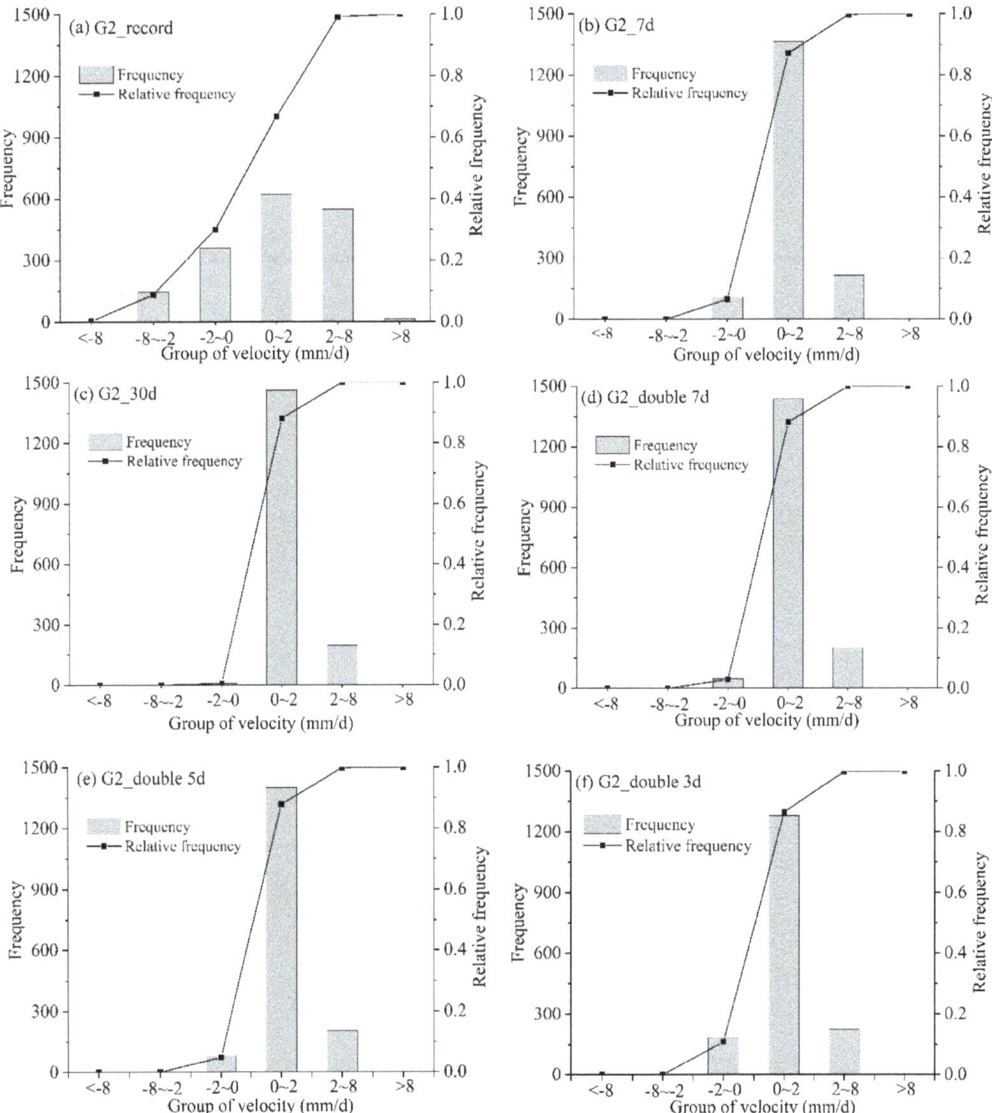

Figure 13. Histogram of rate distribution of G2: (**a**) recorded rates, (**b**) 7−day MA rates, (**c**) 30−day MA rates, (**d**) 7−day DMA rates, (**e**) 5−day DMA rates, (**f**) 3−day DMA rates.

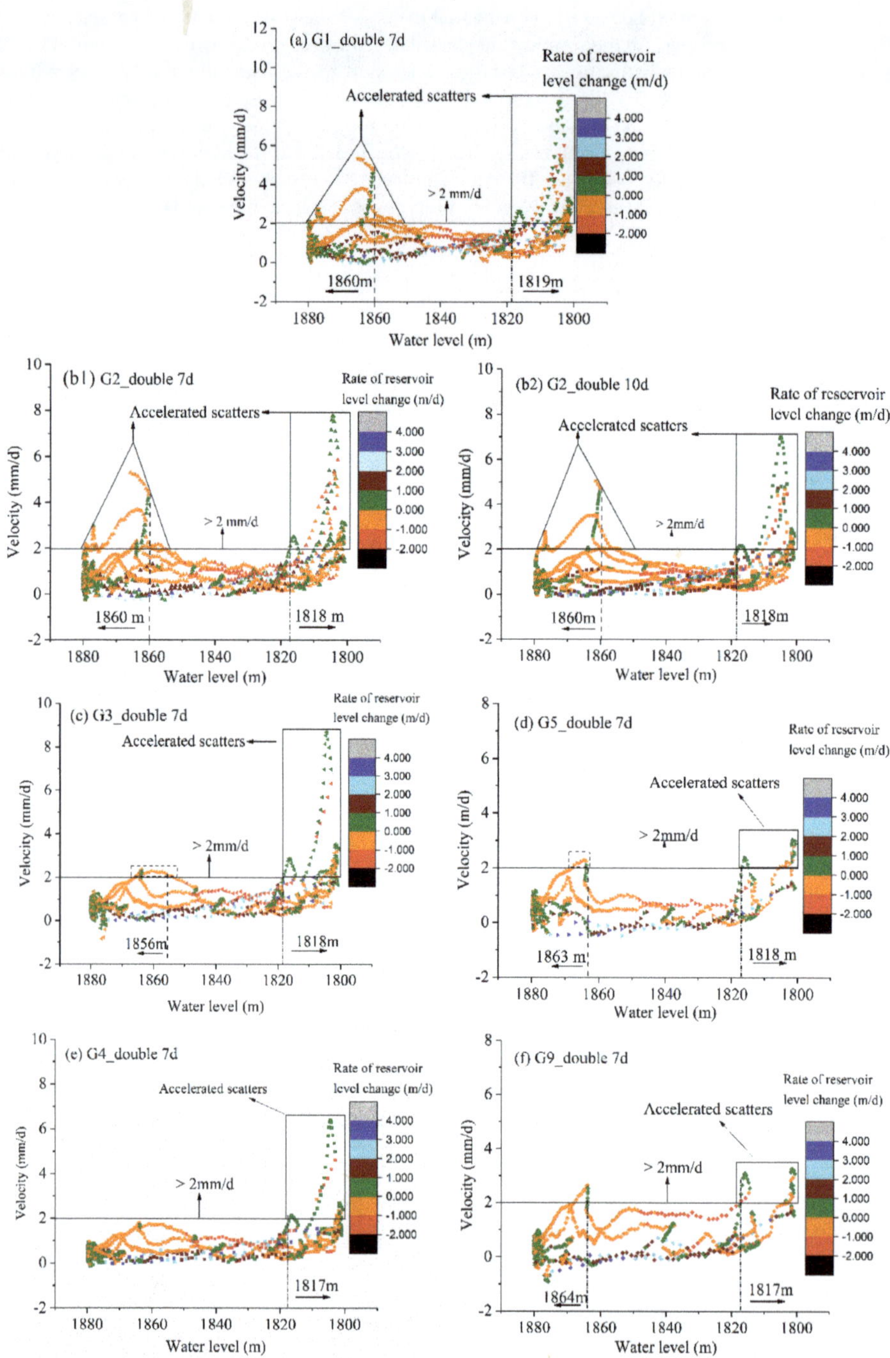

Figure 14. Correlations between the reservoir level and 7−day and 10−day DMA velocities: (**a**) G1_double 7-day, (**b1**) G2_double 7−day, (**b2**) G2_double 10−day, (**c**) G3_double 7−day, (**d**) G4_double 7−day, (**e**) G5_double 7−day, (**f**) G6_double 7−day. In these figures, scatters are colored by different levels of rate of reservoir level change, and the colored levels are shown at the right of every figure.

Water level below 1820 m is a dangerous scenario for the Gapa landslide, which is different to the abovementioned Muyubao landslide in the TGRA for which the warning water level is above a certain water level (172 m) [19]. The warning water level with opposite directions (below or above) of the two landslides can represent the seepage-induced and buoyancy-induced reservoir landslides, respectively, as indicated in the introduction section. Thus, the deformation mechanism controls the positive/negative threshold for critical water level.

In summary, two thresholds can be drawn from the displacement monitoring and reservoir levels: (1) the 2 mm/d of seven-day DMA velocity; and (2) the 1820 m water level. DMA velocity greater than 2 mm/d and water level below 1820 m asl are prone to trigger acceleration and be adverse to the Gapa landslide.

4.3. The Recommended Thresholds for EWS

In this case, Figure 11 has pointed out that the distribution of daily velocity data is extremely discrete with high noise. The daily recorded displacement velocity cannot be directly compared to the extracted threshold (2 mm/d), otherwise it would easily cause false warnings. For example, as Figure 13a shows, there are 564 velocity points greater than 2 mm/d, 33% overall over the G2 monitoring period. Therefore, a threshold for practical applications should be amplified to a certain extent, compared to the moving average threshold.

How can we reasonably and accurately redefine the abovementioned velocity threshold? Recognizing that the alarm should be triggered in an EWS's procedure if the preset thresholds are reached or exceeded is important. Accordingly, the maximum allowable rate in the EWS needs to be found; in other words, the displacement increment needs to be found as the lowest when reaching the DMA velocity of 2 mm/d by a running window. The smallest displacement increment means that the velocity series of the landslide reach the DMA threshold of 2 mm/d most easily, thus indicating that the maximum velocity in the smallest displacement increment is the maximum allowable velocity.

For this purpose, we assume that the time series of the double moving average velocity monotonically increases to 2 mm/d in 13 days. It is easy to find that this situation causes the smallest displacement increment if we set the starting point at zero in the linear/non-linear relationship. Within the non-linear relationship, this is because we can always set the starting point at zero and associate it with the endpoint and the local minimum value of the rate in the non-linear increase for time integrating. Thus, the velocity monotonically increasing with time obtains the smallest displacement increment among linear, concave, and convex curves.

Let us set v_{13}'' equal to 2 mm/d at time t_{13} and have it linearly increased from t_1, as Figure 15 shows. According to Equations (1) and (2) and initial conditions ($v_{13}'' = 2, v_1 = 0$), and the divisions of every term used in the equations in Figure 15c, the value of v_7 can be mathematically derived to be equal to 2 mm/d. Giving v_1 equals zero, v_{13} can be determined as a minimum of 4 mm/d. Similarly, in the case of a concave monotonic curve, v_{13} is greater than 4 mm/d, whereas in the case of a convex monotonic curve, v_{13} is less than 4 mm/d. However, daily velocity in the scatter plot is more concentrated on the onset of an acceleration episode, instead of later or final stages (Figure 11), which is more likely to be consistent with a convex curve. Consequently, 4 mm/d is recommended in an ideal way for the linear type, and, meanwhile, 4 mm/d is on the safe side for the convex type.

The velocity points beyond 4 mm/d are collected under two scenarios covering the monitoring period. Based on the recommended thresholds, scenario (1) solely meets recorded velocity >4 mm/d, and scenario (2) delimits that recorded velocity of >4 mm/d and a water level <1820 m. The number of warning times declines as the number of annual fluctuations increases (Figure 16). If the limiting condition (water level <1820 m) is combined, the number of warning times in scenario (2) sharply reduces compared to scenario (1). The annual relative frequency of all monitoring sites in scenario (2) is less than 10% and is almost less than 3% during the last three years. Since the original records are in chaos in this reservoir-induced landslide case, the water level threshold can reduce false warning frequency. This is because velocity points above the 1820 m water level do not belong to the main accelerated zone (Figure 14).

The warning dates in scenario (2) during the last two years were gathered to analyze the temporal features of warnings. The water levels dropping to below 1820 m asl always happened in May and June, based on statistics. On the other hand, an episode that the water levels fill to 1820 m after dropping to 1800 m asl is also applicable to "water level <1820 m". It always finishes in a short episode (fewer than ten days) during late June to early July. The amount of data during this reservoir filling episode in June is quite small (about five each year), so the data collection in June is a mix of reservoir release and filling.

According to the warning times in May and June, as illustrated on the monthly calendars (Figure 17), the month of June 2019 had the most days of warning times. Over half of the monitoring units released warnings on 9, 13, 15, and 18 June 2019, on 22 and 24 May 2020, and on 14 June 2020; all units released warnings on 9 and 15 June 2019 and on 14 June 2020. From this point of view, the threshold is always discretely exceeded at present, and it is most frequently exceeded in June, especially in mid-June over the last two years.

If the thresholds of velocity and water level are both reached or exceeded, this means that short accelerations are likely to occur. Furthermore, if the thresholds are both continuously exceeded in monitoring days, the Gapa landslide may locally or completely enter the tertiary deformation. This suggests that at least upgrading the warning level to a "caution" level is necessary to allow dam operators to improve attention and surveillance on the landslide. Thus, the thresholds are efficient and effective for judging accelerations, otherwise accelerating behaviors can only be detected by the cumulative displacement, which takes more time and easily causes false warnings.

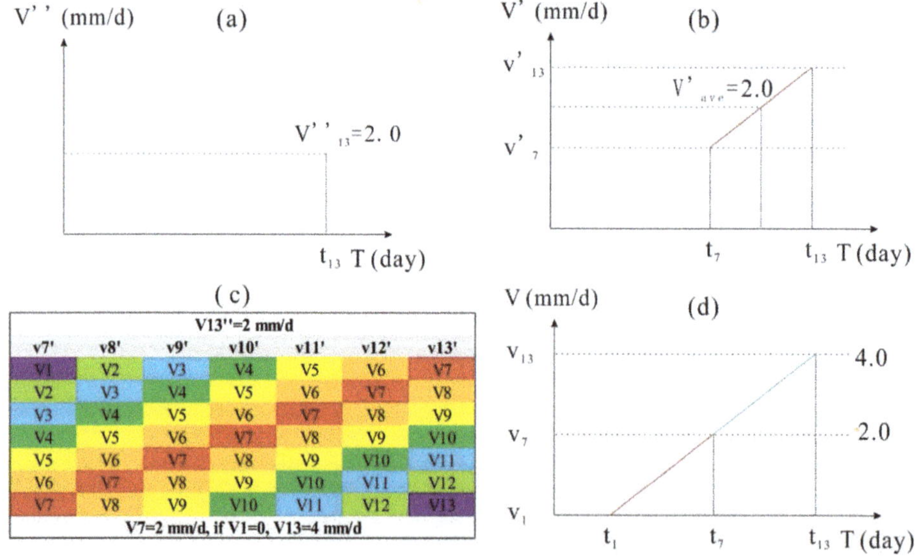

V13''=2 mm/d						
v7'	v8'	v9'	v10'	v11'	v12'	v13'
V1	V2	V3	V4	V5	V6	V7
V2	V3	V4	V5	V6	V7	V8
V3	V4	V5	V6	V7	V8	V9
V4	V5	V6	V7	V8	V9	V10
V5	V6	V7	V8	V9	V10	V11
V6	V7	V8	V9	V10	V11	V12
V7	V8	V9	V10	V11	V12	V13
V7=2 mm/d, if V1=0, V13=4 mm/d						

Figure 15. The process of defining a recommended velocity threshold using the reverse double moving average method in displacement vs. time in a linear relationship: (**a**) the DMA threshold plotted on DMA velocity vs. time; (**b**) the MA velocity series from t_7 to t_{13} plotted on MA velocity vs. time; (**c**) the polynomial of every MA velocity from t_7 to t_{13} expressed in columns (according to this table, the value of V_7 is found to equal 2 mm/d); and (**d**) the original velocity series from t_1 to t_{13} plotted on velocity vs. time.

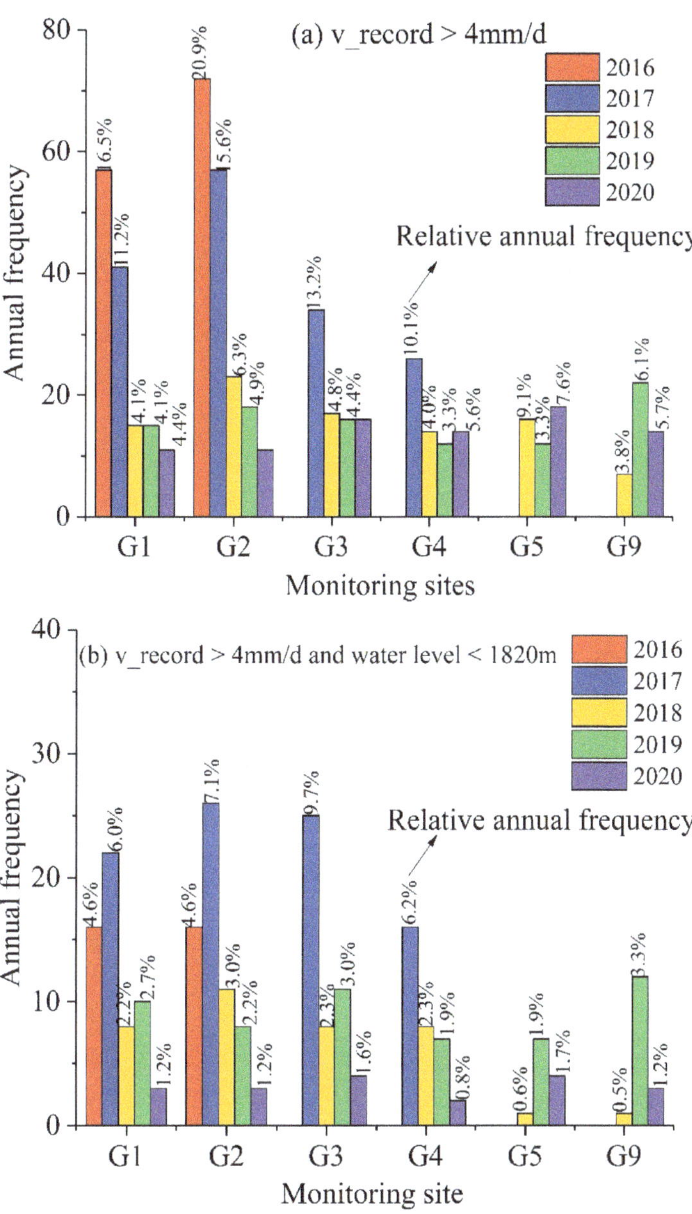

Figure 16. The frequency and annual relative frequency (given in percentage) exceeding the thresholds over monitoring years: (**a**) recorded daily velocity greater than 4 mm/d; (**b**) recorded daily velocity greater than 4 mm/d and water level less than 1820 m asl.

Number of warning sites								Number of warning sites						
0	1	2	3	4	5	6		0	1	2	3	4	5	6
May-19								Jun-19						
		1	2	3	4	5							1	2
6	7	8	9	10	11	12		3	4	5	6	7	8	9
13	14	15	16	17	18	19		10	11	12	13	14	15	16
20	21	22	23	24	25	26		17	18	19	20	21	22	23
27	28	29	30	31				24	25	26	27	28	29	30
May-20								Jun-20						
			1	2	3			1	2	3	4	5	6	7
4	5	6	7	8	9	10		8	9	10	11	12	13	14
11	12	13	14	15	16	17		15	16	17	18	19	20	21
18	19	20	21	22	23	24		22	23	24	25	26	27	28
25	26	27	28	29	30	31		29	30					

Figure 17. Early warning state in May and June 2019/2020. Number of GNSS units releasing warning spans from 0 to 6, 0 and 6 means none of the GNSS units to all GNSS units releasing warnings, respectively.

5. Discussion

Accelerations of the Gapa landslide movement often occur during reservoir fluctuation each year. Analysis of velocity time series based on the moving average method was conducted for warning of periodic acceleration phases with a velocity threshold. Meanwhile, a water level threshold was also assigned to reduce false warnings for GNSS units. The forward and reverse double moving average method is recommended for the time series of GNSS data to define thresholds in EWSs. The proposed method is beneficial in reducing the data noise level, extracting stable pre-thresholds, and defining a final threshold for better performance. Additionally, the water level threshold is crucial as well for warning of possible accelerations of the landslide, as it is mainly influenced by reservoir water fluctuations.

Although the current work provides a foundation for an EWS for the Gapa landslide, the improved Saito's model applied is still limited in different deformation stages, which means that it cannot release a time failure alarm for the landslide based on the current deformation stage. This is because the landslide is believed to enter or continue the constant deformation stage using existing GNSS units and data. Therefore, novel EWS models are urgently needed for landslides with periodic acceleration episodes. Another factor limiting the present EWS is that supplementing and installation of new GNSS units are not cost-effective (the cost of a single GNSS unit is close to 2000 EUR) [10]. The monitoring cost delays overall and detailed recognition of the landslide kinematics and, meanwhile, lowers the coving area of the EWS for the whole landslide. Some other remote sensing techniques could possibly be applied to this target. The Maoxian landslide occurred in June 2017 and killed more than 100 people. Intrieri et al. reported this case study using generated ground deformation maps and displacement time series for detecting precursors of failure with InSAR data [56]. Furthermore, techniques combining GNSS, satellite InSAR (interferometric synthetic aperture radar), ground-based InSAR, and aerial photogrammetry can enhance the understanding of large active landslides at the slope scale and the performance of EWSs, as two cases in Alpine and Apennines, Italy documented [57,58]. Therefore, InSAR techniques could potentially be employed in the Gapa landslide, because it is west-facing and has little effect of snow or vegetation coving in winter. Accessing InSAR data combined with GNSS units for the EWS is believed to enhance its scope and accuracy for the Gapa landslide. On the other hand, thresholds should be calibrated as soon as new data are available in order to guarantee that they are designed flexibly. The comprehensive development of an EWS with threshold update and InSAR technique supplements warrants future studies.

6. Conclusions

Synthesizing the above study, a method highlighted by forward and reverse moving averages for threshold definition is preliminarily proposed. The velocity and water level thresholds were successfully applied to an EWS of the Gapa landslide. The conclusions can be drawn as follows:

(1) The Gapa landslide was formed by bending and topping between the late Pleistocene and early Holocene. The landslide had exhibited very slow movement for the decade before the reservoir impoundment. However, it was reactivated after the impoundment, and its movement was strongly related to the annual reservoir fluctuations, although the annual velocity has now gradually decreased during recent years. Two possibilities of the landslide's evolution—that it enters the acceleration stage or recovers from the periodic fluctuations to a slow-moving/stable state in the future—have been inferred. However, considering the uncertainty and limited monitoring period, the EWS for the Gapa landslide was developed using the improved Saito's model. The warning level is believed to be set at "attention level", according to the landslide's current motion.

(2) In this EWS, the velocity and water level are employed as combined warning indicators. The velocity threshold was defined by the forward and reverse moving average method, and, meanwhile, the seven-day window was decided as the short-term moving window. The recommended velocity threshold is 4 mm/d for current monitoring units. Additionally, a water level threshold with a specific elevation of 1820 m is also used for warning of possible accelerations, based on the distribution characteristics of the DMA velocity time series.

(3) If the daily recorded velocity in each monitoring site exceeds 4 mm/d and, simultaneously, the water level is below 1820 m elevation asl, the warning of likely accelerations will be released in this EWS, allowing users and managers to give more attention and surveillance to the potential hazard. Practical application showed that with the aid of the water level threshold, false warnings are effectively reduced, compared to the sole velocity threshold. It also pointed out that the most alert-prone time zone was mid-June during the last two years. As the thresholds were successfully defined and verified in this typical case, both the velocity and water level thresholds are recommended to be simultaneously used in EWSs for reservoir-induced large-scale landslides in southwestern China and other similar parts of the world.

Author Contributions: S.W. organized and analyzed the data and wrote the paper, X.H. supervised the work, W.Z. and M.B. reviewed and edited the manuscript, Z.Q. and W.S. validated landslide data. All authors have read and agreed to the published version of the manuscript.

Funding: This study was funded by the National Key Research and Development Program of China (No. 2017YFC1501302), the Key Program of the National Natural Science Foundation of China (No. 41630643), the Fundamental Research Funds for the Central Universities, and China University of Geosciences (Wuhan) (No. CUGCJ1701).

Institutional Review Board Statement: Not applicable for studies not involving humans or animals.

Informed Consent Statement: Not applicable for studies not involving humans.

Data Availability Statement: Restrictions apply to the availability of these data. Data were obtained from the Chengdu Engineering Corporation Limited and are available from the authors with the permission of the Chengdu Engineering Corporation Limited.

Acknowledgments: The first author appreciates the support for field monitoring and data provided by Yalong River Hydropower Development Company Limited and Chengdu Engineering Corporation Limited. He is also grateful to the China Scholarship Council for providing a scholarship for this research, which was conducted while he was a visiting PhD student (CSC202006410019) at the University of Bologna, Italy.

Conflicts of Interest: The authors declare no conflict of interest.

References

1. Liang, X.; Gui, L.; Wang, W.; Du, J.; Ma, F.; Yin, K. Characterizing the Development Pattern of a Colluvial Landslide Based on Long-Term Monitoring in the Three Gorges Reservoir. *Remote Sens.* **2021**, *13*, 224. [CrossRef]
2. Tang, H.; Wasowski, J.; Juang, C.H. Geohazards in the three Gorges Reservoir Area, China—Lessons learned from decades of research. *Eng. Geol.* **2019**, *261*, 105267. [CrossRef]
3. Lin, C.; Tseng, C.; Tseng, Y.; Fei, L.; Hsieh, Y.; Tarolli, P. Recognition of large scale deep-seated landslides in forest areas of Taiwan using high resolution topography. *J. Asian Earth Sci.* **2013**, *62*, 389–400. [CrossRef]
4. Intrieri, E.; Gigli, G.; Mugnai, F.; Fanti, R.; Casagli, N. Design and implementation of a landslide early warning system. *Eng. Geol.* **2012**, *147*, 124–136. [CrossRef]
5. Calvello, M.; Piciullo, L. Assessing the performance of regional landslide early warning models: The EDuMaP method. *Nat. Hazard. Earth Syst.* **2016**, *16*, 103–122. [CrossRef]
6. Lagomarsino, D.; Segoni, S.; Fanti, R.; Catani, F. Updating and tuning a regional-scale landslide early warning system. *Landslides* **2013**, *10*, 91–97. [CrossRef]
7. Bao, L.; Zhang, G.; Hu, X.; Wu, S.; Liu, X. Stage Division of Landslide Deformation and Prediction of Critical Sliding Based on Inverse Logistic Function. *Energies* **2021**, *14*, 1091. [CrossRef]
8. Piciullo, L.; Gariano, S.L.; Melillo, M.; Brunetti, M.T.; Peruccacci, S.; Guzzetti, F.; Calvello, M. Definition and performance of a threshold-based early warning model for rainfall-induced Landslides. *Landslides* **2017**, *14*, 995–1008. [CrossRef]
9. Manconi, A.; Giordan, D. Landslide failure forecast in near-real-time. *Geomat. Nat. Hazards Risk* **2016**, *7*, 639–648. [CrossRef]
10. Šegina, E.; Peternel, T.; Urbančič, T.; Realini, E.; Zupan, M.; Jež, J.; Caldera, S.; Gatti, A.; Tagliaferro, G.; Consoli, A.; et al. Monitoring Surface Displacement of a Deep-Seated Landslide by a Low-Cost and near Real-Time GNSS System. *Remote Sens.* **2020**, *12*, 3375. [CrossRef]
11. Macciotta, R.; Hendry, M.; Martin, C.D. Developing an early warning system for a very slow landslide based on displacement monitoring. *Nat. Hazards* **2016**, *81*, 887–907. [CrossRef]
12. Barla, M.; Antolini, F. An integrated methodology for landslides' early warning systems. *Landslides* **2016**, *13*, 215–228. [CrossRef]
13. Giordan, D.; Wrzesniak, A. Automatized Dissemination of Landslide Monitoring Bulletins for Early Warning Applications. In *Understanding and Reducing Landslide Disaster Risk*; Arbanas, Z., Bobrowsky, P.T., Konagai, K., Sassa, K., Takara, K., Eds.; Springer International Publishing: Berlin/Heidelberg, Germany, 2021; pp. 231–235.
14. Guzzetti, F.; Peruccacci, S.; Rossi, M.; Stark, C.P. The rainfall intensity–duration control of shallow landslides and debris flows: An update. *Landslides* **2008**, *5*, 3–17. [CrossRef]
15. Berti, M.; Bernard, M.; Gregoretti, C.; Simoni, A. Physical Interpretation of Rainfall Thresholds or Runoff-Generated Debris Flows. *J. Geophys. Res. Earth Surface* **2020**, *125*. [CrossRef]
16. Rosi, A.; Segoni, S.; Canavesi, V.; Monni, A.; Gallucci, A.; Casagli, N. Definition of 3D rainfall thresholds to increase operative landslide early warning system performances. *Landslides* **2020**, *18*, 1045–1057. [CrossRef]
17. Guzzetti, F.; Gariano, S.L.; Peruccacci, S.; Brunetti, M.T.; Marchesini, I.; Rossi, M.; Melillo, M. Geographical landslide early warning systems. *Earth-Sci. Rev.* **2020**, *200*, 102973. [CrossRef]
18. Zhou, C.; Cao, Y.; Yin, K.; Wang, Y.; Shi, X.; Catani, F.; Ahmed, B. Landslide Characterization Applying Sentinel-1 Images and InSAR Technique: The Muyubao Landslide in the Three Gorges Reservoir Area, China. *Remote Sens.* **2020**, *12*, 3385. [CrossRef]
19. Huang, X.; Guo, F.; Deng, M.; Yi, W.; Huang, H. Understanding the deformation mechanism and threshold reservoir level of the floating weight-reducing landslide in the Three Gorges Reservoir Area, China. *Landslides* **2020**, *17*, 2879–2894. [CrossRef]
20. He, C.; Hu, X.; Xu, C.; Wu, S.; Zhang, H.; Liu, C. Model test of the influence of cyclic water level fluctuations on a landslide. *J. Mt. Sci.-Engl.* **2020**, *17*, 191–202. [CrossRef]
21. Bernardie, S.; Desramaut, N.; Malet, J.P.; Gourlay, M.; Grandjean, G. Prediction of changes in landslide rates induced by rainfall. *Landslides* **2015**, *12*, 481–494. [CrossRef]
22. Cao, Y.; Yin, K.; Zhou, C.; Ahmed, B. Establishment of Landslide Groundwater Level Prediction Model Based on GA-SVM and Influencing Factor Analysis. *Sensors* **2020**, *20*, 845. [CrossRef]
23. Krkač, M.; Špoljarić, D.; Bernat, S.; Arbanas, S.M. Method for prediction of landslide movements based on random forests. *Landslides* **2017**, *14*, 947–960. [CrossRef]
24. Zhao, B.; Dai, Q.; Han, D.; Zhang, J.; Zhuo, L.; Berti, M. Application of hydrological model simulations in landslide predictions. *Landslides* **2020**, *17*, 877–891. [CrossRef]
25. Troncone, A.; Pugliese, L.; Lamanna, G.; Conte, E. Prediction of rainfall-induced landslide movements in the presence of stabilizing piles. *Eng. Geol.* **2021**, *288*, 106143. [CrossRef]
26. Segalini, A.; Valletta, A.; Carri, A. Landslide time-of-failure forecast and alert threshold assessment: A generalized criterion. *Eng. Geol.* **2018**, *245*, 72–80. [CrossRef]
27. Pecoraro, G.; Calvello, M.; Piciullo, L. Monitoring strategies for local landslide early warning systems. *Landslides* **2019**, *16*, 213–231. [CrossRef]
28. Carlà, T.; Farina, P.; Intrieri, E.; Botsialas, K.; Casagli, N. On the monitoring and early-warning of brittle slope failures in hard rock masses: Examples from an open-pit mine. *Eng. Geol.* **2017**, *228*, 71–81. [CrossRef]
29. Huang, F.; Huang, J.; Jiang, S.; Zhou, C. Landslide displacement prediction based on multivariate chaotic model and extreme learning machine. *Eng. Geol.* **2017**, *218*, 173–186. [CrossRef]

30. Li, L.; Zhang, S.X.; Qiang, Y.; Zheng, Z.; Li, S.H.; Xia, C.S.; Xie, X. A Landslide Displacement Prediction Method with Iteration-Based Combined Strategy. *Math. Probl. Eng.* **2021**, *2021*, 6692503. [CrossRef]

31. Intrieri, E.; Carlà, T.; Gigli, G. Forecasting the time of failure of landslides at slope-scale: A literature review. *Earth-Sci. Rev.* **2019**, *193*, 333–349. [CrossRef]

32. Berti, M.; Martina, M.; Franceschini, S.; Pignone, S.; Simoni, A.; Pizziolo, M. Probabilistic rainfall thresholds for landslide occurrence using a Bayesian approach. *J. Geophys. Res. Earth Surface* **2012**, *117*. [CrossRef]

33. Saito, M. Forecasting time of slope failure by tertiary creep. In Proceedings of the Soil Mechanics and Foundation Engineering, Mexico City, Mexico, 1 January 1969; Volume 2, pp. 677–683.

34. Rose, N.D.; Hungr, O. Forecasting potential rock slope failure in open pit mines using the inverse-velocity method. *Int. J. Rock Mech. Min. Sci.* **2007**, *44*, 308–320. [CrossRef]

35. Xu, Q.; Yuan, Y.; Zeng, Y.; Hack, R. Some new pre-warning criteria for creep slope failure. *Sci. China Technol. Sci.* **2011**, *54*, 210–220. [CrossRef]

36. Xu, Q.; Peng, D.; Zhang, S.; Zhu, X.; He, C.; Qi, X.; Zhao, K.; Xiu, D.; Ju, N. Successful implementations of a real-time and intelligent early warning system for loess landslides on the Heifangtai terrace, China. *Eng. Geol.* **2020**, *278*, 105817. [CrossRef]

37. Fan, X.; Xu, Q.; Liu, J.; Subramanian, S.S.; He, C.; Zhu, X.; Zhou, L. Successful early warning and emergency response of a disastrous rockslide in Guizhou province, China. *Landslides* **2019**, *16*, 2445–2457. [CrossRef]

38. Ju, N.; Huang, J.; He, C.; Van Asch, T.W.J.; Huang, R.; Fan, X.; Xu, Q.; Xiao, Y.; Wang, J. Landslide early warning, case studies from Southwest China. *Eng. Geol.* **2020**, *279*, 105917. [CrossRef]

39. Li, C.; Fu, Z.; Wang, Y.; Tang, H.; Yan, J.; Gong, W.; Yao, W.; Criss, R.E. Susceptibility of reservoir-induced landslides and strategies for increasing the slope stability in the Three Gorges Reservoir Area: Zigui Basin as an example. *Eng. Geol.* **2019**, *261*, 105279. [CrossRef]

40. Cruden, D.M.; Varnes, D.J. Landslide Types and Processes. In *Landslides: Investigation and Mitigation*; Transportation Research Board, National Academy of Sciences: Denver, CO, USA, 1996; pp. 36–75.

41. Xu, Q.; Tang, M.; Xu, K.; Huang, X. Research on space-time evolution laws and early warning-prediction of Landslides. *J. Rock Mech. Eng.* **2008**, *27*, 1104–1112. (In Chinese)

42. Kilburn, C.R.J.; Petley, D.N. Forecasting giant, catastrophic slope collapse: Lessons from Vajont, Northern Italy. *Geomorphology* **2003**, *54*, 21–32. [CrossRef]

43. Wu, S.; Hu, X.; Zheng, W.; Zhang, G.; Liu, C.; Xu, C.; Zhang, H.; Liu, Z. Displacement behaviour and potential impulse waves of the Gapa landslide subjected to the Jinping Reservoir fluctuations in Southwest China. *Geomorphology* **2022**, *397*, 108013. [CrossRef]

44. Hu, X.; Wu, S.; Zhang, G.; Zheng, W.; Liu, C.; He, C.; Liu, Z.; Guo, X.; Zhang, H. Landslide displacement prediction using kinematics-based random forests method: A case study in Jinping Reservoir Area, China. *Eng. Geol.* **2021**, *283*, 105975. [CrossRef]

45. Wu, S.; Hu, X.; Zheng, W.; He, C.; Zhang, G.; Zhang, H.; Wang, X. Effects of reservoir water level fluctuations and rainfall on a landslide by two-way ANOVA and K-means clustering. *Bull. Eng. Geol. Environ.* **2021**, *80*, 5405–5421. [CrossRef]

46. Wang, F.; Zhang, Y.; Huo, Z.; Matsumoto, T.; Huang, B. The July 14, 2003 Qianjiangping landslide, Three Gorges Reservoir, China. *Landslides* **2004**, *1*, 157–162. [CrossRef]

47. Semenza, E.; Ghirotti, M. History of the 1963 Vaiont slide: The importance of geological factors. *Bull. Eng. Geol. Environ.* **2000**, *59*, 87–97. [CrossRef]

48. Fan, X.; Xu, Q.; Alonso-Rodriguez, A.; Subramanian, S.S.; Li, W.; Zheng, G.; Dong, X.; Huang, R. Successive landsliding and damming of the Jinsha River in eastern Tibet, China: Prime investigation, early warning, and emergency response. *Landslides* **2019**, *16*, 1003–1020. [CrossRef]

49. Qi, S.; Yan, F.; Wang, S.; Xu, R. Characteristics, mechanism and development tendency of deformation of Maoping landslide after commission of Geheyan reservoir on the Qingjiang River, Hubei Province, China. *Eng. Geol.* **2006**, *86*, 37–51. [CrossRef]

50. Wu, Q.; Tang, H.; Ma, X.; Wu, Y.; Hu, X.; Wang, L.; Criss, R.; Yuan, Y.; Xu, Y. Identification of movement characteristics and causal factors of the Shuping landslide based on monitored displacements. *Bull. Eng. Geol. Environ.* **2019**, *78*, 2093–2106. [CrossRef]

51. Huang, Q.X.; Wang, J.L.; Xue, X. Interpreting the influence of rainfall and reservoir infilling on a landslide. *Landslides* **2016**, *13*, 1139–1149. [CrossRef]

52. Chen, Y.; Irfan, M.; Uchimura, T.; Wu, Y.; Yu, F. Development of elastic wave velocity threshold for rainfall-induced landslide prediction and early warning. *Landslides* **2019**, *16*, 955–968. [CrossRef]

53. Liao, K.; Wu, Y.; Miao, F.; Li, L.; Xue, Y. Using a kernel extreme learning machine with grey wolf optimization to predict the displacement of step-like landslide. *Bull. Eng. Geol. Environ.* **2019**, *79*, 673–685. [CrossRef]

54. Loew, S.; Gschwind, S.; Gischig, V.; Keller-Signer, A.; Valenti, G. Monitoring and early warning of the 2012 Preonzo catastrophic rockslope failure. *Landslides* **2017**, *14*, 141–154. [CrossRef]

55. Criss, R.E.; Yao, W.; Li, C.; Tang, H. A Predictive, Two-Parameter Model for the Movement of Reservoir Landslides. *J. Earth Sci.-China* **2020**, *31*, 1051–1057. [CrossRef]

56. Intrieri, E.; Raspini, F.; Fumagalli, A.; Lu, P.; Del Conte, S.; Farina, P.; Allievi, J.; Ferretti, A.; Casagli, N. The Maoxian landslide as seen from space: Detecting precursors of failure with Sentinel-1 data. *Landslides* **2018**, *15*, 123–133. [CrossRef]

57. Carlà, T.; Tofani, V.; Lombardi, L.; Raspini, F.; Bianchini, S.; Bertolo, D.; Thuegaz, P.; Casagli, N. Combination of GNSS, satellite InSAR, and GBInSAR remote sensing monitoring to improve the understanding of a large landslide in high alpine environment. *Geomorphology* **2019**, *335*, 62–75. [CrossRef]
58. Cenni, N.; Fiaschi, S.; Fabris, M. Integrated use of archival aerial photogrammetry, GNSS, and InSAR data for the monitoring of the Patigno landslide (Northern Apennines, Italy). *Landslides* **2021**, *18*, 2247–2263. [CrossRef]

Article

Use of Multiplatform SAR Imagery in Mining Deformation Monitoring with Dense Vegetation Coverage: A Case Study in the Fengfeng Mining Area, China

Bochen Zhang [1,2], Songbo Wu [3,*], Xiaoli Ding [3], Chisheng Wang [1,4], Jiasong Zhu [1,2] and Qingquan Li [1,2]

[1] MNR Key Laboratory for Geo-Environmental Monitoring of Great Bay Area & Guangdong Key Laboratory of Urban Informatics & Shenzhen Key Laboratory of Spatial Smart Sensing and Services, Shenzhen University, Shenzhen 518060, China; zhangbc@szu.edu.cn (B.Z.); wangchisheng@szu.edu.cn (C.W.); zjsong@szu.edu.cn (J.Z.); liqq@szu.edu.cn (Q.L.)
[2] College of Civil and Transportation Engineering, Shenzhen University, Shenzhen 518060, China
[3] Department of Land Surveying and Geo-Informatics, The Hong Kong Polytechnic University, Hung Hom, KLN, Hong Kong 999077, China; xl.ding@polyu.edu.hk
[4] School of Architecture & Urban Planning, Shenzhen University, Shenzhen 518060, China
* Correspondence: song.bo.wu@connect.polyu.hk

Abstract: Ground deformation related to mining activities may occur immediately or many years later, leading to a series of mine geological disasters, such as ground fissures, collapses, and even mining earthquakes. Deformation monitoring has been carried out with techniques, such as multitemporal interferometric synthetic aperture radar (MTInSAR). Over the past decade, MTInSAR has been widely used in monitoring mining deformation, and it is still difficult to retrieve mining deformation over dense vegetation areas. In this study, we use multiple-platform SAR images to retrieve mining deformation over dense vegetation areas. The high-quality interferograms are selected by the coherence map, and the mining deformation is retrieved by the MSBAS-InSAR technique. SAR images from TerraSAR-X, Sentinel-1A, Radarsat-2, and PALSAR-2 over the Fengfeng mining area, Heibei, China, are used to retrieve the deformation of mining activities covered with dense vegetation. The results show that the subsidence in the Fengfeng mining area reaches up to 90 cm over the period from July 2015 to April 2016. The root-mean-square error (RMSE) between the results from InSAR and leveling is 83.5 mm/yr at two mining sites, i.e., Wannian and Jiulong Mines.

Keywords: Fengfeng mine; mining deformation monitoring; MSBAS; multiplatform SAR data; dense vegetation

Citation: Zhang, B.; Wu, S.; Ding, X.; Wang, C.; Zhu, J.; Li, Q. Use of Multiplatform SAR Imagery in Mining Deformation Monitoring with Dense Vegetation Coverage: A Case Study in the Fengfeng Mining Area, China. *Remote Sens.* **2021**, *13*, 3091. https://doi.org/10.3390/rs13163091

Academic Editors: Oriol Monserrat, Daniele Giordan, Guido Luzi and Niccolò Dematteis

Received: 2 June 2021
Accepted: 2 August 2021
Published: 5 August 2021

Publisher's Note: MDPI stays neutral with regard to jurisdictional claims in published maps and institutional affiliations.

1. Introduction

Mining activities can break the original stress balance of the overlying rock layer, and lead to a series of geological and environmental problems, e.g., aquifer destruction, landslides, and ground structural damage [1–3], which inevitably lead to ground deformation. Thus, detecting and predicting such mining deformation is essential for analyzing the mechanism of mining activities and potential geological hazards. In general, mining deformation can be assessed quantitatively, by empirical and theoretical prediction models [3]. The accuracy of the predicted mining deformation of the models is limited by the input parameters related to the geological condition, geotechnical information, and deformation observed at sparse points. For example, the accuracy of the predicted deformation could range from 48% to 773%, without actual ground deformation information [4]. Therefore, monitoring ground deformation plays an essential role in exploring the deformation mechanism of the mining area.

Ground deformation can be measured by in situ observations from traditional geodetic surveying techniques, such as leveling and global navigation satellite system (GNSS).

These techniques can monitor ground deformation with high accuracy; however, the spatial resolutions are normally poor, and they are often labor intensive. Mining prediction models highly rely on the resolution, sampling rate, and coverage of measurement points. Spaceborne interferometric synthetic aperture radar (InSAR) is an active remote sensing technique, and has become one of the most useful geodetic techniques for ground deformation monitoring [5–9]. Compared with traditional techniques, InSAR has proven its unique advantages through broad coverage and high spatial resolution [10,11]. To further improve the accuracy of traditional InSAR, multitemporal InSAR (MTInSAR) has been developed for analyzing the time series phase of point targets by a series of interferograms. For example, persistent scatter InSAR (PSInSAR) and small baseline subset (SBAS) methods were proposed to reduce the effect of unwanted phase contributions, such as atmosphere, decorrelation noise, and digital elevation model (DEM) phase residual, and to retrieve more accurate ground deformation [12,13]. Currently, MTInSAR has been widely applied in monitoring the time series of deformation in different scenarios, such as seismic motion, landslides, and ground subsidence [14–22]. In addition to mining activities, MTInSAR offers great potential for promoting studies of mining deformation, such as mechanism interpretation, parameter inversion, and deformation prediction [3].

In 1996, Carnec et al. [23] generated the first deformation map caused by underground mining, using InSAR. Then, many applications have been carried out to monitor mining deformation with InSAR in recent decades. For example, in 1999, Wright et al. [4] obtained elevation changes induced by mining activity with two ERS SAR datasets. In 2010, Milan et al. [24] retrieved a series of mining deformations in the area of the Ostrava-Karviná Revier in the Czech Republic using the Stanford method for persistent scatterers (StaMPS). Since mining deformation often occurs in three-dimensional (3D) space, studies with individual SAR datasets using the SAR techniques, such as multi-aperture InSAR [25], and pixel offset tracking [26], can only be dedicated to measuring one-/two-dimensional (1D/2D) deformation along the radar line of sight (LOS) and/or azimuth directions, by which the real pattern of mining deformation is difficult to elucidate. To solve this problem, several methods have been proposed to reconstruct 3D mining displacements, by combining InSAR observations from multitrack or multiplatform systems. For example, in 2011, Hay-Man Ng et al. [27] retrieved the 3D deformation of the southern highland coalfield in New South Wales, Australia, with multiplatform SAR datasets. In 2013, Samsonov et al. [2] proposed the MSBAS method to retrieve the 2D time series of ground deformation of postmining activities along the French-German border.

MTInSAR has been widely used to monitor mining deformation, but it is still limited by environmental issues, such as dense vegetation and extreme weather. Mining areas are often located in rural regions, which are usually covered with dense vegetation, which raises the significant decorrelation of InSAR interferograms and leads to spatial and temporal gaps for mining deformation monitoring [28]. In this article, we carried out a case study of deformation monitoring in a mining area with dense vegetation coverage. We use multiplatform SAR images and MSBAS-InSAR technology to retrieve the mining deformation. The dataset from different SAR platforms is integrated into the MTSBAS-InSAR to estimate the mining deformation. High-quality interferograms are selected according to their coherence maps and weather conditions. The generalized deformation maps of 36 mining sites in the Fengfeng mining area, Hebei, China are obtained by combining the SAR images of TerraSAR-X, Sentinel-1A, Radarsat-2, and PALSAR-2. The deformation results are verified with the measurements from leveling and indicate that the high accuracy of the deformation can be achieved with multiplatform SAR images.

The remainder of this paper is structured as follows. In Section 2, the background of the Fengfeng mining area and SAR datasets used are introduced. Section 3 gives the strategy of data processing and the MSBAS-InSAR method. The result and discussion are shown in Sections 4 and 5 respectively. Section 6 gives the conclusions of this study.

2. Study Area and SAR Datasets

2.1. Background of Fengfeng Mining Area

The Fengfeng mining area is located in the southwestern part of Handan city, China, as shown in Figure 1. Approximately 36 sites and significant subsidence (see Figure 1b) have been reported in and around this area [29,30]. According to the field work photo of Fengfeng mining area in Figure 1b, the ground surface is covered with dense vegetation, e.g., crops and weeds, leading the InSAR interferogram to be easily decorrelated. Especially in the summer period, vegetation grows quickly, and some vegetation is over the height of a man. In addition, the ongoing mining activities has generated large gradient subsidence that cause difficulties to InSAR for monitoring of deformation.

Figure 1. (**a**) Geolocation of the Fengfeng mining area, framed by the white polygon; (**b**) the picture of the field mining area indicates large subsidence and dense ground vegetation.

2.2. SAR Datasets

Multisensor spaceborne SAR satellite datasets from both ascending and descending tracks over the Fengfeng area were collected to measure the mining deformation, as outlined in Table 1. The coverage of these data and the distribution of the main mining regions are shown in Figure 2. A digital elevation model (DEM) from shuttle radar topography missions (SRTM), provided by the United States Geological Survey (USGS) with a spatial resolution of 30 m, is adopted for topography compensation and geocoding. We also collected measurements of 152 leveling stations in the 6 phases at two mining sites, i.e., Jiulong and Wannian Mines, to validate the InSAR results.

Table 1. SAR datasets used in this study.

Satellite	TerraSAR-X	Sentinel-1A	Radarsat-2	PALSAR-2
Orbital	descending	ascending	ascending	ascending
Period	29 July 2015–7 April 2016	2 December 2015–31 March 2016	30 November 2015–5 March 2016	19 July 2015–6 December 2015
No. of image	14	7	4	3
No. of selected image	8	7	4	3

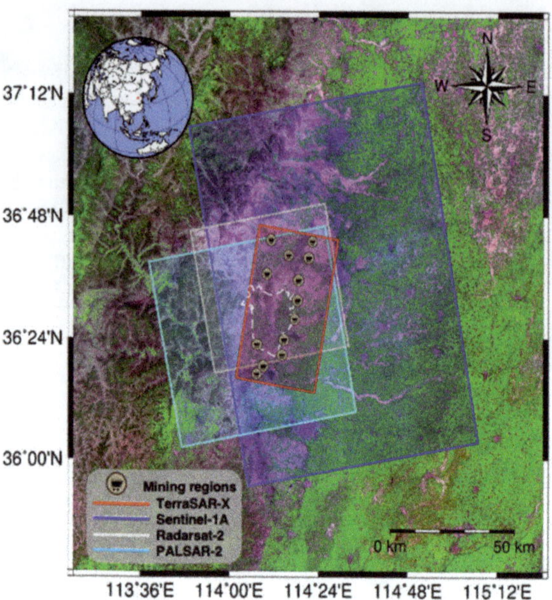

Figure 2. Coverage of SAR satellite datasets used in this study. The main mining regions are marked with symbols. The background image is a false color composite from Landsat-8.

3. Methodology

3.1. Interferometric Processing

The data preprocessing of each SAR dataset used in this study was carried out using GAMMA software [31], including single look complex (SLC) image extraction and coregistration, interferogram generation, flat-earth phase removal, phase noise filtering, phase unwrapping, and geocoding.

3.2. Retrieval of Time-Series Deformation

In this study, the MSBAS technique in [2] was used to retrieve the time-series deformation of the Fengfeng mining area. MSBAS is an advanced extension of the traditional SBAS [13], that extends the deformation retrieval from the line-of-sight (LOS) direction to the two-dimensional direction, i.e., the horizontal (east-west direction) and vertical (up-down direction) deformation. In addition to the advantage of 2D deformation, MSBAS can significantly improve the temporal sampling of deformation time series with multiplatform SAR images. The difference from the SBAS method is that before time-series analyzing, all unwrapped interferograms must be geocoded and interpolated into a common grid with the same spatial resolution, as shown in Figure 3. The lower resolution of the SAR dataset is usually chosen as the resolution of the common grid, which is the final spatial resolution of the deformation result. Both SBAS and MSBAS are briefly introduced in the following sections.

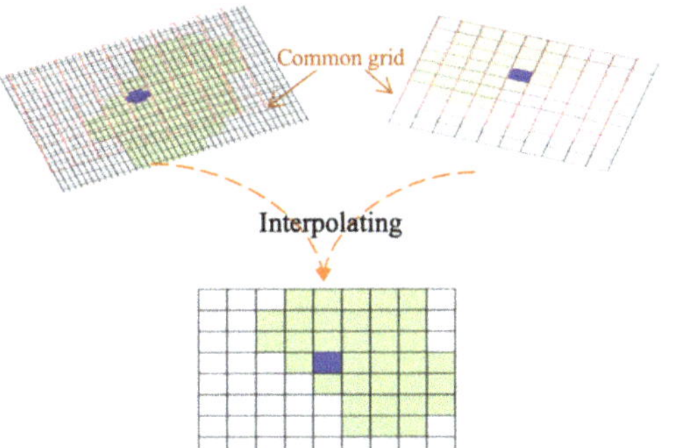

Figure 3. Schematic diagram of interpolating the multiplatform SAR interferogram to a common georeferenced grid.

3.2.1. SBAS Technique

For each SAR data processing, it is assumed that N SAR images from a single platform are obtained at the time order of t_1, t_2, t_3, $\cdots$, t_N. M differential interferograms (where $\frac{N}{2} \leq M \leq \frac{N(N-1)}{2}$) with a short multimaster baseline are generated and well phase unwrapped. The unwrapped phase ($\Delta\varphi$) of one pixel on an interferogram can be written as:

$$\Delta\varphi = \Delta\varphi_{defo} + \Delta\varphi_{DEMeror} + \Delta\varphi_{orb} + \Delta\varphi_{atm} + \Delta\varphi_{dec} = \frac{4\pi}{\lambda} D_{LOS} + \frac{4\pi}{\lambda} \frac{B_\perp H}{R sin(\theta)} + \Delta\varphi_{orb} + \Delta\varphi_{atm} + \Delta\varphi_{dec} \qquad (1)$$

where $\Delta\varphi_{defo}$ is the deformation phase in the *LOS* direction between the period of two SAR acquisitions; $\Delta\varphi_{DEMerror}$ is the topographic residual phase, caused by the imperfection of the external DEM; $\Delta\varphi_{orb}$ is the phase related to the orbital error that can be compensated by the orbital model and external information [32,33]; $\Delta\varphi_{atm}$ and $\Delta\varphi_{dec}$ are the atmospheric phase screen (APS) and decorrelation noise phase, respectively; and λ, D_{LOS}, θ, $B_\perp$, and R are the radar wavelength, deformation, incident angle, perpendicular baseline, and slant range distance between the radar sensor and ground target, respectively.

After compensating for the orbital error ($\Delta\varphi_{orb}$) and part of the atmospheric phase (i.e., altitude-related), Equation (1) can be expressed in the form of matrices when all interferograms are taken into consideration:

$$\Delta\boldsymbol{\varphi} = \boldsymbol{AX} + \boldsymbol{e} \qquad (2)$$

where $\Delta\boldsymbol{\varphi}$ is the unwrapped phase vector of a pixel in all interferograms with the dimension of $M \times 1$, $\boldsymbol{X}$ is the $P \times 1$ unknown parameter vector that is used to model the deformation phase (D_{LOS}) and DEM error (H) [13], $\boldsymbol{e}$ is the noise phase component mainly including the atmospheric phase and decorrelation noise, A is a design matrix linking the unknows and unwrapped phase, which has M rows and P columns. Generally, the solution can be estimated using least-square inversion, i.e., $\boldsymbol{X} = \left(\mathbf{A}^T\mathbf{A}\right)^{-1}\left(\mathbf{A}^T\boldsymbol{\varphi}\right)$. Regularization methods, such as Tikhonov, can also be used to solve the problem when design matrix A is rank deficient [34].

The LOS deformation from an individual SAR geometry or platform can be expressed as [35],

$$D_{LOS} = D_N \sin(\theta)\sin(\beta) - D_E \cos(\beta)\sin(\theta) + D_U \cos(\theta) \qquad (3)$$

where β is the azimuth angle of the flight direction. D_U, D_N, and D_E are the deformation vectors along the up-down, north-south, and east-west directions, respectively. Equation (3) can be rewritten in the form of a matrix as:

$$D_{LOS} = [S_N \ S_E \ S_U] \cdot [D_N \ D_E \ D_U]^{\mathrm{T}} \tag{4}$$

where S is the *LOS* unit vector with north, east, and up components, i.e., $S_N = sin(\beta)sin(\theta)$, $S_E = -cos(\beta)sin(\theta)$, and $S_U = cos(\theta)$. For individual SAR geometry, the result in Equations (3) or (4) cannot be decomposed into 3D displacements. The SBAS methodology can be modified to generate an approximate solution of time series displacement in two or three dimensions, when SAR images are obtained from more than one acquisition geometry [36].

3.2.2. MSBAS Technique

In MSBAS method, each SAR dataset should have an overlap period in time and with different observation geometries. After the data processing with SBAS method, the DEM errors, orbital errors, and atmospheric errors related to altitude are compensated from the unwrapped interferograms for each SAR dataset to achieve the deformation interferograms. Then, the unwrapped deformation interferograms of each SAR dataset are interpolated into the common grid, as shown in Figure 3. The 3D displacement time series on the common grid point can then be resolved as:

$$\Delta\varphi = \begin{bmatrix} S_N A & S_E A & S_U A \end{bmatrix} \begin{bmatrix} D_N \\ D_E \\ D_U \end{bmatrix} \tag{5}$$

where $\Delta\varphi$ is the unwrapped phase vector generated by the multiplatform SAR images, A is the coefficient matrix, and depending on the interferograms generated, S is the LOS unit vector containing geometric information. Currently, most spaceborne SAR satellites have near-polar sun-synchronous orbits, the SAR sensor is relatively insensitive to displacement in the north-south direction, compared to the vertical and east-west directions. The deformation in the north-south direction can be often ignored [36], and Equation (5) can be then rewritten as:

$$\begin{bmatrix} \Delta\varphi \\ 0 \end{bmatrix} = \begin{bmatrix} \hat{A} \\ \zeta I \end{bmatrix} \begin{bmatrix} D_E \\ D_U \end{bmatrix} \tag{6}$$

where the coefficient matrix is updated to $\hat{A} = \{S_E A \ S_U A\}$, ζ is the regularization factor, and I is the unit matrix. Similar to the SBAS method, equation system (6) may be a rank-deficient problem. Transactive singular value decomposition (TSVD) and Tikhonov regularization can be used to find the least square solution in the sense of the minimum norm. In MSBAS, Tikhonov regularization is applied. However, due to the effect of atmospheric phase and decorrelation noise, the solution of Equation (6) can be biased when the time period between each SAR dataset does not overlap or match very well [36]. Therefore, the displacement can be modelled by the deformation model with unknown parameters. For example, an interval deformation velocity model is used to model the 3D displacement, Equation (5) can be rewritten as

$$\begin{bmatrix} \Delta\varphi \\ 0 \end{bmatrix} = \begin{bmatrix} \hat{A} \\ \zeta I \end{bmatrix} \begin{bmatrix} V_E \\ V_U \end{bmatrix} \tag{7}$$

where V_E and V_U are the interval deformation velocities in consecutive SAR acquisitions along the east-west and vertical directions, respectively. $\hat{A}$ is the coefficient matrix, and contains the time intervals between consecutive SAR acquisition. The solution of Equation (7) is solved by applying Tikhonov regularization of a second order. In the regularization parameter, $\zeta = 0.1$ is used to produce a smooth solution. The cumulative displacement of the 2D time series can be recovered by the deformation model according to time. For the interval linear deformation velocity model, the 2D displacement time series can be recovered by Equation (8).

$$D_E^i = \sum_{i=1}^{K} V_E^i \Delta t_i, \, D_U^i = \sum_{i=1}^{K} V_U^i \Delta t_i \tag{8}$$

The accuracy of MSBAS is highly dependent on the quality of the observation, i.e., $\Delta\varphi$. The phase quality of an interferogram is easily affected by decorrelation and atmospheric artifacts. The selection of interferograms is therefore essential in MSBAS data processing. For this reason, all interferograms in which the spatiotemporal baselines were within the given thresholds (see the detailed threshold value in Table A1 in Appendix A) were generated from the available SAR images in this case study. A multilook operation was carried out to suppress the phase noise. Due to the dense vegetation cover during the observation period, we removed interferograms with low coherence in the mining area (mean coherence smaller than 0.28). Phase unwrapping was performed using the MCF method in GAMMA software. In addition, since the decorrelation effect caused by dense vegetation heavily affected the usage of SAR images with different wavelengths, in this case study, we attempted to gradually integrate the multiplatform SAR interferograms with high phase quality into the MSBAS estimator, to reduce the effect of decorrelation for retrieval of mining deformation. Detailed results are presented in Section 4.

4. Results

4.1. Deformation with Single SAR Platform

The mining area suffering ground deformation was identified first, before InSAR time-series data processing. We used two TerraSAR-X images with a 22-day revisit period to generate the deformation interferogram. The results in Figure 4 show that many de-formation areas probably induced by mining activities were identified. Five segmentation areas (i.e., A, B, C, D, and E, as shown in Figure 4 with white rectangles) were divided to clarify the mining regions, which was used to identify the phase quality of the interferograms in the mining area.

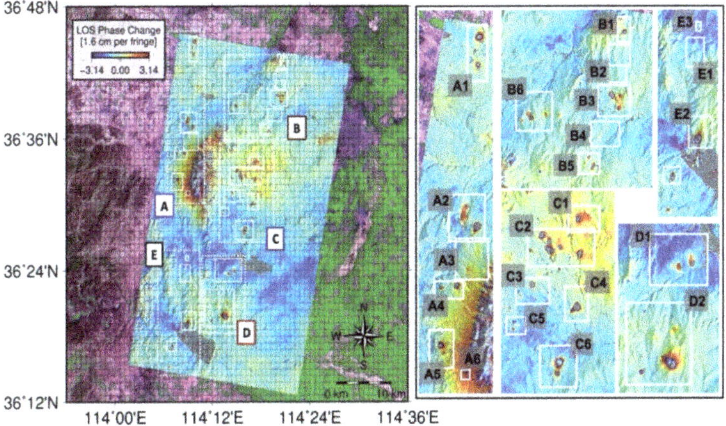

Figure 4. The deformation map of the mining areas estimated by two TerraSAR-X images with a 22-day revisit period.

After the identification of the mining site, we used 14 TerraSAR-X images from June 2015 to April 2016 to estimate the ground deformation over the Fengfeng mining area. Thirteen interferograms with short baseline information were generated, as shown in Figure 5. The coherence maps of each segmentation area were used to select the high-quality interferograms for the MSBAS method, and some examples of coherence maps are given in Figure A1 of Appendix A. According to the coherence maps, the interferogram with 11-day and 22-day time intervals was more effective in monitoring mining deformation with dense vegetation. In addition, due to the heavy weather conditions shown in Figure A2, the surface crops may have changed rapidly during the observation period, leading the backscattering coefficient of the object to vary greatly, and some interferograms with short spatiotemporal baselines may still be affected. If interferograms with longer temporal baselines were included, we could not obtain many coherent pints in the central area of mines, due to the decorrelation caused by dense vegetation. Therefore, we have made corresponding adjustments to only use the data from December 2015 to April 2016 with mainly 11-day baseline interference pairs, to eliminate severe decorrelation in large areas. Seven interferograms with good phase quality were generated according to the spatiotemporal baseline in Figure 5 with a red network.

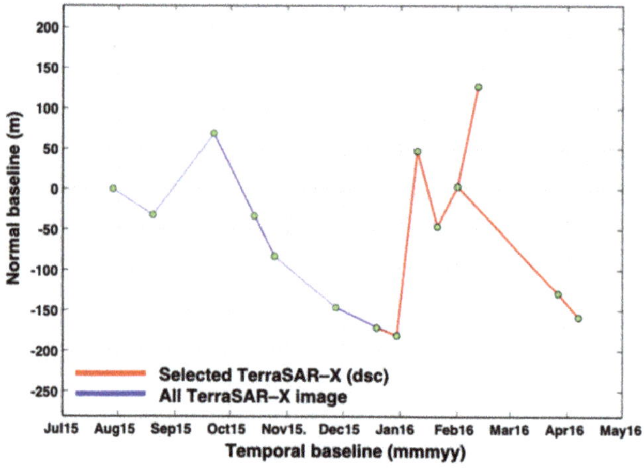

Figure 5. The baseline configuration of the TerraSAR–X dataset.

Since only the TerraSAR–X SAR dataset was used in this section, the SBAS method was used to retrieve the mining deformation. Figure 6 displays the LOS deformation velocity map of the investigated area. The whole area remained stable over the observed period, except for the significant mining deformation over the mine area. The mining deformation of five segmentation regions was well retrieved. Denser mining areas were identified separately, and the deformation boundary of two closer mining areas were also identified, indicating the applicability of the TerraSAR–X in mining deformation areas during the winter period. The maximum deformation among the mining sites reached approximately 100 cm/yr. Two mining sites, Wannian and Jiulong, are displayed in the subfigure of Figure 6. The maximum LOS deformations of these two sites were approximately 23 cm and 16 cm, respectively.

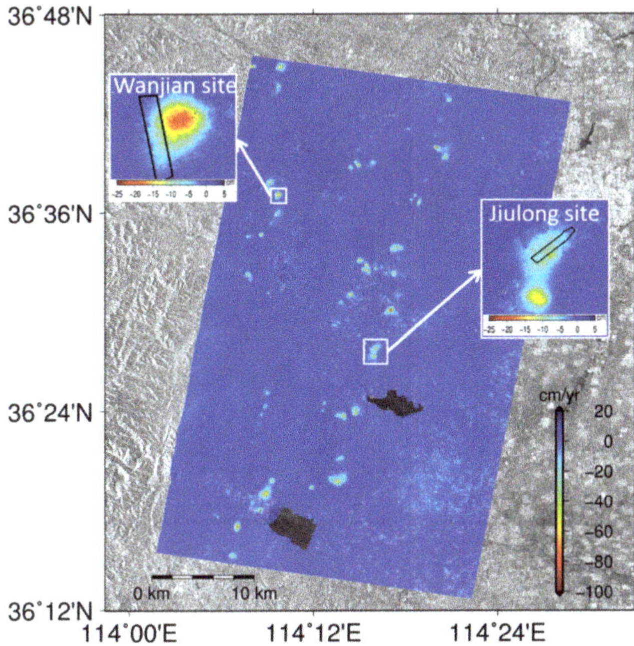

Figure 6. The deformation velocity between December 2015 and April 2016 was estimated with TerraSAR–X data.

The time-series deformation of the Fengfeng mine is shown in Figure 7. During the observed period, all the mining sites suffered significant deformation, and the deformation of each mining site started at different moments, which indicates that mining activities were carried out in all mining areas during the observation period, and the scale of mining activities and working hours varied. The maximum LOS deformation related to mining activities was approximately 30 cm.

4.2. Deformation with Two SAR Platforms

To retrieve the 2D displacement of the Fengfeng mine, the Sentinel-1A dataset from the ascending orbital direction between December 2015 and April 2016 was used for estimation, together with the TerraSAR–X dataset. The spatiotemporal baseline configuration of the two SAR datasets is shown in Figure 8a, in which the temporal baseline of both datasets was well matched. Thus, the MSBAS technique in Section 3.2 was used to retrieve the 2D mining displacement. Since the coverage of Sentinel-1A is larger than the coverage of TerraSAR–X, the common area of the displacement has the same coverage as the TerraSAR–X dataset. The retrieved 2D displacement of the Fengfeng mine is shown in Figure 9. The maximum accumulated subsidence in Figure 9a reached 33 cm during the observed period. Due to the surface subsidence of the mining area, the edges of the subsidence area collapsed toward the center area, resulting in a horizontal east-west displacement that moved to the center area. Therefore, each mining site suffered from not only subsidence but also horizontal displacement. Figure 9b shows that the relative horizontal displacement of the Fengfeng mining area reached a maximum of 20 cm. Since InSAR is insensitive to horizontal deformation in the north-south direction, the horizontal displacement corresponding to the north-south direction was difficult to retrieve.

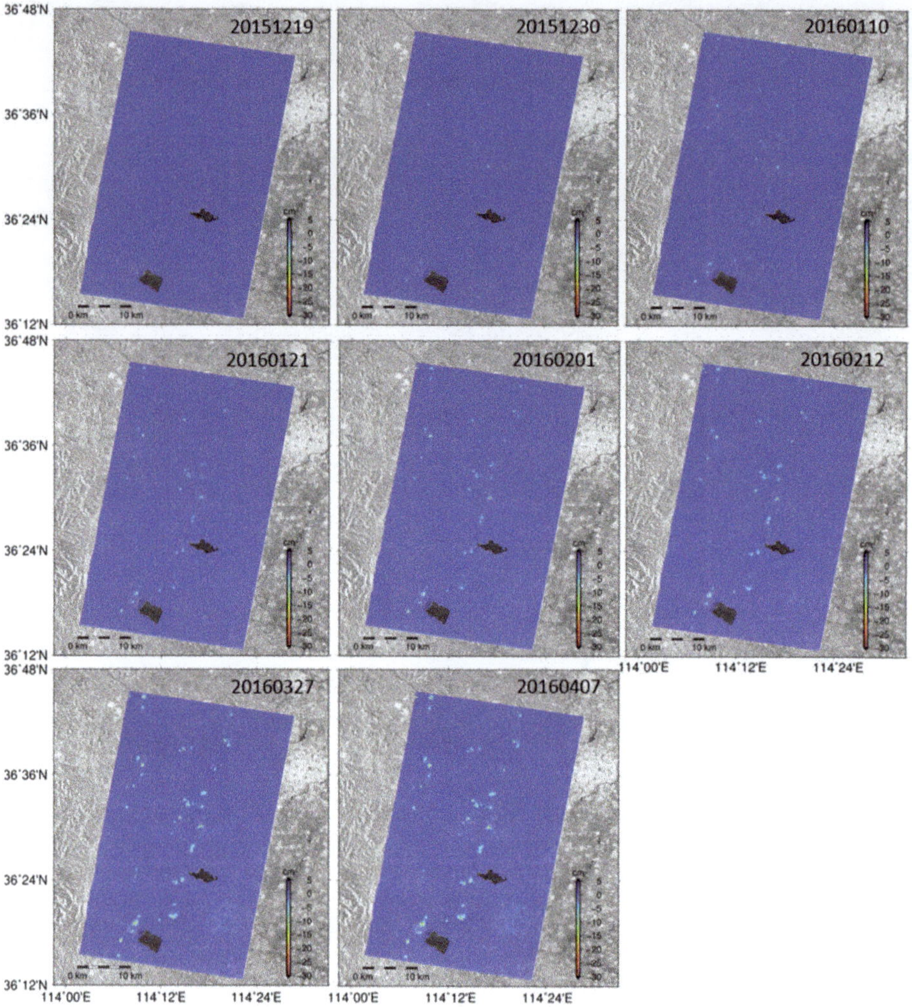

Figure 7. The deformation time series between September 2015 and April 2016, estimated with TerraSAR–Xdata.

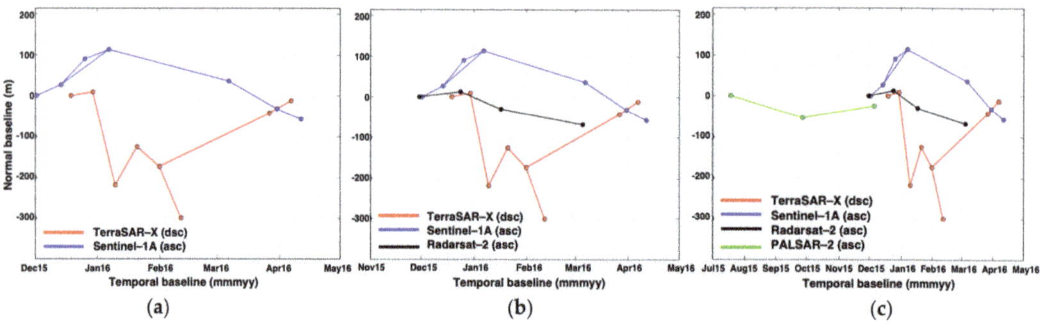

Figure 8. The baseline configuration of (**a**) the TerraSAR–X and Sentinel-1A datasets, (**b**) the TerraSAR–X and Sentinel-1A datasets and Radarsat-2, and (**c**) the TerraSAR–X and Sentinel-1A datasets, Radarsat-2, and PALSAR-2.

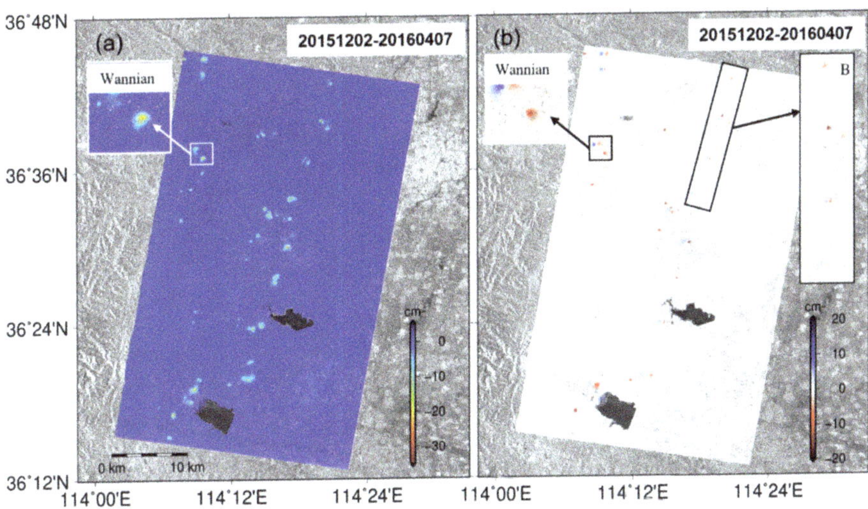

Figure 9. (**a**) The subsidence map and (**b**) horizontal displacement map of the Fengfeng mine area with TerraSAR–X and Sentinel-1A from 20151219 to 20160407.

4.3. Deformation with Three SAR Platforms

Since Sentinel-1A is susceptible to the influence of the atmosphere, we added a set of Radarsat-2 images with the C-band to detect the corresponding 2D displacement results. The baseline configuration is displayed in Figure 8b. The coverage of Radarsat-2 could not fully cover TerraSAR–X data, as shown in Figure 2, and the 2D displacement of some research areas could not be retrieved. The final 2D displacement is shown in Figure 10. The maximum accumulated subsidence in Figure 10a reached 33 cm during the observed period. The vertical subsidence of the Wannian mining site was enlarged, and the horizontal displacement of the B mining region was more significant.

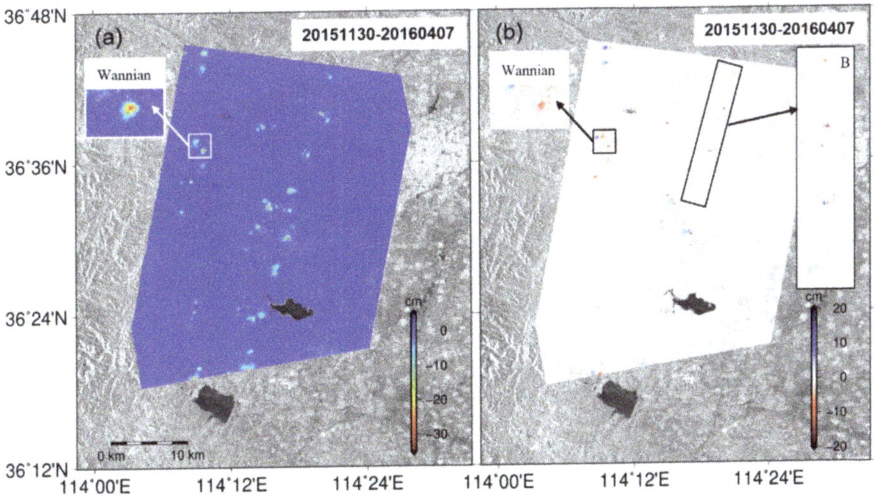

Figure 10. (**a**) The subsidence map and (**b**) horizontal displacement map of the Fengfeng mine area with TSX, Sentinel-1A and Radarsat-2 from 20151219 to 20160407.

4.4. Deformation with All SAR Platforms

PALSAR-2 data use a long radar wavelength in L-band, and can penetrate part of the vegetation. We therefore used three available PALSAR-2 data to retrieve the mining deformation in the summer period, and extended the deformation time series for MSBAS. The ground deformation in mining areas often produced regional subsidence that funnel along its dig working face, as sketched in Figure 11. Largest vertical subsidence suffered in the regions near the center axis of the dig working face, while horizontal displacements of these regions was close to zero, as the results of the Wannian mining area show in Figures 9 and 10. For ascending and descending orbital geometry, the horizontal deformation of the area near the mining center caused completely opposite effects, as shown by points A and C in the Figure 11. With multiplatform SAR images from opposite observed geometries in a common period, the influence on the combined deformation results will be weakened. Therefore, we ignored the horizontal displacement and converted all the LOS direction deformations into vertical, to solve the longer time series of mining subsidence.

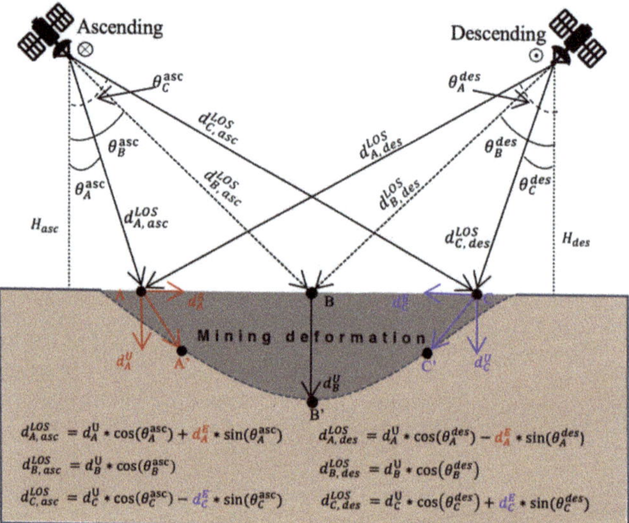

Figure 11. Schematic diagram of mining deformation monitoring with multiplatform SAR from ascending and descending orbital directions.

The subsidence of the Fengfeng mine from July 2015 to April 2016 is shown in Figure 12a. Due to the difference between the coverage of each SAR dataset, the deformation map was reshaped again. The subsidence of most of the Fengfeng mining regions was retrieved, and the maximum subsidence was approximately 90 cm. The subsidence map larger than 5 cm is displayed in Figure 12b. In total, 36 deformed mining sites were revealed, and their time-series subsidence is shown in Figures A3 and A4 of the Appendix A. The maximum subsidence of each mining site is listed in Table A2 of the Appendix A. The maximum subsidence occurred in the D2 mining region and reached 88.8 cm. The ground coverage of the subsidence with different intervals is listed in Table A2. Compared with the official report investigation in 2016 [37], the deformation areas from the InSAR results were larger than the report provided by the government, indicating that the subsidence area has increased, due to underground mining activities. It also demonstrated the usability of the InSAR method in the mining area with denser vegetation coverage.

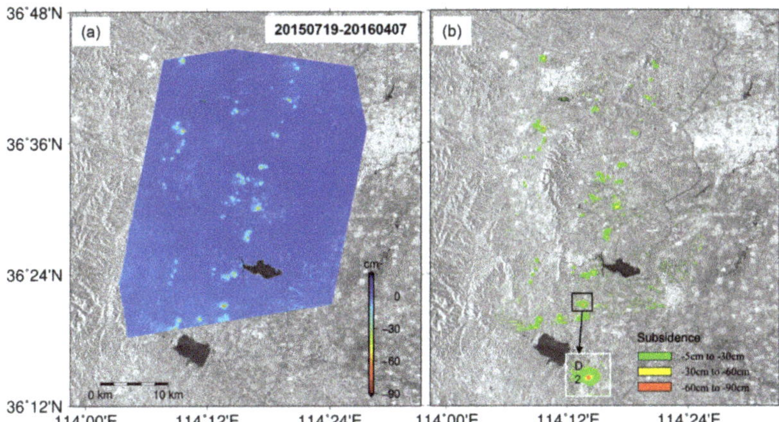

Figure 12. (**a**) Vertical deformation of the Fengfeng mining area from 20150719 to 20160407. (**b**) The subsidence of mining sites larger than 5 cm.

5. Discussion

5.1. Accuracy Validation

We used the subsidence results of ground leveling measurements at two mining sites, Wanjian and Jiulong Mines, to validate the results derived from the multiplatform SAR datasets. The location of the leveling points is shown with the white dots on in Figure 13. The path of the leveling stations was along the side part of the mining area, to obtain ground subsidence. We compared the time-series subsidence between the MSBAS-derived results and the leveling on twelve selected points, which were distributed uniformly over the mining area, as labeled in Figure 13. Because the resolution of the MSBAS-derived results depends on the common reference grid of the multiplatform SAR dataset, we selected MSBAS measurements with distances to leveling stations smaller than 100 m, to compare with the leveling data of the two sites. All qualified MSBAS measurements were used to calculate the uncertainty and are displayed in Figures 14 and 15. The results of MSBAS agreed well with the leveling measurements. The differences between them might relate to the uncertainty of the atmospheric delays, decorrelation effect, and differences in location and time between the InSAR and leveling data, such as the unmatched data in the last period of the two leveling stations at the Jiulong site, i.e., JL03 and JLQ39.

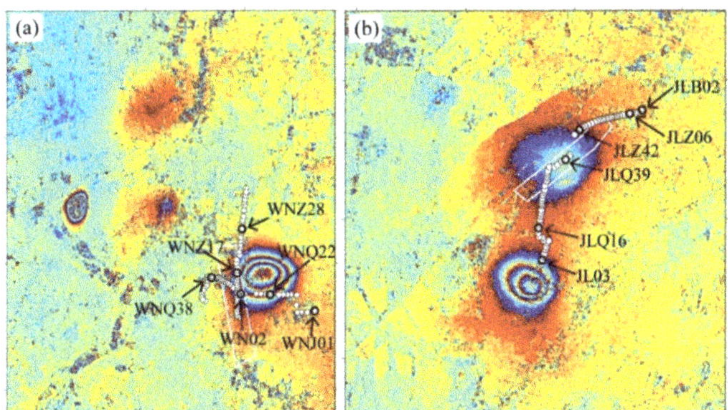

Figure 13. The distribution of the leveling stations marked with white dots of the (**a**) Wannian site and (**b**) Jiulong site. The background is one of the wrapped TerraSAR–X interferograms.

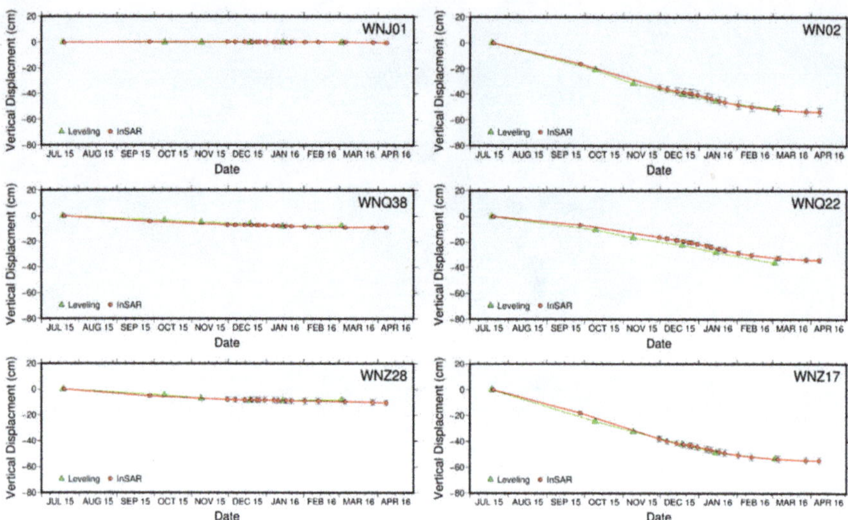

Figure 14. Comparison of time-series subsidence between leveling data and MSBAS results at the selected points of the Wannian site.

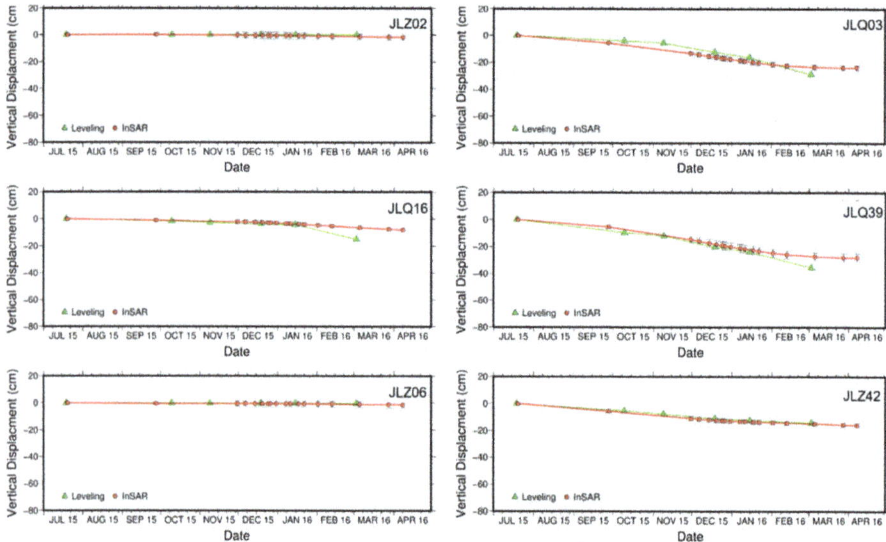

Figure 15. Comparison of time-series subsidence between leveling data and MSBAS results at the selected points of the Jiulong site.

Since the date of the acquisition of leveling and the SAR images were quite different, it was difficult to compare the exact value between them, thus, the deformation velocity was used to make the cross validation. We therefore calculated the subsidence velocity with the leveling data. The points of MSBAS measurements with a distance smaller than 100 m were compared with the leveling data, to reduce the effect of the shift in geolocation. Figure 16 gives the scatterplots between the leveling and the results of InSAR at the two mining sites. The corresponding RMSE is 83.51 mm/year. Due to the limited number of interferograms used in retrieving the deformation, the RMSE between InSAR and leveling results is a little bit large. This value could be further reduced if more high-quality interferograms with short temporal baseline can be obtained from multiplatform SAR, however, it is difficult for the monitoring scenario like the region in this study with the state-of-the-art SAR missions. The significant difference of leveling stations that is framed with a dashed black rectangle, may be caused by the large gradient deformation in the center of the mining site that is leading to the unwrapping error [3]. According to the presented studies in [4,38], it has been suggested that the accuracy of mining subsidence measured by InSAR are from millimeters to few centimeters. Therefore, the accuracy of mining deformation detected by the MSBAS and multiplatform SAR datasets is acceptable for this study.

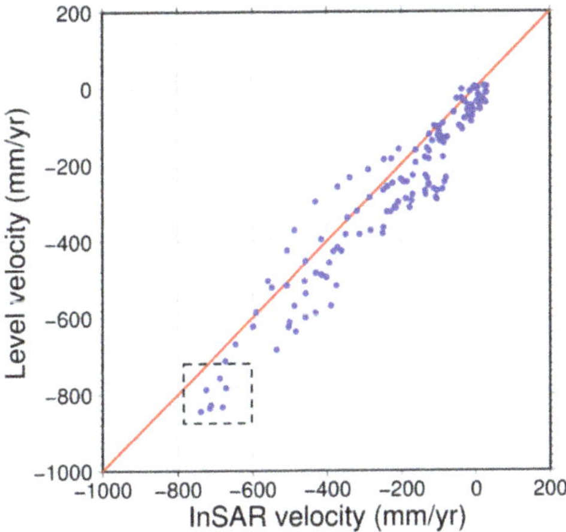

Figure 16. Comparison of the subsidence velocity between the MSBAS-InSAR results and the leveling measurements at the two mining sites of Jiulong and Wannian.

5.2. Advantages of Multiplatform SAR Data in Mining Deformation Monitoring

Because mining areas are often located in rural areas covered with dense vegetation, ground coverage, extreme weather, and large gradients of deformation will raise the decorrelation of InSAR interferograms and lead to incomplete spatial coverage with information gaps for mining deformation monitoring. It is worth noting from this study that the L-band PALSAR-2 data were effective for mining deformation monitoring, even in summer, and pairs with long baselines can be included in the interferometric analysis; therefore, it would be good to use the L-band data for mining deformation monitoring. However, the revisit period (i.e., from 40 to 90 days) of PALSAR-2 is too long to achieve enough temporal samples of the deformation for further analysis of the mechanism of mining activities. The objective of our contribution was to show that important improvements to these limitations are still possible. In denser vegetation areas, the mining deformation can be retrieved by properly selecting the high-quality interferograms from multiplatform SAR datasets. The fusion processing of multiplatform data can not only obtain the vertical subsidence of the mining area but also obtain the horizontal displacement. Meanwhile, a more satisfactory coverage of spatial monitoring and the increased temporal sampling of the deformation can be obtained, compared with the single SAR platform dataset.

6. Conclusions

Deformation associated with mining activities is usually difficult to precisely monitor in regard to both time and space because of the diverse topography, large-gradient ground subsidence, and dense vegetation coverage. These problems possibly result in decorrelation in InSAR, and therefore limit the application of InSAR in mining areas. In this study, we used multiplatform SAR data and MSBAS technique to monitor mining deformation over the Fengfeng mining area, China, from July 2015 to April 2016. The derived results indicated that the Fengfeng mining area experienced serious ground subsidence caused by mining activities during the observed period. The largest subsidence reached 88.8 cm. Joint analysis of the InSAR-derived subsidence and the ground true leveling data proved the high accuracy of the MSBAS results. The study suggests that using multiplatform SAR datasets helps to retrieve the ground deformation in mining areas with dense vegetation.

Author Contributions: Conceptualization, B.Z., S.W. and C.W.; methodology, B.Z.; software, B.Z. and S.W.; validation, B.Z. and S.W.; formal analysis, S.W., B.Z. and X.D.; investigation, C.W.; resources, B.Z. and X.D.; data curation, B.Z. and S.W.; writing—original draft preparation, S.W. and B.Z.; writing—review and editing, S.W. and B.Z.; supervision, S.W., X.D., J.Z. and Q.L.; project administration, B.Z. and X.D.; funding acquisition, B.Z. and X.D. All authors have read and agreed to the published version of the manuscript.

Funding: This research was supported the National Key Research and Development Program of China (2018YFB2101005), the Guangdong Basic and Applied Basic Research Foundation (No. 2021A1515011427), the Research Grants Council (RGC) of Hong Kong (PolyU 152232/17E, PolyU 152164/18E, and PolyU 15223319/19E), the Research Institute for Sustainable Urban Development (RISUD) (1-BBWB), the National Natural Science Foundation of China (41974006), the Shenzhen Scientific Research and Development Funding Program (JCYJ20200807110745001, KQJSCX20180328093 453763, and JCYJ201803305125101282), the Research Program from the Department of Education of Guangdong Province (grant No. 2018KTSCX196), and the National Administration of Surveying, Mapping and Geoinformation of China (Grant 201412016).

Data Availability Statement: Not applicable.

Acknowledgments: The authors would like to thank the DLR for providing the TerraSAR-X dataset under the science proposal LAN2979 and GEO3603, and the ESA/Copernicus for providing the Sentinel-1A SAR images, respectively. The Radarsat-2 was from CSA. The PALSAR-2 was provided by the JAXA under the science proposal PI3380. The authors would like to thank the China University of Mining and Technology for providing leveling data. The authors would also like to thank the anonymous reviewers for their constructive comments and suggestions.

Conflicts of Interest: The authors declare no conflict of interest.

Appendix A

This appendix has shown the details and data supplemental to the main text.

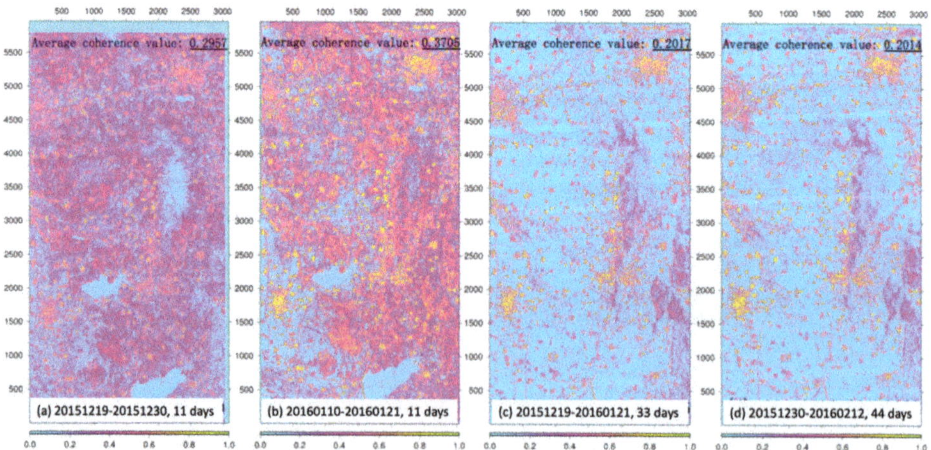

Figure A1. The examples of the coherence maps of TSX datasets.

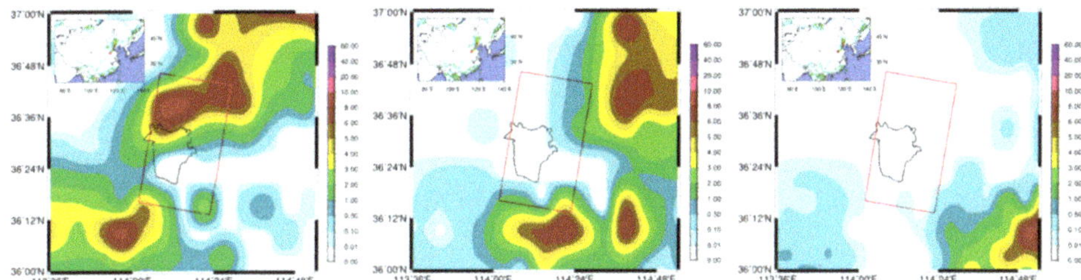

Figure A2. The examples of the weather condition on three TSX acquisition dates i.e., 20150729, 20150820, and 20160201.

Table A1. The threshold of spatiotemporal baseline for each SAR data processing.

SAR Data	Temporal Baseline (Days)	Perpendicular Baseline (m)
TSX	60	300
Sentinel-1A	100	100
Radarsat-2	100	300
PALSAR-2	300	200

Table A2. The detailed information of displacement in 36 mining sites.

Regions	No.	No. of Deformed Areas	Ground Coverage [km^2]			Maximum Displacement [cm]
			−5 cm to −30 cm	−30 cm to −60 cm	−60 cm to −90 cm	
A1	1	1	0.861	0.1436	0.000038	−61.56
A2	2	1	0.6036	0.1074	0.0371	−81.81
	3	1	1.1483	0.0368	0	−47.99
A3	4	1	0.2919	0.0306	0	−51.83
A4	5	1	0.1973	0.0187	0	−46.68
	6	1	0.1617	0	0	−29.37
A5	7	1	0.1415	0	0	−13.58
	8	1	0.1217	0	0	−19.49
A6	9	2	0.0874	0	0	−24.56
B1	10	2	0.7195	0.0001	0	−33.93
B2	11	1	0.2147	0	0	−21.29
	12	1	0.1072	0.0014	0	−41.66
B3	13	3	0.7566	0.0869	0	−58.38
B4	14	1	0.0563	0	0	−23.59
B5	15	3	0.5722	0.03	0	−49.31
	16	2	0.6993	0.0581	0	−56.46
B6	17	1	0.1465	0	0	−28.01
C1	18	1	0.9503	0.0673	0	−49.08
	19	1	0.0928	0	0	−22.41
C2	20	4	1.4109	0.0884	0	−54.42
	21	2	0.2545	0.00068	0	−35.59
	22	1	0.4426	0.0166	0	−36.82
C3	23	1	0.4461	0.0112	0	−38.34
C4	24	1	1.0695	0.2351	0.00084	−66.5
C5	25	2	0.3035	0.00003	0	−30.35
C6	26	2	1.6334	0.0739	0	−52.55
D1	27	3	0.3516	0.0012	0	−36.45
	28	4	2.0048	0.1332	0	−55.03
	29	1	1.4196	0.1339	0.0427	−88.8
D2	30	1	0.7794	0.112	0	−51.34
	31	1	0.1956	0.0054	0	−36.9
	32	2	1.9831	0	0	−28.35
E1	33	1	0.2088	0.0153	0	−39.63
	34	1	0.2129	0.0002	0	−30.73
E2	35	2	2.2508	0.1558	0.0001	−60.84
E3	36	1	0.2996	0.0003	0	−32.64

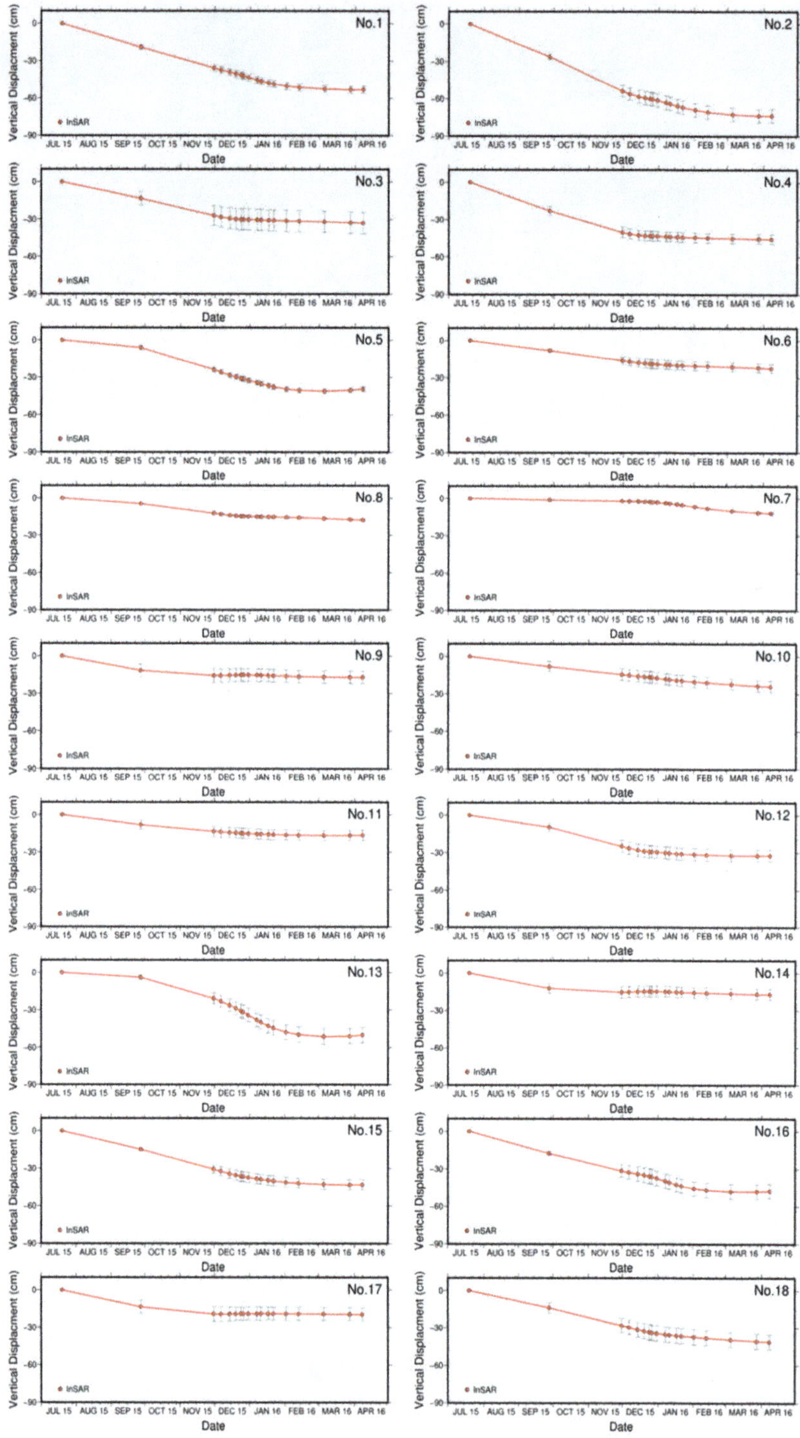

Figure A3. The subsidence time series of mining sites from No.1 to No.18.

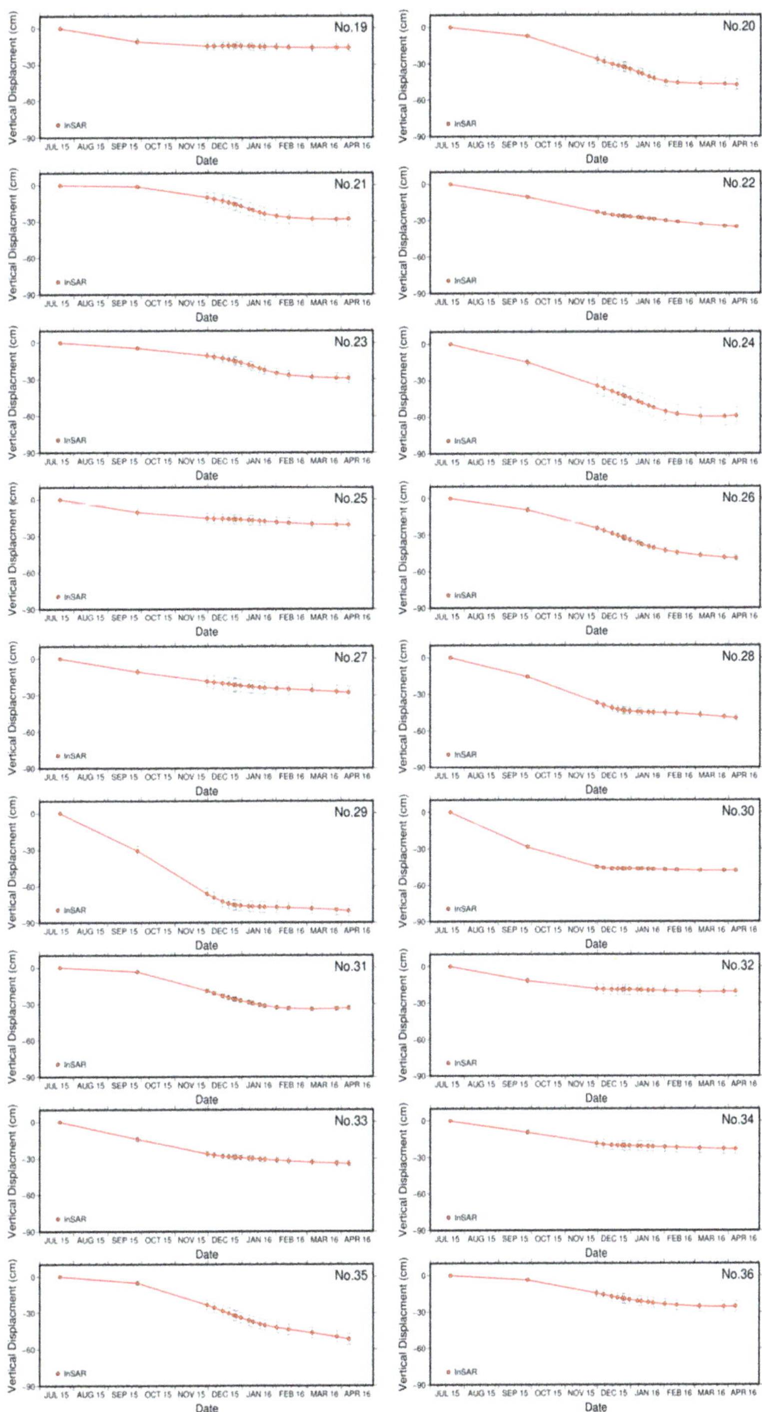

Figure A4. The subsidence time series of mining sites from No.19 to No.36.

References

1. Liu, G.; Guo, H.D.; Hanssen, R.; Perski, Z.; Li, X.W.; Yue, H.Y.; Fan, J.H. The application of InSAR technology to mining area subsidence monitoring. *Remote Sens. Land Res.* **2008**, *20*, 51–55.
2. Samsonov, S.; D'Oreye, N.; Smets, B. Ground deformation associated with post-mining activity at the French-German border revealed by novel InSAR time series method. *Int. J. Appl. Earth Obs. Geoinf.* **2013**, *23*, 142–154. [CrossRef]
3. Yang, Z.; Li, Z.; Zhu, J.; Wang, Y.; Wu, L. Use of SAR/InSAR in mining deformation monitoring, parameter inversion, and forward predictions: A review. *IEEE Geosci. Remote Sens. Mag.* **2020**, *8*, 71–90. [CrossRef]
4. Wright, P.; Stow, R. Detecting mining subsidence from space. *Int. J. Remote Sens.* **1999**, *20*, 1183–1188. [CrossRef]
5. Bamler, R.; Hartl, P. Synthetic aperture radar interferometry. *Inverse Probl.* **1998**, *14*, R1. [CrossRef]
6. Rosen, P.A.; Hensley, S.; Joughin, I.R.; Li, F.K.; Madsen, S.N.; Rodriguez, E.; Goldstein, R.M. Synthetic aperture radar inter-ferometry. *Proc. IEEE* **2000**, *88*, 333–382. [CrossRef]
7. Massonnet, D.; Feigl, K.L. Radar interferometry and its application to changes in the Earth's surface. *Rev. Geophys.* **1998**, *36*, 441–500. [CrossRef]
8. Eldhuset, K.; Andersen, P.H.; Hauge, S.; Isaksson, E.; Weydahl, D.J. ERS tandem InSAR processing for DEM generation, glacier motion estimation and coherence analysis on Svalbard. *Int. J. Remote Sens.* **2003**, *24*, 1415–1437. [CrossRef]
9. Ding, X.L.; Liu, G.X.; Li, Z.W.; Li, Z.L.; Chen, Y.Q. Ground subsidence monitoring in Hong Kong with satellite SAR inter-ferometry. *Photogramm. Eng. Remote Sens.* **2004**, *70*, 1151–1156. [CrossRef]
10. Burgmann, R.; Rosen, P.A.; Fielding, E.J. Synthetic aperture radar interferometry to measure earth's surface topography and its deformation. *Annu. Rev. Earth Planet. Sci.* **2000**, *28*, 169–209. [CrossRef]
11. Hanssen, R.F. *Radar Interferometry: Data Interpretation and Error Analysis*; Kluwer Academic Publishers: Dordrecht, The Netherlands, 2001.
12. Ferretti, A.; Prati, C.; Rocca, F. Permanent scatterers in SAR interferometry. *IEEE Trans. Geosci. Remote Sens.* **2001**, *39*, 8–20. [CrossRef]
13. Berardino, P.; Fornaro, G.; Lanari, R.; Sansosti, E. A new algorithm for surface deformation monitoring based on small baseline differential SAR interferograms. *IEEE Trans. Geosci. Remote Sens.* **2002**, *40*, 2375–2383. [CrossRef]
14. Wu, S.; Zhang, L.; Ding, X.; Perissin, D. Pixel-Wise MTInSAR Estimator for integration of coherent point selection and unwrapped phase vector recovery. *IEEE Trans. Geosci. Remote Sens.* **2018**, *57*, 2659–2668. [CrossRef]
15. Xu, W.; Wu, S.; Materna, K.; Nadeau, R.; Floyd, M.; Funning, G.; Chaussard, E.; Johnson, C.W.; Murray, J.R.; Ding, X.; et al. Interseismic ground deformation and fault slip rates in the greater san francisco bay area from two decades of space geodetic data. *J. Geophys. Res. Solid Earth* **2018**, *123*, 8095–8109. [CrossRef]
16. Wu, S.; Yang, Z.; Ding, X.; Zhang, B.; Zhang, L.; Lu, Z. Two decades of settlement of Hong Kong International Airport measured with multi-temporal InSAR. *Remote Sens. Environ.* **2020**, *248*, 111976. [CrossRef]
17. Sun, Q.; Zhang, L.; Ding, X.L.; Hu, J.; Liang, H.Y. Investigation of slow-moving landslides from ALOS/PALSAR images with TCPInSAR: A case study of Oso, USA. *Remote Sens.* **2014**, *7*, 72–88. [CrossRef]
18. Pepe, A.; Bonano, M.; Zhao, Q.; Yang, T.; Wang, H. The use of C-/X-band time-gapped sar data and geotechnical models for the study of shanghai's ocean-reclaimed lands through the SBAS-DInSAR technique. *Remote Sens.* **2016**, *8*, 911. [CrossRef]
19. Zhang, B.; Wang, C.; Ding, X.; Zhu, W.; Wu, S. Correction of ionospheric artifacts in SAR data: Application to fault slip inversion of 2009 southern Sumatra earthquake. *IEEE Geosci. Remote Sens. Lett.* **2018**, *15*, 1327–1331. [CrossRef]
20. Duan, M.; Xu, B.; Li, Z.W.; Wu, W.H.; Cao, Y.M.; Liu, J.H.; Wang, G.Y.; Hou, J.X. A new weighting method by considering the physical characteristics of atmospheric turbulence and decorrelation noise in SBAS-InSAR. *Remote Sens.* **2020**, *12*, 2557. [CrossRef]
21. Xiong, S.T.; Wang, C.S.; Qin, X.Q.; Zhang, B.C.; Li, Q.Q. Time-series analysis on persistent scatter-interferometric synthetic aperture radar (PS-InSAR) derived displacements of the Hong Kong-Zhuhai-Macao bridge (HZMB) from sentinel-1A observations. *Remote Sens.* **2021**, *13*, 546. [CrossRef]
22. Qin, X.Q.; Zhang, L.; Yang, M.S.; Luo, H.; Liao, M.S.; Ding, X.L. Mapping surface deformation and thermal dilation of arch bridges by structure-driven multi-temporal D-InSAR analysis. *Remote Sens. Environ.* **2018**, *216*, 71–90. [CrossRef]
23. Carnec, C.; Massonnet, D.; King, C. Two examples of the use of SAR interferometry on displacement fields of small spatial extent. *Geophys. Res. Lett.* **1996**, *23*, 3579–3582. [CrossRef]
24. Lazecký, M. InSAR used for subsidence monitoring of mining area OKR, Czech Republic. In *Fringe 2009, Proceedings of the Workshop, Frascati, Italy, 30 November–4 December 2009*; ESA Communications: Noordwijk, The Netherlands, 2010.
25. Bechor, N.B.D.; Zebker, H.A. Measuring two-dimensional movements using a single InSAR pair. *Geophys. Res. Lett.* **2006**, *33*. [CrossRef]
26. Michel, R.; Avouac, J.-P.; Taboury, J. Measuring ground displacements from SAR amplitude images: Application to the landers earthquake. *Geophys. Res. Lett.* **1999**, *26*, 875–878. [CrossRef]
27. Ng, A.H.-M.; Ge, L.; Zhang, K.; Chang, H.-C.; Li, X.; Rizos, C.; Omura, M. Deformation mapping in three dimensions for underground mining using InSAR—Southern highland coalfield in New South Wales, Australia. *Int. J. Remote Sens.* **2011**, *32*, 7227–7256. [CrossRef]
28. Wegmuller, U.; Werner, C.; Strozzi, T.; Wiesmann, A. Monitoring mining induced surface deformation. *IEEE Int. Geosci. Remote Sens.* **2004**, *3*, 1933–1935.

29. Diao, X.; Wu, K.; Hu, D.; Li, L.; Zhou, D. Combining differential SAR interferometry and the probability integral method for three-dimensional deformation monitoring of mining areas. *Int. J. Remote Sens.* **2016**, *37*, 5196–5212. [CrossRef]
30. Xia, Y.; Wang, Y. InSAR- and PIM-based inclined goaf determination for illegal mining detection. *Remote Sens.* **2020**, *12*, 3884. [CrossRef]
31. Werner, C.; Wegmüller, U.; Strozzi, T.; Wiesmann, A. Gamma SAR and interferometric processing software. *Remote Sens.* **2000**, *1620*, 1620.
32. Zhang, L.; Ding, X.L.; Lu, Z.; Jung, H.S.; Hu, J.; Feng, G.C. A novel multitemporal InSAR model for joint estimation of de-formation rates and orbital errors. *IEEE Int. Geosci. Remote Sens.* **2014**, *52*, 3529–3540. [CrossRef]
33. Baehr, H. *Orbital Effects in Spaceborne Synthetic Aperture Radar Interferometry*; KIT Scientific Publishing: Karlsruhe, Germany, 2013.
34. Schmidt, D.A.; Bürgmann, R. Time-dependent land uplift and subsidence in the Santa Clara valley, California, from a large interferometric synthetic aperture radar data set. *J. Geophys. Res. Space Phys.* **2003**, *108*, 9. [CrossRef]
35. Hu, J.; Li, Z.-W.; Ding, X.; Zhu, J.; Zhang, L.; Sun, Q. Resolving three-dimensional surface displacements from InSAR measure-ments: A review. *Earth-Sci. Rev.* **2014**, *133*, 1–17. [CrossRef]
36. Samsonov, S.; D'Oreye, N. Multidimensional time-series analysis of ground deformation from multiple InSAR data sets applied to Virunga Volcanic province. *Geophys. J. Int.* **2012**, *191*, 1095–1108. [CrossRef]
37. General Office of the Municipal Government. Notice of the General Office of Handan Municipal People's Government on Issuing Handan's 2016 Geological Disaster Prevention Plan. Available online: https://www.hd.gov.cn/hdzfxxgk/gszbm/auto23692/201808/t20180802_799421.html?keywords=%E5%9C%B0%E8%B4%A8%E7%81%BE%E5%AE%B3.2016 (accessed on 15 May 2021).
38. Raucoules, D.; Maisons, C.; Carnec, C.; Le Mouelic, S.; King, C.; Hosford, S. Monitoring of slow ground deformation by ERS radar interferometry on the Vauvert salt mine (France): Comparison with ground-based measurement. *Remote Sens. Environ.* **2003**, *88*, 468–478. [CrossRef]

remote sensing

Article

Multi-Temporal Satellite Interferometry for Fast-Motion Detection: An Application to Salt Solution Mining

Lorenzo Solari [1,*], Roberto Montalti [2], Anna Barra [1], Oriol Monserrat [1], Silvia Bianchini [3] and Michele Crosetto [1]

[1] Centre Tecnològic de Telecomunicacions de Catalunya (CTTC/CERCA), Division of Geomatics, Avenida Gauss, 7, E-08860 Castelldefels, Spain; anna.barra@cttc.cat (A.B.); omonserrat@cttc.cat (O.M.); michele.crosetto@cttc.cat (M.C.)

[2] TRE Altamira, Carrer de Còrsega, 381, 08037 Barcelona, Spain; roberto.montalti@tre-altamira.com

[3] Department of Earth Sciences, University of Firenze, Via La Pira, 4 50121 Firenze, Italy; silvia.bianchini@unifi.it

[*] Correspondence: lorenzo.solari@cttc.cat; Tel.: +34-93-645-29-00

Received: 19 October 2020; Accepted: 25 November 2020; Published: 29 November 2020

Abstract: Underground mining is one of the human activities with the highest impact in terms of induced ground motion. The excavation of the mining levels creates pillars, rooms and cavities that can evolve in chimney collapses and sinkholes. This is a major threat where the mining activity is carried out in an urban context. Thus, there is a clear need for tools and instruments able to precisely quantify mining-induced deformation. Topographic measurements certainly offer very high spatial accuracy and temporal repeatability, but they lack in spatial distribution of measurement points. In the past decades, Multi-Temporal Satellite Interferometry (MTInSAR) has become one of the most reliable techniques for monitoring ground motion, including mining-induced deformation. Although with well-known limitations when high deformation rates and frequently changing land surfaces are involved, MTInSAR has been exploited to evaluate the surface motion in several mining area worldwide. In this paper, a detailed scale MTInSAR approach was designed to characterize ground deformation in the salt solution mining area of Saline di Volterra (Tuscany Region, central Italy). This mining activity has a relevant environmental impact, depleting the water resource and inducing ground motion; sinkholes are a common consequence. The MTInSAR processing approach is based on the direct integration of interferograms derived from Sentinel-1 images and on the phase splitting between low (LF) and high (HF) frequency components. Phase unwrapping is performed for the LF and HF components on a set of points selected through a "triplets closure" method. The final deformation map is derived by combining again the components to avoid error accumulation and by applying a classical atmospheric phase filtering to remove the remaining low frequency signal. The results obtained reveal the presence of several subsidence bowls, sometimes corresponding to sinkholes formed in the recent past. Very high deformation rates, up to −250 mm/yr, and time series with clear trend changes are registered. In addition, the spatial and temporal distribution of velocities and time series is analyzed, with a focus on the correlation with sinkhole occurrence.

Keywords: multi-temporal interferometry; mining; salt dissolution; MTInSAR; sinkholes

---

1. Introduction

Subsidence is a common ground deformation where underground mining activities are carried out. The surface displacements, usually characterized by high deformation rates, are triggered by the excavation of new mining levels which creates cavities, redistributes the stresses, or modifies the groundwater circulation [1]. Among the many different types of underground mining activities, salt solution is one of the most impactful on the environment [2]. The extraction of saturated brines through the dissolution of deep salt levels commonly produces high subsidence rates [2–5]. The forced circulation of water in the salt levels leaves back cavities that in the long-term evolve in surface effects as subsidence bowls, collapse chimneys and sinkholes. Some examples of such processes include: the Old Belvedere Spinello brinefield in Southern Italy [6,7], the Barycz and Lezkowice salt mines in Poland [8,9], the Tuzla brinefield in Bosnia Herzegovina [10–12], the Gallenoncourt and Vauvert salt mines in France [13,14], the Ocnele Mari mining area in Romania [15–17], and various sites in the United States of America [2,18,19].

Remote sensing techniques can offer a useful support to studies on the ground effects of natural or anthropogenic subsidence, including mining-related motion. In the past decade, remote sensing techniques have become a reliable alternative to classical geotechnical investigations and measurements. Among many, Differential and Multi-temporal satellite interferometry (DInSAR and MTInSAR) techniques (e.g., [20–22]), terrestrial and airborne laser scanning (e.g., [23–25]), global navigation satellite system data (e.g., [26–28]) provided the best results. Focusing on satellite interferometry (InSAR), mining-induced subsidence is one of the common fields of application of this technique. Some of the most recent works have been presented by various authors, as [28–34]. InSAR can also provide useful information for sinkholes detection and monitoring, as demonstrated in the case of the Wink sinkholes in Texas (USA, [35]), along the Dead Sea shorelines in Israel and Jordan [36] or in the Ebro Valley (Spain, [37]). A review of sinkhole hazard assessment using satellite interferometry has been presented by Theron and Engelbrecht [38]. A review of MTInSAR techniques is given in Crosetto et al. [39].

Salt dissolution mining is not a common target for interferometric analysis; just a few bibliographic examples exist. The first example is provided by Raucoles et al. [14]. These authors analyzed ERS 1/2 radar images by means of a DInSAR approach to measure the subsidence induced by the dissolution of deep salt levels (1900 to 2800 m b.s.l.) in the Vauvert mine (southern France). The presence of cavities that tend to close, induce the creation of a network of fracture propagating to the surface. This results in a large subsidence bowl, 8-km wide, with velocities up to 2 cm/yr. Samieie-Esfahan et al. [40] performed a MTInSAR analysis to characterize a subsidence area in the Friesland Region (northern Netherlands). Subsidence was related to solution mining activities in addition to the presence of a gas field. The author demonstrated the usefulness of velocity decomposition methods to detect the presence of horizontal components of motions and avoid under-/over-estimation of the results. Recently, Gee et al. [41] confirmed the presence of a subsidence bowl in the same area, with maximum velocities up to 13 mm/yr. Another example is provided by Liu et al. [42] who investigated land subsidence in the solution mine of Changde, China. The interferometric analysis of Sentinel-1 images between June 2015 and January 2017 highlighted the presence of a maximum accumulated deformation of 200 mm. It is worth mentioning the attempt made by these authors to correlate the atmospheric conditions and the solvent temperature with the seasonality of the InSAR-derived time series.

InSAR is nowadays a well-recognized ground monitoring tool in mining areas, thanks to the millimeter accuracy, the relatively short revisiting time and the ability to provide a high number of measurement points, higher than any other ground-based point-like technique (e.g., levelling). Despite the clear advantages, InSAR suffers some well-known limitations when analyzing mining areas. There are two common sources of decorrelation that prevent the detection of reliable measurement points: phase aliasing, due to non-linear motions with displacements higher than a quarter of wavelength between two acquisitions, and coherence loss, related to frequent surface changes produced by the mining activity. The latter is particularly relevant in case of open pit mines. Such limitations can be partly overcome by adopting ad hoc local scale processing strategies, e.g., applying a non-linear of hybrid phase unwrapping approach (e.g., [43]) or using specific coherence thresholding (e.g., [44]).

In this paper, the results of a detailed scale MTInSAR approach applied in the Saline di Volterra brine field in the Volterra municipality (Tuscany, Central Italy) are presented. The area is characterized by a long-term solution mining activity that has a considerable environmental impact in terms of ground motion, i.e., creation of hundreds of sinkholes, and depletion of the water resource, e.g., salinization. A target-oriented MTInSAR processing method is used to squeeze as much as possible the information contained in the Sentinel-1 images. The analysis revealed the presence of several subsidence bowls, sometimes corresponding to sinkholes formed in the recent past. The subsidence bowls have a common deformation pattern, with line of sight (LOS) velocities increasing forward the center of the bowl and reaching, in some cases, peaks of over -200 mm/yr. A detailed velocity- and time series-based analysis was performed to highlight the differences in the temporal evolution of sinkholes formed at various stages of the mining activity. According to this analysis, different deformation regimes were identified for the sinkholes of two high deformation areas in the neighboring of Saline di Volterra. This is the first time ground deformation is quantified in this area; thus, the results obtained can increase the awareness of local entities on the ground effects induced by the mining activity. Moreover, the evaluation of the motion triggered by salt dissolution mining has been rarely investigated by means of satellite interferometry.

2. Study Area

The area of interest is located a few kilometers south of the cities of Volterra and Montecatini Val di Cecina in the Tuscany Region (central Italy) within a hilly landscape dominated by gentle slopes and moderate reliefs. The village of Saline di Volterra is surrounded by five different mining concessions where the two deformation targets of this research are located (red contours in Figure 1). The name of the village "Saline" (salt mine in English) reflects the long-term mining activity of this area. The mining concessions cover a total of 28.5 km^2; the biggest one, Buriano, spreads for 10 km^2, whereas the smallest one, Casanova, has a surface of only 1 km^2 (black contours in Figure 1). An overview on the solution mining activity in this area is provided in Section 2.1.

In addition to mining-related phenomena (i.e., sinkholes) the slopes surrounding the study area are characterized by intense denudation processes and superficial erosion phenomena that reach their maximum evolution forming badlands areas whose presence creates a peculiar landscape [45].

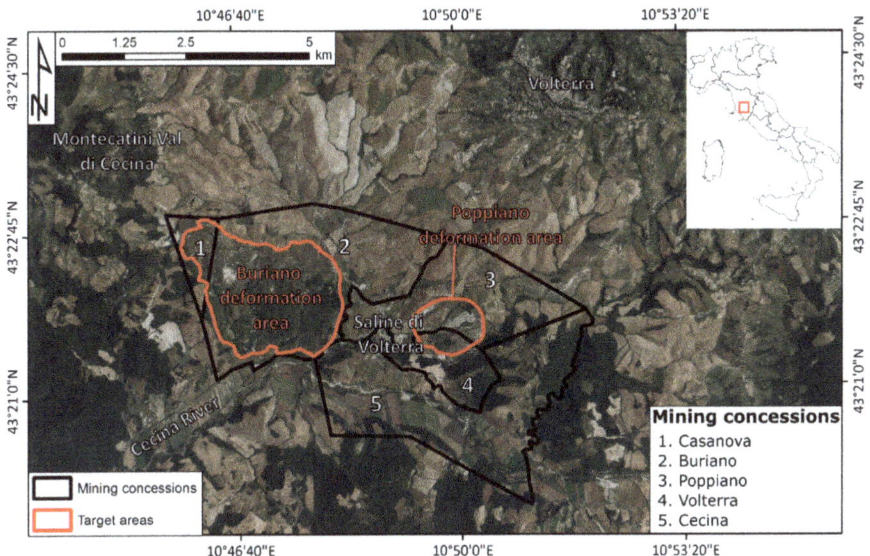

Figure 1. Localization of the area of interest and of the mining concessions in the surroundings of Saline di Volterra in the Volterra Municipality. The background image is a 20 cm AGEA orthophoto (from [46]).

The following three chapters introduce (i) the local geological context, which facilitated and promoted the mining activities, (ii) the history of the Saline di Volterra brine fields with an overview on the mining activity, and (iii) the ground effects registered in the area.

2.1. Geological Context

On a geological point of view, the area of interest is located in the Volterra basin, a NW-SE-oriented Miocene-Pliocene extensional basin developed in response to the eastward migration of the Apennine thrust-belt [47]. The basin is filled by a sedimentary succession ~2500 m thick [48] that includes an evaporitic Messinian succession composed of gypsum and halite layers alternated with conglomerates, sandstone, and claystone [49]. The upper part of this Messinian sequence, known as "Saline di Volterra formation," hosts laminitic gypsum, turbiditic gypsarenite, microcrystalline gypsum, and halite with intercalations of clay, silt, and sandstones [48]. The Saline di Volterra formation has a variable thickness ranging from a few meters to 200 m, depending on the position within the Volterra basin [49]. The whole sequence is unconformably overlaid by Pliocene marine clays ("Blue Clays"), which represent the closing sedimentation of the basin and widely outcrop in the investigated area (Figure 2).

Little information is available to reconstruct the subsurface setting of the Messinian sequence in the area of interest. Nicolich et al. [50] investigated the area of the Buriano mining concession through high resolution 3D seismic surveys, discovering the presence of cavities in the saline layer at a depth of 150 m to 300 m. Burgalassi et al. [51] stated that the productive salt levels are 10 to 40 m thick and are found at a depth ranging from 60 to 400 m. Speranza et al. [49,52] analyzed the rheological properties of the salt levels in the mining area of Saline di Volterra. The salt samplings were collected from deep cores which intersected the salt deposit between 122 and 165 m below surface, for a total thickness of 43 m. These authors recognized four salt facies, with variable percentage of halite and gypsum layers and variable grain size of the crystals. Speranza et al. [49] estimated that halite represents the 40% of the total volume of the Messinian series.

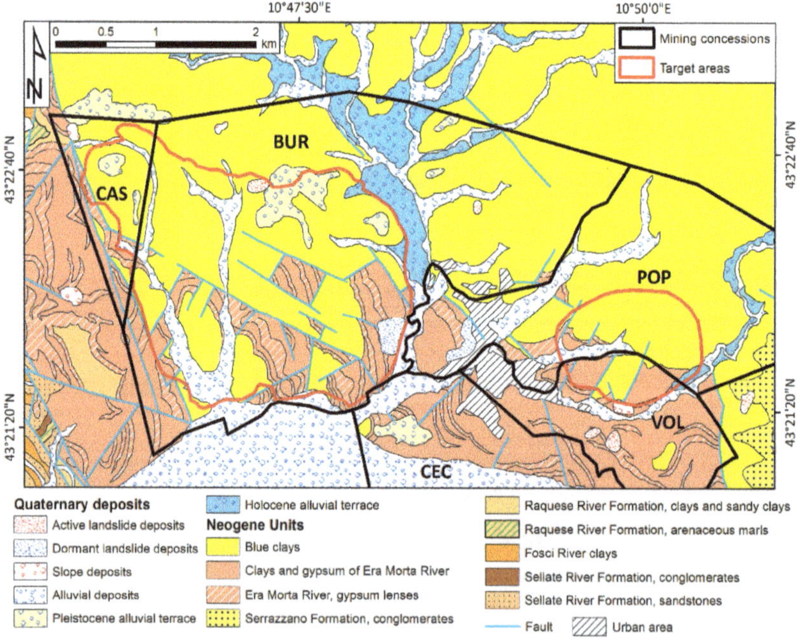

Figure 2. Geological map of the study area. From the Geological Map of Italy, 1:50.000 scale, Sheet 295-Pomarance [53].

2.2. Mining Activity in the Area of Interest

The presence of exploitable salt levels in Saline di Volterra was already known during the Roman Empire. The Romans called the area "moje" from the Latin word "muriae" (brine in English). The real exploitation of the salt layers started in the Middle Ages when the saturated water that gushed out from several springs was left to evaporate naturally [54]. The intensive industrial extraction of the salt levels started at the beginning of the twentieth century (1910) by the Saline di Stato, a company ruled by the central government, and by the private company Solvay [54]. In the area of interest, five mining concessions are present (Figure 1): Buriano and Casanova in the Montecatini Val di Cecina municipality, Cecina, Poppiano, and Volterra in the Volterra municipality.

Salt solution is the mining technique applied here, following the Trump method. This method is rather simple: A dissolving fluid, typically water, is pumped into the depth of interest to dissolve the salt level; the saturated brine is then pumped back to the surface [2,55]. The process is subdivided in two phases. First, the water is pumped through the boreholes into the depth of interest and starts to dissolve the salt level. Compressed air is used to constrict and contain the fluid within the salt level, creating a flow that allows connecting the multiple cavities created by adjacent boreholes. Then, some boreholes are equipped with pumps to extract the produced semi-saturated brines; other boreholes continue to reinject new fresh water. After multiple cycles, the final saturation value of 300 g/L is reached [50] and the brine is extracted and channeled, through a system of pipes, to the treatment plant of Rosignano, along the coast (38 km western than Saline di Volterra). In the treatment plant, the sodium chloride is finally extracted from the brine trough a chemical process. The injection/extraction boreholes are disposed in the mining concessions in a regular grid with spacing between the boreholes of 40–45 m. Every mining panel counts 60 to 360 boreholes [51]. The actual production rate requires to drill 50–60 additional boreholes per year [56].

Burglassi et al. [51] estimated that the total amount of salt deposits in the subsurface of this area is 466 million tons; however, considering an extraction ratio of 65% [57], the effective amount of mineable material is 300 million tons. The time needed for the total consumption of this resource has been estimated in about 200 years. The main environmental impacts of the mining activity are three: (i) overexploitation of fresh water, (ii) salinization of streams and aquifers, and (iii) subsidence/sinkholes [58].

2.3. Sinkhole Database

As previously said, subsidence is one of the main environmental impacts of the mining activity in Saline di Volterra. Sinkholes are the clearest evidence of this process. In this work, aerial orthophotos are exploited to create a multi-temporal inventory of sinkhole occurrence for the two target areas. Digital orthophotos were made available through the web map service of the Tuscany Region [46]. Sinkholes were manually mapped using a database of 9 orthophotos acquired in different years, i.e., ranging between 1954 and 2018 (Table 1). The manual mapping of sinkholes in the area is straightforward; the vegetation cover is sparse, and the sinkholes can be easily detected once formed because of their contrast with the surroundings. Such activity has been carried out by a remote sensing expert using a geographical information system.

Figure 3 presents the results of the orthophoto analysis for the mining concessions of Buriano-Casanova (Figure 3A) and Poppiano-Volterra (Figure 3B). It is worth noting that the definition of the sinkhole contours can be affected by the quality of the oldest images and by the not perfect alignment of some of the WMS services used. This is acceptable since the contours are only used to testify the surface effects of the mining activity.

Table 1. Digital orthophotos used to create the multi-temporal sinkhole database.

Year	Nominal Scale	Image Acquisition	Data Owner
1954	1:10′000	Italian Aeronautical Group (GAI)	Italian Geographic Military Institute (IGM)
1978	1:10′000	Rossi Brescia Aerophotogrammetry	Tuscany Region
1988	1:10′000	CGR Parma	Tuscany Region
1999	1:10′000	CGR Parma	Agricultural Payments Agency (AGEA)
2003	1:10′000	CGR Parma	Agricultural Payments Agency (AGEA)
2007	1:10′000	BLOM-C.G.R.	BLOM-C.G.R.
2010	1:10′000	Rossi Brescia Aerophotogrammetry	Agricultural Payments Agency (AGEA)
2013	1:10′000	BLOM-C.G.R.	Agricultural Payments Agency (AGEA)
2018	1:5′000	BLOM-C.G.R.	Agricultural Payments Agency (AGEA)

The total number of mapped sinkholes is 105, with a maximum of 22 and 19 phenomena for the periods 1954–1978 and 1978–1988, respectively. The age of the sinkholes is clearly connected to the activity of the salt dissolution wells. Buriano-Casanova was the oldest area where the mining activity took place; thus, it is the area where the oldest sinkholes are recorded. On the contrary, Poppiano-Volterra is the most recent concession to be exploited for salt extraction. Figure S1 presents the contours of the mining sectors active in different times, as derived from the interpretation of the set of digital orthophotos. As said before, salt dissolution mining began in 1910 and, until 1978, it was mainly localized in the south-western sector of the Buriano concession. Then, the mining activity expanded to the north, involving the Casanova concession as well. In Poppiano-Volterra a small sector of the Volterra concession was active already before 1978. The expansion of the mining activity to the Poppiano concession took place at the beginning of the 2000s. The oldest sector seems to be nowadays inactive, as testified by the migration of the sinkhole occurrence in the two areas of interest. Two graphic interchange format (gif) files are proposed as supplementary files to highlight the sinkholes' occurrence over time in the mining areas of Buriano-Casanova and Poppiano-Volterra (Figures S2 and S3).

The raise and decrease of the number of sinkholes for comparable time periods, e.g., 1978–1988 and 1988–1999 or 2007–2010 and 2010–2013, is probably related to different rates of extraction. Of course, the development of sinkholes is not instantaneous and there can be a lag time of years between their formation and the mining exploration of a salt level in a certain area. Moreover, considering the fact that not the entire water pumped underground is extracted, the dissolution of the underground holes and chambers can continue for years [2].

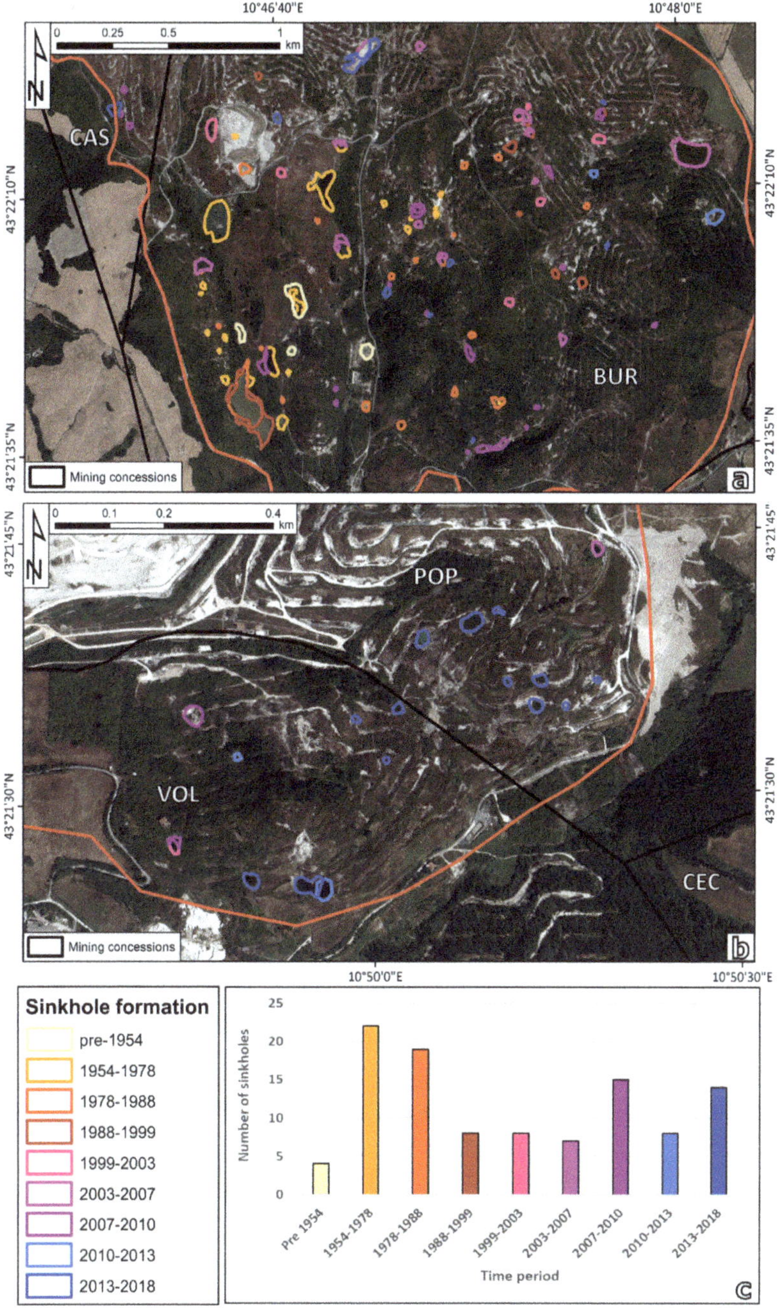

Figure 3. Sinkholes catalogue for the mining concessions of Buriano-Casanova (**a**) and Poppiano-Volterra (**b**). The background image is an AGEA 2018 orthophoto (Table 1). The distribution of boreholes for temporal periods is presented in inset **c**. Sinkholes with contour variations are counted as their first appearance.

3. InSAR Processing

A detailed scale processing approach, based on the direct integration of interferograms, is carried out to retrieve the deformation occurred in the mining area of Saline di Volterra. The workflow of the methodology is presented in Figure 4. The processing chain developed by the Geomatics Division of the Centre Tecnològic de Telecomunicacions de Catalunya was exploited to perform the analysis [59].

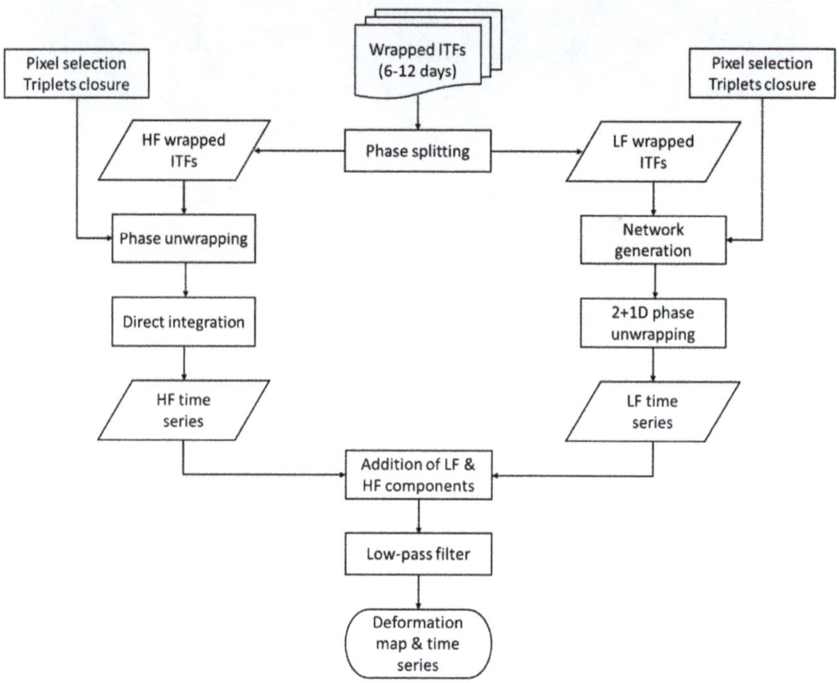

Figure 4. Workflow of the processing approach. ITFs, interferograms; HF, high frequency; LF, low frequency.

Data processing starts with the generation of the full-resolution interferogram stack, following the classical differential interferometry approach [60]. Interferograms (ITFs) are generated starting from a stack of 127 coregistered Sentinel-1 images (both 1A and 1B), acquired in descending orbit and C-band (wavelength of ~5.6 cm). The first image of the stack is set as master (19/09/2016). The images cover the period between September 2016 and November 2018. ITFs are generated with the shorter temporal baseline possible, i.e., 6 or 12 days. A 30-m Shuttle Radar Topographic Mission Digital Elevation Model (SRTM 1 Arc-Second Global – version 3, [61]) is used to remove the topographic contribution from the interferograms. The precise orbits provided by the European Space Agency are used as well [62].

Phase splitting is performed on the stack of wrapped ITFs to extract the low (LF) and high (HF) frequency components. A spatial filter is applied to every ITF to do so. A Butterworth filter is used in this work, in the form:

$$G(\Omega) = \frac{1}{\sqrt{1 + \left(\frac{\Omega}{\Omega_C}\right)^{2N}}} \tag{1}$$

where G is the gain, Ω is the frequency, Ω_C is the cut-off frequency, and N is the order. The spatial filtering generates the LF interferograms, whose residuals represent the HF component. The LF component is assumed to be almost completely formed by the atmospheric contribution [63] although some fast movements with high spatial correlation can still be present. The HF component contains the slow movements (few mm/yr to several cm/yr), the fast local movements (below 2 mm/day), and the residual topographic errors [64]. The analysis of LF and HF ITFs is performed by following two parallel paths (Figure 4).

The HF path is composed of three steps: (i) pixel selection, (ii) phase unwrapping, and (iii) direct integration. Pixel selection is based on an approach called "triplets closure." As the name suggests, the method is based on triplets of radar images (i, j, and k) and interferograms generated with different temporal baselines, in this case 6 and 12 days ($\Delta\Phi_{i-j}$, $\Delta\Phi_{j-k}$, $\Delta\Phi_{i-k}$). The error (ε) is:

$$\varepsilon = \Delta\Phi_{i-j} + \Delta\Phi_{j-k} - \Delta\Phi_{i-k} \tag{2}$$

In case of low noise ITFS, ε is close to zero. The error increases along to the level of noise. The standard deviation of ε calculated with (2) over the whole interferograms stack for each pixel is used to select the points reliable enough to unwrap the phase. The pixels with standard deviation values below the threshold are selected as phase unwrapping candidates. Phase unwrapping is performed by means of the minimum cost flow method [65]. Finally, a point-wise direct integration of the unwrapped interferograms is performed. In a second step, the unwrapped interferograms and thus the final result of direct integration, are all referred to a stable reference point. The direct integration is expressed in the form:

$$n = 0 \div N \left\{ \begin{array}{c} \Phi_n = \Phi_{n-1} + \Delta\Phi_{(n-1)n} \\ \Phi_0 = 0 \end{array} \right. \tag{3}$$

where Φ_n is the accumulated phase at the acquisition time n, Φ_{n-1} is the accumulated phase at the acquisition time $n-1$, and $\Phi_{(n-1)n}$ is the unwrapped interferometric phase of the interferogram calculated using the images acquired at the times $n-1$ and n. The first image is set to zero (Φ_0). The output of the direct integration is, for each point, the temporal evolution of the phase with respect to the first acquisition. Each point is then referred to a stable reference point. The phase is converted in LOS displacement by following:

$$Disp_n = \frac{\Phi_n}{4\pi}\lambda \tag{4}$$

where $Disp_n$ is the accumulated displacement of the point at the acquisition time n, Φ_n is the accumulated phase at the acquisition time n (in radians), and λ is the wavelength of the satellite. Each point is then characterized by n values of displacement, forming the time series, and by a value of velocity calculated over the reference period.

The LF path is composed by three steps: (i) network generation, (ii) pixel selection, (iii) 2+1D phase unwrapping. The LF analysis begins with the set of short baseline LF ITFs which were previously generated. In order to increase the redundancy of the network, the LF ITFs are crossed to derive synthetic ITFs with longer temporal baselines. The measurement point candidates are then selected by applying the triplets closure concept (2) to the newest LF network of interferograms. The 2+1D unwrapping method is finally applied to derive the time series of deformation [59]. Through this method it is possible to check the consistency of the spatial unwrapping in time, owing the redundant ITFs network, and on a pixel-wise basis, i.e., on the points selected with the triplets closure method.

The final products of the two parallel processing paths are time series and deformation maps for the low and high frequency components. The two products are then combined again in order to avoid error accumulation related to the LF-HF split. This error accumulation results in residual trends, with opposite sign between the LF and HF results. Finally, a classical atmospheric phase filtering is applied to remove the remaining low frequency signal not due to the ground deformation. The final deformation map is then georeferenced in a geographic reference system.

4. Results

The obtained deformation map is presented in Figure 5. A total of ~24,000 points cover the mining area of Saline di Volterra with an average density of ~800 measurement points/km^2 (MP/km^2). This high density is due to the fact that the thresholding applied during the InSAR processing allowed maximizing the MP coverage within the two target areas, where the sinkholes are present and where the highest deformation rates are likely to be found. This implies to have a noisier deformation map outside of the high displacement areas. This is acceptable since the targets of this paper are exactly these "fast" moving pixels.

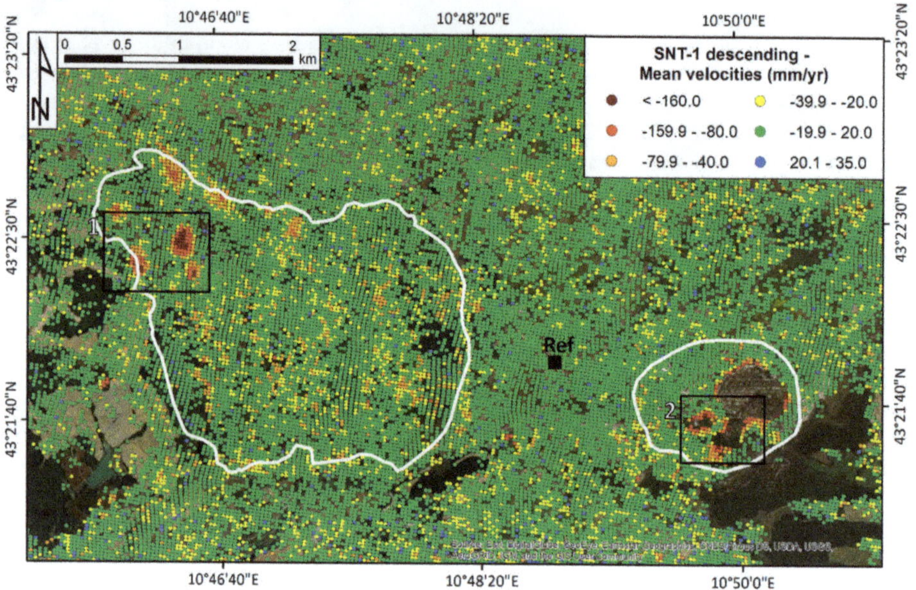

Figure 5. Sentinel-1-derived deformation map for the area of interest. The black dot is the reference point, located in the stable area of the Saline di Volterra village. The white contours are the two target areas. The two black square represent the two areas with the highest deformation rates (1, Buriano-Casanova, part of the Buriano deformation area; 2, Poppiano-Volterra, part of the Poppiano deformation area) which are the focus of the next two sections. The background image is an ESRI World Image.

The deformation map is classified on the basis of the average velocity values, expressed in mm/yr. A stability threshold of ±20 mm/yr was defined to highlight the subsidence bowls within the two mining areas. This threshold is defined depending on the standard deviation of the velocities, here equal to 2.5 times such value. A classical red-green-blue scale bar is used to classify the deformation map; yellow to dark red points refer to movements away from the sensor with negative sign, light blue points refer to movements through the sensor whereas green points fall within the stability threshold. All the measurements are referred to the LOS and are differential values, i.e., referred to a reference point assumed as stable. The reference point was selected in the urban area of Volterra in correspondence of the main road crossing the hamlet, where the oldest edifices are built. To the knowledge of the authors, the area is stable as confirmed also by other interferometric data available for the whole Tuscany Region [46,66].

The deformation map clearly highlights the presence of several subsidence bowls within the contours of the two areas of interest that are part of the mining concessions around Saline di Volterra. Some of the bowl reach in their central portion LOS velocities higher than −200 mm/yr.

The characterization of the two mining areas, Buriano-Casanova and Poppiano-Volterra, is presented in the following sections. It is worth noting that there are some other moving areas (with positive or negative LOS velocity values) outside the mining areas that may be linked to the motion of complex landslides and small earthflows, which characterize the clayey slopes around Volterra [67].

4.1. Buriano-Casanova Mining Area

The northwestern portion of the Buriano-Casanova mining area presents at least five well-defined subsidence bowls with LOS velocities higher than 80 mm/yr. Among these, the bowl A and B (Figure 6a) are the ones with the highest deformation rates, up to −250 and −230 mm/yr, respectively. This implies that the two subsidence bowls reach in their central portion an accumulated displacement higher than 500 mm in a period a little longer than two years. The isolated points showing an uplift are far from the subsidence bowls and from the mapped sinkholes; they are supposed to be related to pixel-scale errors (e.g., LF residuals) that have to be taken into account in this kind of target-oriented processing approach.

The subsidence bowls are located in the most recent mining expansion of this area, at the border between the Buriano and Casanova concessions, where some of the most recent sinkholes are found (Figure 3). Bowls A and B occupy a surface of ~0.04 km^2 and ~0.03 km^2, respectively. In the first case, the orthophotographic analysis did not highlight the presence of any formed sinkholes, whereas in the area of bowl B a sinkhole is evident from the 2013' orthophoto and it is expanding in the last years (see Figure 6b).

Figure 6c shows the time series of deformation for the bowl A. The time series extracted from the center of the bowl (point 1) has a purely linear trend with minor oscillations of the displacement values. The deformation rate for point 1 is equal to −210 mm/yr. Point 2 is located in the southern border of the subsidence bowl and it is characterized by a lower deformation rate (−122 mm/yr), as expected moving away from the center of the subsidence area. The time series is again linear with a minor acceleration event between May and July 2018 (point 2). During this period, the velocity is ~200 mm/yr. This behavior is shared by several points in the same sector of the bowl. The northern portion of the bowl presents the lowest velocities, up to 40 mm/yr and linear trends with no significant accelerations (point 3).

Figure 6d presents the time series of deformation for the bowl B. The time series analysis is focused on the points around the mapped sinkholes. Points 4 and 5 are extracted from the southern and northern border of the sinkhole, respectively. The time series referred to as point 4 shows a linear trend with a strong acceleration between August and November 2017, with LOS velocities passing from ~−200 mm/yr to ~−420 mm/yr. Such acceleration is not evident in the time series of point 5 whose deformation rate (−137 mm/yr) is constant over time with minor order oscillations. It is interesting to notice that moving away from the sinkhole the time series tend to stabilize in the last 6 months (point 6 is an example of this trend).

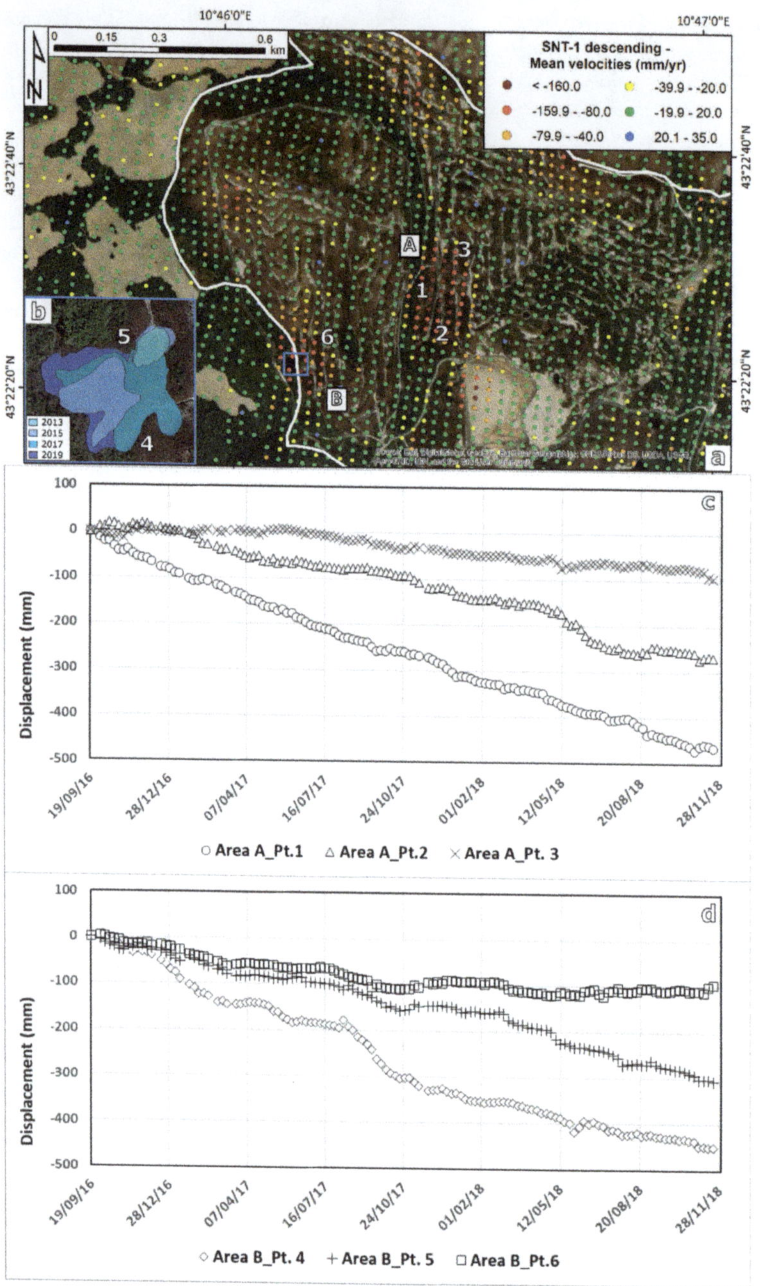

Figure 6. (**a**) Deformation map for the high deformation area of Buriano-Casanova. A and B are the two subsidence bowls with the highest deformation rates: the first with no signs of sinkhole development, the second with a well-developed sinkhole expanding over time (see inset b, whose location is indicated by the blue square). Numbers 1 to 6 indicate areas where the time series of insets (**c**) and (**d**) have been extracted. The background image of inset (**a**) is an ESRI World Image. Inset (**b**) image is a Google Earth imagery.

4.2. Poppiano-Volterra Mining Area

The central portion of the Poppiano-Volterra area presents a distribution of the measurement points within the subsidence bowls slightly different than the previous example. In fact, only the external border of the moving areas is well-defined. In fact, the loss of coherence is so strong that a large part of what it is supposed to be the main subsidence area is not covered by any measurement point, even considering the specific data processing applied. The loss of coherence can have a triple interpretation: (i) land cover characteristics; (ii) high surface changes between the images of the stack, and (iii) presence of fast motions that create decorrelation. It is worth noting that the land cover is basically the same as the Buriano-Casanova mining concession and surface changes induced by salt dissolution mining are not as big as in the case of e.g., open pit mining. Considering this, the hypothesis of the presence of fast deformation rates exceeding the ~2 mm/day unwrapping threshold seems the most reasonable.

Because of the lack of measurement points, the subsidence bowls are less defined as it is for the Buriano-Casanova area. One clear bowl can be identified in the area of point 1 (Figure 7a). The low point density prevents further visual interpretation of the deformation map. In general terms, LOS velocities greatly increase forward the low coherence areas, passing from less than ±10 mm/yr to more than 100 mm/yr, in some cases exceeding −220 mm/yr.

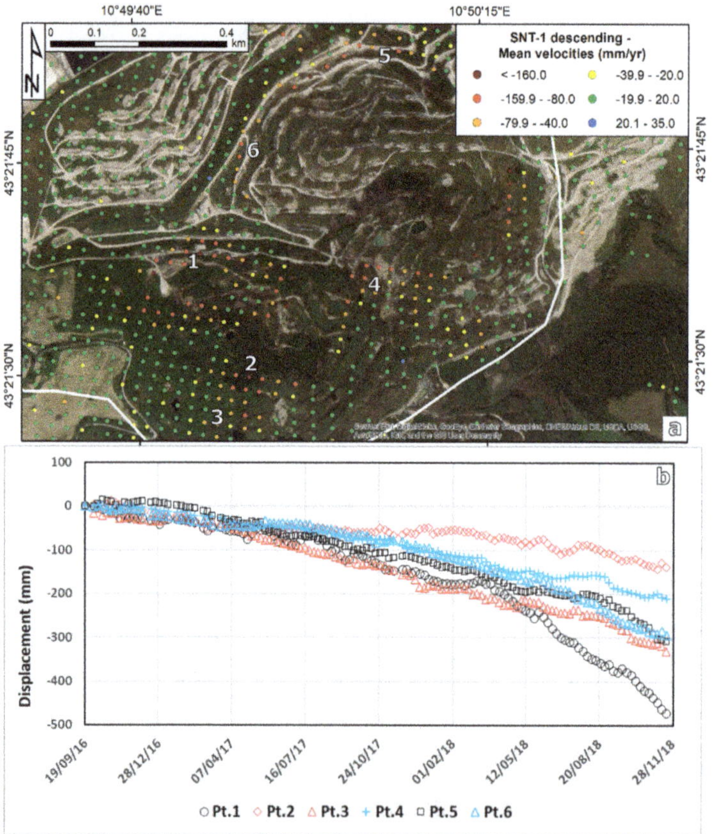

Figure 7. (**a**) Deformation map for the mining area of Poppiano-Volterra. Numbers 1 to 6 indicate areas where the time series of inset (**b**) have been extracted. The background image of inset (**a**) is an ESRI World Image.

The time series of deformation extracted for the points with the highest deformation rates along the internal border of the high deformation areas are presented in Figure 7b (points 1,2,4,5, and 6). All these time series highlight an acceleration with different magnitudes in the last months of the monitored period. The breaking point is located between June and August 2018.

Point 1 is the one showing the most evident acceleration. Although having already an accumulated deformation of 250 mm in less than 2 years (LOS velocity of -152 mm/yr), this point triplicated its deformation trend after the beginning of June 2018. In the second part of the time series the accumulated deformation is equal to 220 mm in a ~6 months period (the estimated LOS velocity is equal to ~450 mm/yr). Such acceleration is not as evident in the other time series, but the velocity change is still relevant, ranging between 120 and 180 mm/yr. The behavior of the time series in the last six months supports the previous hypothesis, i.e., the presence of very high deformation rates where measurement points could not be obtained.

4.3. Comparison between the Spatial Distribution of Sinkholes and the Deformation Map

A hot-spot like analysis is performed to extract the biggest subsidence bowls from the deformation map presented in Figure 5 and to compare their spatial distribution with the sinkhole database. Such analysis is based on the concept of active deformation areas (ADA) defined by Barra et al. [68]. An ADA defines a moving area composed of a certain number of moving measurement points that are aggregated on the basis of neighboring criteria. Moving means that the LOS velocity of a measurement point is above a threshold defined by the user. The software tools developed by Navarro et al. [69] are used to perform the ADA extraction.

In this paper, a LOS velocity threshold equal to 2.5 times the standard deviation of the deformation map (±20 mm/yr) is used to select the moving measurement points. Every ADA is composed of a minimum of 10 points and the clustering distance is set to 28 m. The analysis allows highlighting 25 ADA in the two mining areas, 21 for Buriano-Casanova, and 4 for Poppiano-Volterra. Average LOS velocity values range between -25 mm/yr and -109 mm/yr.

The visual comparison between the ADA distribution and the sinkhole database is shown in Figure 8. Considering a search radius of 50 m, the ADA associated to one or more mapped sinkholes are 18, 15 for Buriano-Casanova and 3 for Poppiano-Volterra. The number of sinkholes found within a subsidence bowl is 64 (almost half of the total).

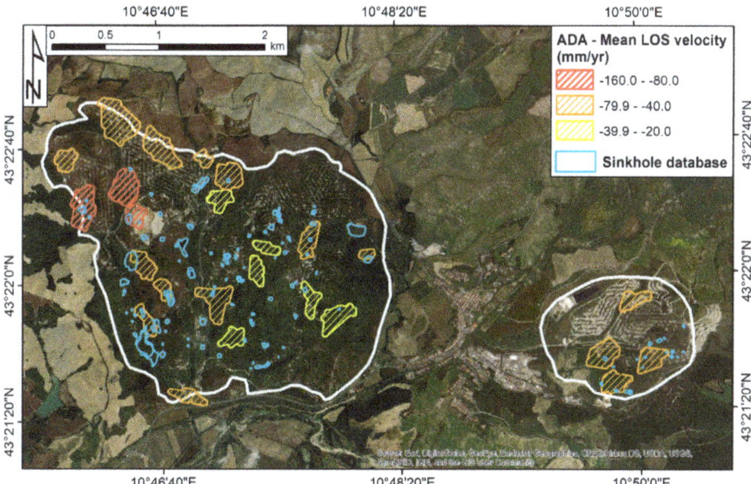

Figure 8. Spatial distribution of sinkholes with respect to the deformation areas extracted from the deformation map presented in Figure 5. The background image is an ESRI World Image.

4.4. Sinkhole-Scale Analysis of the LOS Velocity Values

A further analysis is performed at sinkhole scale. The deformation map is filtered by using the polygons of the 105 sinkholes mapped in the areas of Buriano-Casanova and Poppiano-Volterra. A 10-m buff is created around each polygon to define an area of influence. The average, maximum, minimum, and standard deviation values of the LOS velocities of the measurement points within each sinkhole are extracted. This information is available for more than half of the sinkholes, 58 in the Buriano-Casanova area and 9 in the Poppiano-Volterra concession. For the remaining sinkholes, no measurement points are available.

The analysis of the velocity values allows discovering that sixty percent of the sinkholes (including the areas of influence) present average velocities within the ±10 mm/yr interval. Twenty-eight percent of the sinkholes have average LOS velocities smaller than −20 mm/yr. Only 10% of the sinkholes show average deformation rates smaller than −40 mm/yr; 3 of them exceed −80 mm/yr.

In order to assist the interpretation of the results, the pure evaluation of the average velocity values is associated to the information regarding the age of formation of the sinkholes. The results are presented in Figure 9. The box plot is showing the median, minimum, and maximum values of the LOS velocities of all the measurement points included in the sinkholes of each temporal interval. The graph shows an evident increase of the deformation rates for the most recent sinkholes. The oldest temporal intervals (before 1988) show very similar velocity distributions, with median values around −10 mm/yr. LOS velocities regularly increase until the period 2010–2013, when a strong increase of the maximum velocities is recorded.

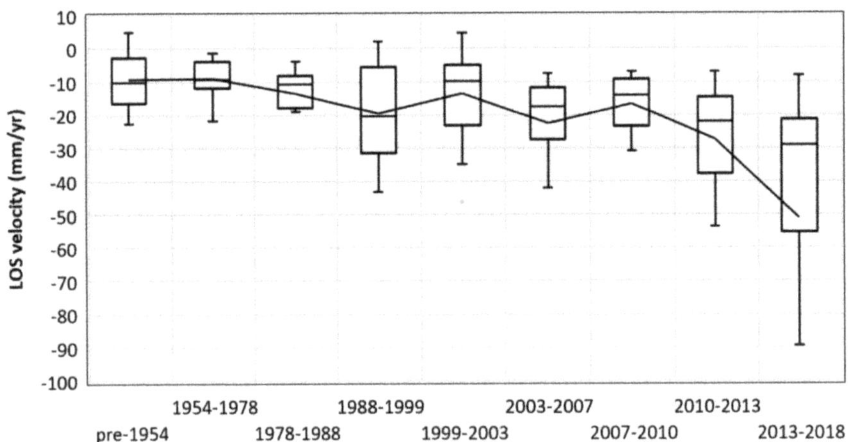

Figure 9. Box plot showing the variability of the velocity values depending on the time of formation of the sinkholes in the mining concession around Saline di Volterra. The solid line represents the mean value, the stroke within each box plot the median value.

5. Discussion

Nowadays, multi-temporal satellite interferometry is a reliable and recognized tool for the assessment of mining-induced ground motion. In this work, MTInSAR was applied for the first time to detect and characterize subsidence in a salt-dissolution mining area in southern Tuscany. Such anthropic activity has a significant impact on the water resources of Saline di Volterra and triggers the formation of several sinkholes that, fortunately, do not affect any built-up area.

Although the impact of ground deformation is, for now, not directly interfering with any human activity, the quantification of the deformation triggered by this mining activity is certainly of interest for the local entities and land planners, considering that only part of the mining concessions is currently

exploited. To do so, a target-oriented MTInSAR approach was used to squeeze as much as possible the information contained in the Sentinel-1 images. The processing methodology is "target-oriented" since its goal is to detect in the best way fast deformation, within the technical and intrinsic limitations of the technique. This assumes that sinkholes are likely connected to very high subsidence rates (e.g., [14]), well recognizable from other slow-moving phenomena that characterize the area (i.e., landslides). It is important to maximize the spatial sampling of measurement points and to reduce the temporal baseline of interferograms in order to decrease the aliasing error. The result is a deformation map involving low-coherence radar target where the fast moving areas are well-detected (see Figure 6) or detected as much as technically possible (see Figure 7). The tradeoff is a higher standard deviation of the velocity estimation and the presence of measurement points with uncorrelated signal. It is worth noting that such local-scale approach is clearly not suitable for small-scale mapping activities, where the high point density would negatively impact on the definition of the moving areas, creating too many false positives.

The availability of data for one single orbit can lead to an underestimation of the real velocity vector since the horizontal component cannot be calculated. Anyway, past experiences indicate that the main component of deformation connected to sinkhole formation is vertical [70,71] although other authors report the presence of anomalous strong horizontal component before the sinkhole collapse [72].

The processing approach allowed deriving a high density of measurement points covering the mining concessions around Saline di Volterra. The analysis of the deformation map revealed the presence of several subsidence bowls which are in some cases correspondent to areas where sinkholes have been already formed in the recent past (Figure 8). The formation of sinkholes is quite common where salt dissolution mining is carried out. The high number of phenomena mapped in this area is probably linked to the fact that the salt level is relatively shallow, with cavities formed 150–200 m below surface. This is consistent with other mining areas worldwide. As stated by Warren [2], the presence of caverns below 150 m can "lead to catastrophic chimneying and stoping in sediments above cavity." The same author adds that "with deeper operations the land surface tends to subside into a bowl of subsidence." The temporal occurrence of sinkholes is certainly linked to the mining expansion over time. Figure 3 and Figure S1 well visualize the relationship between sinkhole formation and mining activity, assuming the existence of a lag time between the end of the salt extraction at a certain depth and the formation of the sinkhole. Such lag time is difficult to estimate without the availability of additional subsurface data.

A detailed analysis of the distribution of LOS velocities and time series of selected measurement points was carried out within some selected subsidence bowls. As it is expected, the spatial pattern of the LOS velocity is common for all the bowl. LOS velocities greatly increase through the center of the subsidence area. What differ on a case basis is the deformation trend of the measurement points within each bowl (Figures 6 and 7). The time series can be classified in (i) linear, (ii) bilinear with an acceleration in the last part of the monitored period, and iii) bilinear with a deceleration in the last part of the monitored period. The first case is consistent with subsidence bowls where no sinkholes have been formed yet (area A in Figure 6), whereas a deceleration in the last part of the monitored period is registered where well-developed sinkholes have been formed in the last 10 years (area B in Figure 6). A bilinear trend with a final (and sometimes strong) acceleration is registered by the measurement points of the Poppiano-Volterra concession, where the mining activity is the most recent and where just a few small sinkholes were mapped. It is worth mentioning that the older the sinkhole, the lower the deformation rates in its surroundings (Figure 9). We postulate that the oldest sinkholes have reached the maximum expansion allowed by the local hydrogeological and geological conditions, also following the abandonment of the oldest mining concessions. Unfortunately, the availability of subsurface data is extremely limited and the access to the area is restricted, preventing the possibility to perform field surveys or to collect ground monitoring data useful for interferometric data validation.

6. Conclusions

This paper presented a local-scale application of multi-temporal satellite interferometry targeting a solution mining area in southern Tuscany. The surroundings of Saline di Volterra host several brine wells that pump water into a salt level at a depth ranging between 60 and 400 m below surface. The mining activity has a relevant environmental impact in terms of depletion of the water resources and in terms of ground motion, creating several sinkholes which were mapped through multi-temporal analysis of orthophotos.

The deformation map, obtained through the analysis of Sentinel-1 images, revealed the presence of several subsidence bowls, sometimes corresponding to sinkholes formed in the recent past. The subsidence bowls have a common deformation pattern, with LOS velocities increasing forward the center of the bowl. The temporal evolution of the measurement points can vary a lot on case basis. Finally, a correlation between LOS velocities and age of formation of sinkholes is found.

Although with some uncertainties, the processing approach presented in this paper was able to detect fast deformation rates that are usually puzzling to solve in mining areas. This detailed scale and target-oriented approach demonstrated its capability to provide useful information in terms of density of measurement points and quality of the time series. A future evolution of this work could be further analyses that involve the decrease of the temporal frequency of data processing and its application in other mining contexts, going forward to the creation of a monitoring system based on satellite interferometric data.

Supplementary Materials: The following are available online at http://www.mdpi.com/2072-4292/12/23/3919/s1, Figure S1: Multi-temporal mining expansion in the area of Saline di Volterra. Figure S2: sinkhole evolution in the area of Buriano-Casanova. Figure S3: sinkhole evolution in the area of Poppiano-Volterra.

Author Contributions: Conceptualization, L.S. and R.M.; methodology, A.B. and O.M.; software, A.B., O.M. and M.C.; writing—original draft preparation, L.S.; writing—review and editing, R.M, A.B., O.M., S.B., M.C.; supervision, M.C. All authors have read and agreed to the published version of the manuscript.

Funding: This research received no external funding.

Acknowledgments: This work was partially funded by the Spanish Ministry of Economy and Competitiveness through the DEMOS project "Deformation monitoring using Sentinel-1 data" (Ref: CGL2017-83,704-P).

Conflicts of Interest: The authors declare no conflict of interest.

References

1. Bell, F.G.; Stacey, T.R.; Genske, D.D. Mining subsidence and its effect on the environment: Some differing examples. *Environ. Geol.* **2000**, *40*, 135–152. [CrossRef]
2. Warren, J.K. Solution mining and salt cavern usage. In *Evaporites*; Warren, J.K., Ed.; Springer International Publishing: Cham, Switzerland, 2016; Volume 1, pp. 1303–1374.
3. Neal, J.T. Prediction of subsidence resulting from creep closure of solutioned-mined caverns in salt domes. In Proceedings of the 4th International Symposium on Land Subsidence, Houston, TX, USA, 12–17 May 1991; Johnson, I.A., Ed.; LAHS Publications: Houston, TX, USA, 1991; pp. 225–234. Available online: http://hydrologie.org/redbooks/a200/iahs_200_0225.pdf (accessed on 28 April 2020).
4. Martínez, J.D.; Johnson, K.S.; Neal, J.T. Sinkholes in evaporite rocks. *Am. Sci.* **1998**, *86*, 39–52. [CrossRef]
5. Whyatt, J.; Varley, F. Catastrophic failures of underground evaporite mines. In Proceedings of the 27th International Conference on Ground Control in Mining, Morgantown, WV, USA, 29–31 July 2008; Peng, S.S., Mark, C., Finfinger, G.L., Tadolini, S.C., Khair, A.W., Heasley, K., Luo, Y., Eds.; West Virginia University: Morgantown, WV, USA, 2008; pp. 29–31.
6. Gisotti, G. Case of induced subsidence for extraction of salt by hydrosolution. In Proceedings of the 4th International Symposium on Land Subsidence, Houston, TX, USA, 12–17 May 1991; Johnson, I.A., Ed.; LAHS Publications: Houston, TX, USA, 1991; pp. 225–234. Available online: http://hydrologie.org/redbooks/a200/iahs_200_0235.pdf (accessed on 28 April 2020).
7. Guarascio, M. Microseismic monitoring of solution mining cavities. In Proceedings of the Twentieth International Symposium on the Application of Computers and Mathematics in the

Mineral Industries, Johannesburg, South Africa, 19–23 October 1987; Lemmer, I.C., Schaum, H., Camisani-Calzolari, F.A.G.M., Eds.; South African Institute of Mining and Metallurgy: Johannesburg, South Africa, 1987; Volume 1, pp. 49–54.

8. Garlicki, A. Solution mining of Miocene salts in Poland and its environmental impact. In Proceedings of the Seventh Symposium on Salt, Kyoto, Japan, 6–9 April 1992; Kakihana, H., Hoshi, T., Eds.; Elsevier Science Publisher BV: Amsterdam, The Netherlands, 1993; Volume 1, pp. 419–424.

9. Perski, Z.; Hanssen, R.; Wojcik, A.; Wojciechowski, T. InSAR analyses of terrain deformation near the Wieliczka Salt Mine, Poland. *Eng. Geol.* **2009**, *106*, 58–67. [CrossRef]

10. Mancini, F.; Stecchi, F.; Gabbianelli, G. GIS-based assessment of risk due to salt mining activities at Tuzla (Bosnia and Herzegovina). *Eng. Geol.* **2009**, *109*, 170–182. [CrossRef]

11. Mancini, F.; Stecchi, F.; Zanni, M.; Gabbianelli, G. Monitoring ground subsidence induced by salt mining in the city of Tuzla (Bosnia and Herzegovina). *Environ. Geol.* **2009**, *58*, 381–389. [CrossRef]

12. Stecchi, F.; Antonellini, M.; Gabbianelli, G. Curvature analysis as a tool for subsidence-related risk zones identification in the city of Tuzla (BiH). *Geomorphology* **2009**, *107*, 316–325. [CrossRef]

13. Buffet, A. The collapse of Compagnie des Salins SG4 and SG5 drillings. In Proceedings of the Solution Mining Research Institute Fall Meeting, Rome, Italy, 4–7 October 1998; pp. 79–105.

14. Raucoules, D.; Maisons, C.; Carnec, C.; Le Mouelic, S.; King, C.; Hosford, S. Monitoring of slow ground deformation by ERS radar interferometry on the Vauvert salt mine (France): Comparison with ground-based measurement. *Remote Sens. Environ.* **2003**, *88*, 468–478. [CrossRef]

15. Zamfirescu, F.; Mocuta, M.; Constantinecu, T.; Medves, E.; Danchiv, A. The main causes of a geomechanical accident of brine caverns at field II of Ocnele Mari-Romania. *Mater. Geoenviron.* **2003**, *50*, 431–434.

16. Balteanu, D.; Enciu, P.; Deak, G. A large scale collapse in the Ocnele Mari salt mine field, Getic Subcarpathians, Romania. *Studia Geomorphol. Carpatho Balc.* **2006**, *40*, 119–126.

17. Poenaru, V.D.; Badea, A.; Savin, E.; Teleaga, D.; Poncos, V. Land degradation monitoring in the Ocnele Mari salt mining area using satellite imagery. In *Earth Resources and Environmental Remote Sensing/GIS Applications II*; Michel, U., Civco, D.L., Eds.; International Society for Optics and Photonics: Bellingham, WA, USA, 2011; Volume 8181, p. 81810.

18. Galloway, D.; Jones, D.R.; Ingebritsen, S.E. *Land Subsidence in the United States*; US Geological Survey: Denver, CO, USA, 1999; pp. 1–177.

19. Johnson, K.S. Subsidence hazards due to evaporite dissolution in the United States. *Environ. Geol.* **2005**, *48*, 395–409. [CrossRef]

20. Amelung, F.; Galloway, D.L.; Bell, J.W.; Zebker, H.A.; Laczniak, R.J. Sensing the ups and downs of Las Vegas: InSAR reveals structural control of land subsidence and aquifer-system deformation. *Geology* **1999**, *27*, 483–486. [CrossRef]

21. Chaussard, E.; Wdowinski, S.; Cabral-Cano, E.; Amelung, F. Land subsidence in central Mexico detected by ALOS InSAR time-series. *Remote Sens. Environ.* **2014**, *140*, 94–106. [CrossRef]

22. Qu, F.; Zhang, Q.; Lu, Z.; Zhao, C.; Yang, C.; Zhang, J. Land subsidence and ground fissures in Xi'an, China 2005–2012 revealed by multi-band InSAR time-series analysis. *Remote Sens. Environ.* **2014**, *155*, 366–376. [CrossRef]

23. Palamara, D.R.; Nicholson, M.; Flentje, P.; Baafi, E.; Brassington, G.M. An evaluation of airborne laser scan data for coalmine subsidence mapping. *Int. J. Remote Sens.* **2007**, *28*, 3181–3203. [CrossRef]

24. Zhou, D.; Wu, K.; Chen, R.; Li, L. GPS/terrestrial 3D laser scanner combined monitoring technology for coal mining subsidence: A case study of a coal mining area in Hebei, China. *Nat. Hazards* **2014**, *70*, 1197–1208. [CrossRef]

25. Ge, Y.; Tang, H.; Gong, X.; Zhao, B.; Lu, Y.; Chen, Y.; Lin, Z.; Chen, H.; Qiu, Y. Deformation monitoring of earth fissure hazards using terrestrial laser scanning. *Sensors* **2019**, *19*, 1463. [CrossRef]

26. Jing-Xiang, G.; Hong, H. Advanced GNSS technology of mining deformation monitoring. *Procedia Earth Planet Sci.* **2009**, *1*, 1081–1088. [CrossRef]

27. Ustun, A.; Tusat, E.; Yalvac, S. Preliminary results of land subsidence monitoring project in Konya Closed Basin between 2006–2009 by means of GNSS observations. *Nat. Hazards Earth Syst. Sci.* **2010**, *10*, 1151. [CrossRef]

28. Argyrakis, P.; Ganas, A.; Valkaniotis, S.; Tsioumas, V.; Sagias, N.; Psiloglou, B. Anthropogenically induced subsidence in Thessaly, central Greece: New evidence from GNSS data. *Nat. Hazards* **2020**, *102*, 1–22. [CrossRef]

29. Przyłucka, M.; Herrera, G.; Graniczny, M.; Colombo, D.; Béjar-Pizarro, M. Combination of conventional and advanced DInSAR to monitor very fast mining subsidence with TerraSAR-X Data: Bytom City (Poland). *Remote Sens.* **2015**, *7*, 5300–5328. [CrossRef]

30. Yang, Z.; Li, Z.; Zhu, J.; Yi, H.; Hu, J.; Feng, G. Deriving dynamic subsidence of coal mining areas using InSAR and logistic model. *Remote Sens.* **2017**, *9*, 125. [CrossRef]

31. Du, Z.; Ge, L.; Ng, A.H.M.; Li, X. Investigation on mining subsidence over Appin–West Cliff Colliery using time-series SAR interferometry. *Int. J. Remote Sens.* **2018**, *39*, 1528–1547. [CrossRef]

32. Malinowska, A.A.; Witkowski, W.T.; Hejmanowski, R.; Chang, L.; Van Leijen, F.J.; Hanssen, R.F. Sinkhole occurrence monitoring over shallow abandoned coal mines with satellite-based persistent scatterer interferometry. *Eng. Geol.* **2019**, *262*, 105336. [CrossRef]

33. López-Vinielles, J.; Ezquerro, P.; Fernández-Merodo, J.A.; Béjar-Pizarro, M.; Monserrat, O.; Barra, A.; Blanco, P.; García-Robles, J.; Filatov, A.; García-Davalillo, J.C.; et al. Remote analysis of an open-pit slope failure: Las Cruces case study, Spain. *Landslides* **2020**, 1–16. [CrossRef]

34. Pawluszek-Filipiak, K.; Borkowski, A. Monitoring mining-induced subsidence by integrating differential radar interferometry and persistent scatterer techniques. *Eur. J. Remote Sens.* **2020**, 1–13. [CrossRef]

35. Kim, J.W.; Lu, Z.; Degrandpre, K. Ongoing deformation of sinkholes in Wink, Texas, observed by time-series Sentinel-1a SAR interferometry (preliminary results). *Remote Sens.* **2016**, *8*, 313. [CrossRef]

36. Baer, G.; Magen, Y.; Nof, R.N.; Raz, E.; Lyakhovsky, V.; Shalev, E. InSAR measurements and viscoelastic modeling of sinkhole precursory subsidence: Implications for sinkhole formation, early warning, and sediment properties. *J. Geophys. Res. Earth Surf.* **2018**, *123*, 678–693. [CrossRef]

37. Galve, J.P.; Castañeda, C.; Gutiérrez, F. Railway deformation detected by DInSAR over active sinkholes in the Ebro Valley evaporite karst, Spain. *Nat. Hazards Earth Syst. Sci.* **2015**, *15*, 2439–2448. [CrossRef]

38. Theron, A.; Engelbrecht, J. The role of Earth observation, with a focus on SAR interferometry, for sinkhole hazard assessment. *Remote Sens.* **2018**, *10*, 1506. [CrossRef]

39. Crosetto, M.; Monserrat, O.; Cuevas-González, M.; Devanthéry, N.; Crippa, B. Persistent scatterer interferometry: A review. *ISPRS J. Photogramm. Remote Sens.* **2016**, *115*, 78–89. [CrossRef]

40. Samieie-Esfahany, S.; Hanssen, R.; Van Thienen-Visser, K.; Muntendam-Bos, A. On the effect of horizontal deformation on InSAR subsidence estimates. In Proceedings of the Fringe 2009 Workshop, Frascati, Italy, 30 November–4 December 2009; Lacoste-Francis, H., Ed.; ESA Communications ESTEC: Noordwijk, The Netherlands, 2009; Volume 30.

41. Gee, D.; Sowter, A.; Grebby, S.; De Lange, G.; Athab, A.; Marsh, S. National geohazards mapping in Europe: Interferometric analysis of The Netherlands. *Eng. Geol.* **2019**, *256*, 1–22. [CrossRef]

42. Liu, X.; Xing, X.; Wen, D.; Chen, L.; Yuan, Z.; Liu, B.; Tan, J. Mining-induced time-series deformation investigation based on SBAS-InSAR technique: A case study of drilling water solution rock salt mine. *Sensors* **2019**, *19*, 5511. [CrossRef] [PubMed]

43. Mura, J.C.; Paradella, W.R.; Gama, F.F.; Silva, G.G.; Galo, M.; Camargo, P.O.; Silva, A.Q.; Silva, A. Monitoring of non-linear ground movement in an open pit iron mine based on an integration of advanced DInSAR techniques using TerraSAR-X data. *Remote Sens.* **2016**, *8*, 409. [CrossRef]

44. Wang, S.; Lu, X.; Chen, Z.; Zhang, G.; Ma, T.; Jia, P.; Li, B. Evaluating the feasibility of illegal open-pit mining identification using insar coherence. *Remote Sens.* **2020**, *12*, 367. [CrossRef]

45. Bianchini, S.; Del Soldato, M.; Solari, L.; Nolesini, T.; Pratesi, F.; Moretti, S. Badland susceptibility assessment in Volterra municipality (Tuscany, Italy) by means of GIS and statistical analysis. *Environ. Earth Sci.* **2016**, *75*, 1–14. [CrossRef]

46. Tuscany Region WMS Service. Available online: http://www502.regione.toscana.it/geoscopio/servizi/wms/OFC.htm (accessed on 8 September 2020).

47. Patacca, E.; Sartori, R.; Scandone, P. Tyrrhenian basin and Apenninic arcs: Kinematic relations since late Tortonian times. *Mem. Soc. Geol. It.* **1990**, *45*, 425–451.

48. Testa, G.; Lugli, S. Gypsum–anhydrite transformations in Messinian evaporites of central Tuscany (Italy). *Sediment Geol.* **2000**, *130*, 249–268. [CrossRef]

49. Speranza, G.; Vona, A.; Vinciguerra, S.; Romano, C. Relating natural heterogeneities and rheological properties of rocksalt: New insights from microstructural observations and petrophyisical parameters on Messinian halites from the Italian Peninsula. *Tectonophysics* **2016**, *666*, 103–120. [CrossRef]
50. Nicolich, R.; Primiero, A.; Zgur, F.; Di Marzo, N. 3D seismic imaging and numerical modelling of subsidence in solution mining of rocksalt. In Proceedings of the 64th EAGE Conference & Exhibition, Florence, Italy, 27–30 May 2002; European Association of Geoscientists & Engineers: Houton, The Netherlands, 2002.
51. Burgalassi, D.; Cheli, B.; Luzzati, T.; Del Soldato, V.; Freschi, E. Analisi delle ricadute ambientali della Solvay sul territorio della Val di Cecina. In *La Solvay e in Val di Cecina: Ricadute Socio-Economiche e Ambientali di Una Grande Industria Chimica sul Suo Territorio*; Cheli, B., Luzzati, T., Eds.; Edizioni Plus srl: Pisa, Italy, 2010; pp. 145–217. (In Italian)
52. Speranza, G.; Cosentino, D.; Tecce, F.; Faccenna, C. Paleoclimate reconstruction during the Messinian evaporative drawdown of the Mediterranean Basin: Insights from microthermometry on halite fluid inclusions. *Geochem. Geophys. Geosyst.* **2013**, *14*, 5054–5077. [CrossRef]
53. Italian Geological Map (1:50000 Nominal Scale). Available online: https://www.isprambiente.gov.it/Media/carg/note_illustrative/295_Pomarance.pdf (accessed on 1 September 2020).
54. Nannoni, R.; Capperi, M. *Miniere e Minerali della Val di Cecina*; Gruppo Mineralogico Cecinese, Tipografie Grafiche Favillini: Livorno, Italy, 1989; p. 68. (In Italian)
55. Johnson, K.S. Land subsidence above man-made salt-dissolution cavities. In *Land Subsidence Case Studies and Current Research: Proceedings of the Dr. Joseph, F. Poland Symposium on Land Subsidence*; Borcher, J.W., Ed.; Association of Engineering Geologists Special Publication: Belmont, CA, USA, 1998; Volume 8, pp. 385–392.
56. Tuscany Region Environmental Agency (ARPAT). *Quadro Conoscitivo Ambientale degli Insediamenti Solvay Nelle Province di Pisa e Livorno (2000–2005)*; Technical Report; Tuscany Region Environmental Agency (ARPAT): Firenze, Italy, 2006. (In Italian)
57. Pinna, S. *Rischi Ambientali e Difesa del Territorio*; Franco Angeli Edizioni: Milan, Italy, 2002; p. 176. (In Italian)
58. Tuscany Region Environmental Agency. The Environmental Impact of Salt Dissolution Mining in Saline di Volterra. Available online: http://www.arpat.toscana.it/notizie/arpatnews/2013/174-13/174-13-gli-impatti-ambientali-nelle-attivita-minerarie-connesse-alla-coltivazione-del-salgemma (accessed on 20 August 2020).
59. Devanthéry, N.; Crosetto, M.; Monserrat, O.; Cuevas-González, M.; Crippa, B. An approach to persistent scatterer interferometry. *Remote Sens.* **2014**, *6*, 6662–6679. [CrossRef]
60. Rosen, P.A.; Hensley, S.; Joughin, I.R.; Li, F.K.; Madsen, S.N.; Rodriguez, E.; Goldstein, R.M. Synthetic aperture radar interferometry. *Proc. IEEE* **2000**, *88*, 333–382. [CrossRef]
61. Shuttle Radar Topography Mission (SRTM) 1 Arc-Second Global, Product Description. Available online: https://www.usgs.gov/centers/eros/science/usgs-eros-archive-digital-elevation-shuttle-radar-topography-mission-srtm-1-arc?qt-science_center_objects=0#qt-science_center_objects (accessed on 6 November 2020).
62. Sentinels-POD-team. *Sentinels POD Service File Format Specifications*; Technical Report n° GMES-GSEGEOPGFS-10-0075; European Space Agency: Paris, France, 2013.
63. Gomba, G.; Parizzi, A.; De Zan, F.; Eineder, M.; Bamler, R. Toward operational compensation of ionospheric effects in SAR interferograms: The split-spectrum method. *IEEE Trans. Geosci. Remote* **2015**, *54*, 1446–1461. [CrossRef]
64. Crosetto, M.; Monserrat, O.; Cuevas, M.; Crippa, B. Spaceborne differential SAR interferometry: Data analysis tools for deformation measurement. *Remote Sens.* **2011**, *3*, 305–318. [CrossRef]
65. Costantini, M. A novel phase unwrapping method based on network programming. *IEEE Trans. Geosci. Remote.* **1998**, *36*, 813–821. [CrossRef]
66. Raspini, F.; Bianchini, S.; Ciampalini, A.; Del Soldato, M.; Solari, L.; Novali, F.; Del Conte, S.; Rucci, A.; Ferretti, A.; Casagli, N. Continuous, semi-automatic monitoring of ground deformation using Sentinel-1 satellites. *Sci Rep.* **2018**, *8*, 1–11. [CrossRef]
67. Bianchini, S.; Solari, L.; Casagli, N. A gis-based procedure for landslide intensity evaluation and specific risk analysis supported by persistent scatterers interferometry (PSI). *Remote Sens.* **2017**, *9*, 1093. [CrossRef]
68. Barra, A.; Solari, L.; Béjar-Pizarro, M.; Monserrat, O.; Bianchini, S.; Herrera, G.; Crosetto, M.; Sarro, R.; González-Alonso, E.; Mateos, R.M.; et al. A methodology to detect and update active deformation areas based on sentinel-1 SAR images. *Remote Sens.* **2017**, *9*, 1002. [CrossRef]

69. Navarro, J.A.; Tomás, R.; Barra, A.; Pagán, J.I.; Reyes-Carmona, C.; Solari, L.; Vinnieles, J.L.; Falco, S.; Crosetto, M. ADAtools: Automatic detection and classification of active deformation areas from PSI displacement maps. *ISPRS Int. J. Geo Inf.* **2020**, *9*, 584. [CrossRef]
70. Paine, J.G.; Buckley, S.M.; Collins, E.W.; Wilson, C.R. Assessing collapse risk in evaporite sinkhole-prone areas using microgravimetry and radar interferometry. *J. Environ. Eng. Geophys.* **2012**, *17*, 75–87. [CrossRef]
71. Nof, R.N.; Baer, G.; Ziv, A.; Raz, E.; Atzori, S.; Salvi, S. Sinkhole precursors along the Dead Sea, Israel, revealed by SAR interferometry. *Geology* **2013**, *41*, 1019–1022. [CrossRef]
72. Jones, C.E.; Blom, R.G. Bayou Corne, Louisiana, sinkhole: Precursory deformation measured by radar interferometry. *Geology* **2014**, *42*, 111–114. [CrossRef]

Publisher's Note: MDPI stays neutral with regard to jurisdictional claims in published maps and institutional affiliations.

remote sensing

MDPI

Article

Detailed Mapping of Lava and Ash Deposits at Indonesian Volcanoes by Means of VHR PlanetScope Change Detection

Moritz Rösch [1,2] and Simon Plank [2,*]

1   Department of Geography and Geology, University of Wuerzburg, Am Hubland, 97074 Wuerzburg, Germany; moritz.roesch@dlr.de
2   German Remote Sensing Data Center (DFD), German Aerospace Center (DLR), Muenchener Str. 20, 82234 Wessling, Germany
*   Correspondence: simon.plank@dlr.de

Abstract: Mapping of lava flows in unvegetated areas of active volcanoes using optical satellite data is challenging due to spectral similarities of volcanic deposits and the surrounding background. Using very high-resolution PlanetScope data, this study introduces a novel object-oriented classification approach for mapping lava flows in both vegetated and unvegetated areas during several eruptive phases of three Indonesian volcanoes (Karangetang 2018/2019, Agung 2017, Krakatau 2018/2019). For this, change detection analysis based on PlanetScope imagery for mapping loss of vegetation due to volcanic activity (e.g., lava flows) is combined with the analysis of changes in texture and brightness, with hydrological runoff modelling and with analysis of thermal anomalies derived from Sentinel-2 or Landsat-8. Qualitative comparison of the mapped lava flows showed good agreement with multispectral false color time series (Sentinel-2 and Landsat-8). Reports of the Global Volcanism Program support the findings, indicating the developed lava mapping approach produces valuable results for monitoring volcanic hazards. Despite the lack of bands in infrared wavelengths, PlanetScope proves beneficial for the assessment of risk and near-real-time monitoring of active volcanoes due to its high spatial (3 m) and temporal resolution (mapping of all subaerial volcanoes on a daily basis).

Keywords: lava; volcanoes; PlanetScope; change detection; object-based image analysis

Citation: Rösch, M.; Plank, S. Detailed Mapping of Lava and Ash Deposits at Indonesian Volcanoes by Means of VHR PlanetScope Change Detection. *Remote Sens.* **2022**, *14*, 1168. https://doi.org/10.3390/rs14051168

Academic Editors: Daniele Giordan, Guido Luzi, Oriol Monserrat and Niccolò Dematteis

Received: 3 February 2022
Accepted: 23 February 2022
Published: 26 February 2022

Publisher's Note: MDPI stays neutral with regard to jurisdictional claims in published maps and institutional affiliations.

1. Introduction

Hazards triggered by volcanic activity have massive environmental, economic, and humanitarian impacts [1,2]. Lava flows can destroy entire settlements and infrastructure [3]. Ash and tephra deposits cause damage to critical infrastructure, buildings, and agricultural land [4] and pose a threat to human health [5]. Furthermore, pyroclastic flows [6,7] and lahars [8] are responsible for a large number of the fatalities due to volcanic hazards [1]. Secondary volcanic hazards, particularly volcano-induced tsunamis, also possess large destructive forces [9,10]. Due to the catastrophic impacts of volcanic hazards, monitoring these processes is an essential component for risk assessment and planning in disaster management [11–13]. For that, satellite-based data are increasingly being integrated into the monitoring process of volcanoes, as they allow large-scale views of areas that are difficult to reach and hazardous due to active eruptions. In addition to the frequently used infrared (IR) (e.g., [14–16]) and synthetic aperture radar (SAR) (e.g., [10,17]) data, optical satellite data are also useful for the monitoring of active volcanoes. Very high-resolution (VHR) optical systems have been used to map ash and tephra deposits [18], lava flows [19], pyroclastic flows, and lahars [20,21] at volcanoes. Lower-resolution optical imagery (e.g., Sentinel-2, Landsat-7/-8) (10–30 m) have also been successfully used to detect volcanic hazards (e.g., [22–24]). A commonly used method for mapping and quantifying volcanic hazard deposits using optical satellite data is the differencing of the Normalized Difference Vegetation Index (NDVI) [18,23]. The results of this methodology can provide information about secondary volcanic hazards (e.g., precipitation-induced lahars) [25]. However, NDVI differentiation (dNDVI) is limited exclusively to the vegetation area of a volcano, leaving land surface changes in the crater area or sparsely vegetated flank areas undetected [18]. Especially in these areas, the differentiation between volcanic deposits (esp. lava flows) and the sparsely vegetated or completely unvegetated background is particularly difficult due to the similar spectral signatures [22,26]. In addition, clear sky conditions are required when using optical systems. Due to the highly dynamic nature of volcanic hazards and the resulting small-scale morphological changes, as well as the frequent cloud cover in volcanic environments, the temporal and geometric resolution of optical systems is critical for a successful classification of volcanic hazards [18,26]. The daily temporal and 3 m spatial resolution of PlanetScope represent significant advantages over previously used remote sensing data, making it a potentially suitable data source for monitoring volcanic hazards. Despite all this, so far only few studies have used PlanetScope to monitor and investigate active volcanoes. Aldeghi et al. [18] successfully used PlanetScope data to determine the spatial extent of ash and tephra deposits following an eruption of Fuego, Guatemala. Other studies used PlanetScope data exclusively for visual analysis of geomorphological changes at Krakatau, Indonesia, [19] or to generate multi-temporal digital elevation models (DEMs) from which lava flow deposits at Nabro, Eritrea, could be derived [14].

This work therefore aims to investigate the potential of PlanetScope for detecting land surface changes due to volcanic hazards. The focus is on the development of new change detection methods for lava flow classification that take advantage of PlanetScope data. For this purpose, existing change analysis methods for identifying vegetation changes due to volcanic deposition were further developed and combined with hydrological runoff modelling from digital elevation models. Multi-sensor analysis can compensate for limitations of individual data sources and thus exploit the full potential of satellite data for monitoring active volcanoes (e.g., [13,14,16,17,27,28]). For this reason, multispectral Sentinel-2 and Landsat-8 data were implemented into the methodology to analyze thermal anomalies. This enabled us to map lava flows during eruptive phases at two active volcanoes (Karangetang 2018/2019 (I) and 2019/2020 (II), Krakatau 2018 (I)) in Indonesia. Furthermore, ash and tephra deposits following eruptions at Krakatau 2018 (II) and Agung 2017 (I) were mapped based on vegetation changes.

2. Study Area

Karangetang (2.781°N; 125.407°E) (Figure 1) is an active stratovolcano located on Siau Island on the Sangihe Island Arc in NE Indonesia [29]. The volcanic cone (1797 m a. s. l.) is characterized by andesitic to basaltic-andesitic rocks and is one of the most active volcanoes in Indonesia, with more than 40 eruptions since 1675. For the most part, the explosive eruptive behavior of Karangetang results in pyroclastic flows, lahars, as well as lava flows, and thus poses an immediate risk to the population [12,30].

In this work, lava flow deposits were mapped during two different eruptive phases at Karangetang (Table 1). The first eruptive phase (Karangetang I) began in November 2018 and continued until March 2019. During this period, large lava effusions developed at the N crater, resulting in a lava flow towards NNW. A second lava flow also flowed WNW from the N crater in late November 2018 [31]. Effusive activity continued throughout December 2018 into January 2019 and was repeatedly accompanied by avalanches and pyroclastic flows. On 2 March 2019, the lava flow on the NNW flank crossed a road near the coast. Two days later, the lava flow reached the coastline in the NNW of the island and ended in the ocean. During the activity in February 2019, 190 people had to be evacuated. From 12 February 2019 onwards activity decreased and the eruption phase ended in March 2019 [31].

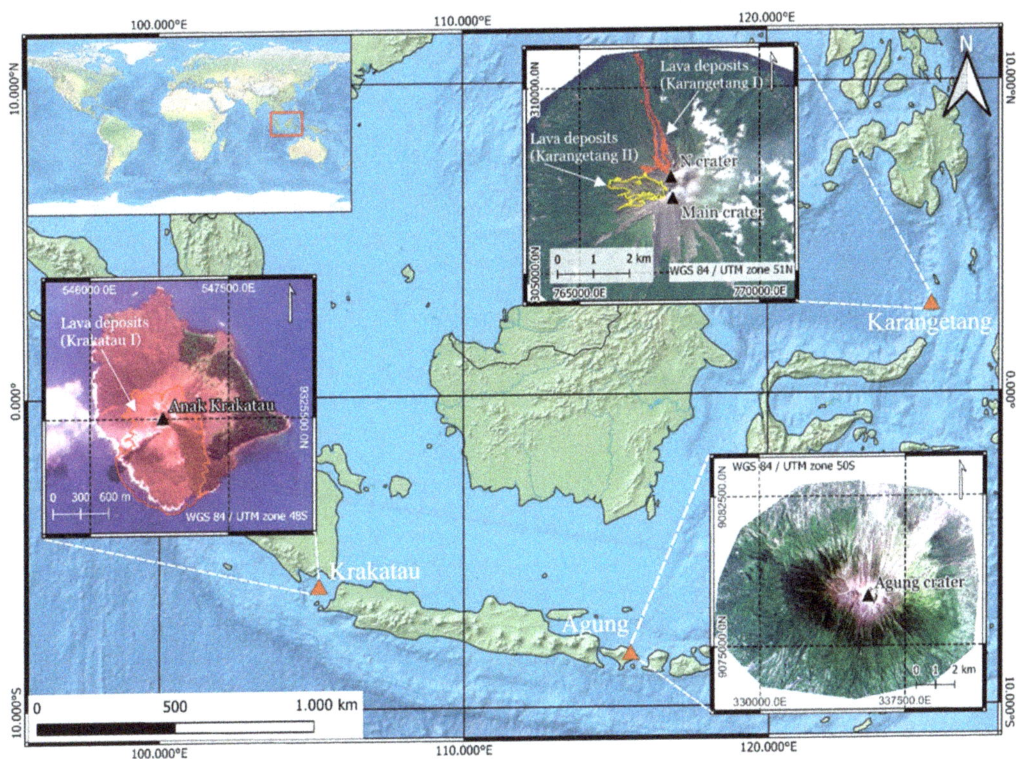

Figure 1. Study area in Indonesia with analyzed volcanoes Karangetang (2.781°N; 125.407°E), Agung (8.343°S; 115.508°E), and Krakatau (6.102°S; 105.423°E). For Karangetang and Krakatau, the derived lava areas are displayed. Zoom-in maps of volcanoes based on Sentinel-2 imagery (Karangetang: 20 November 2019; Agung: 18 September 2017; Krakatau: 17 September 2018). Basemap made with Natural Earth (www.naturalearthdata.com).

Table 1. Analyzed volcanoes and eruption events.

Volcano	Eruption Event	Time Period	Description
Karangetang	Karangetang I	November 2018–March 2019	Lava flow from the N crater towards NNW and WNW
	Karangetang II	July 2019–January 2020	Lava flow from the main crater towards W
Agung	Agung I	21 November 2017–30 November 2017	Multiple ash eruptions with large ash and tephra deposits
Krakatau	Krakatau I	June 2018–December 2018	Strombolian activity with lava flows towards S
	Krakatau II	22 December 2018	Explosive eruption with flank collapse triggering tsunami

The second eruptive phase at Karangetang (Karangetang II) occurred between July 2019 and January 2020 (Table 1) and was characterized by a variety of incandescent block debris avalanches, lava flows, frequent gas and steam emissions, and ash clouds [32,33]. In addition to block debris avalanches with incandescent material, a WNW lava flow developed from the main crater in late July 2019. Lava effusion increased in the following weeks and continued to develop downslope in August and September 2019. Lava effusion was also observed at the N crater during this period. Following a large Strombolian eruption in late November 2019, steady lava effusion intensified, creating a large-scale lava field on the W flank of Karangetang [32]. During December 2019, lava effusion gradually decreased and incandescent block debris avalanches became dominant on the W flank. This activity continued until mid-January 2020. Thereafter, volcanic activity abruptly stopped, and only isolated hotspots were sighted in both craters [33].

Mount Agung (8.343°S; 115.508°E) (Figure 1) is an active stratovolcano (2997 m a. s. l.) located in the east of the island of Bali on the Sunda Island Arc [34]. The volcano is composed mainly of basaltic-andesitic rocks and has a partly explosive Plinian or Sub-Plinian eruptive form. This often results in pyroclastic flows and lahars [34]. Due to the high population density of the island of Bali, Agung is a particularly dangerous volcano [35].

In mid-August 2017, a new eruptive phase began (Agung I) after a prolonged period of inactivity [34,36]. This intensified in November 2017 (Table 1), with massive ash eruptions occurring on 21 November 2017 and the subsequent days. In the process, ash columns rose to heights ranging from 4.8 to 9.3 km a. s. l. between 25 and 26 November 2017 [37]. During this period, the ash cloud drifted mainly W and SW [36,37]. The eruptions resulted in the closure of the international airport in Bali (60 km SW of Agung) for several days between 26 and 29 November 2017. In addition, approximately 25,000 people were evacuated in the immediate vicinity of Agung. The largest eruptions continued until 30 November 2017. After that, only smaller ash and gas eruptions were observed [36].

Krakatau (6.102°S; 105.423°E) (Figure 1) is a volcanic complex in the Sunda Strait that consists of a caldera with a stratovolcano inside. The volcanic complex is composed primarily of andesites or basaltic andesites [38]. It is characterized by frequent alternation between Strombolian or Vulcanian eruptions [10]. However, violent phreatomagmatic and Surtseyan eruptions are also common [9,10]. Mainly lava flows and pyroclastic flows occur during eruptions of Krakatau [10]. A caldera collapse during the catastrophic 1883 eruption destroyed the majority of the pre-1883 Krakatau Island. Volcanism continued after the 1883 events, eventually producing Anak Krakatau, "the son of Krakatau" [39]. The islands located in the immediate surroundings of Anak Krakatau are all uninhabited [38], which is why there is no direct threat to humans from the volcanic hazards of Anak Krakatau. However, a major secondary threat to the densely populated coastal regions of Java and Sumatra are volcano-induced tsunamis [9,40], which led to catastrophic consequences following major eruptions of Krakatau in 1883 [39,40] and in 2018 [9,10].

In this work, two volcanic events at Krakatau were analyzed (Table 1). The first eruptive phase (Krakatau I) began in late June 2018, with lava flows developing on the S and SE flanks of Anak Krakatau accompanied with Strombolian activity. In addition, movement occurred on the SW flank due to the volcanic activity [10]. Lava effusion and Strombolian activity continued into September 2018, whereby an increase in eruptive activity was observed [41]. As a result, steady lava effusion flowed into the sea off the S coast of Anak Krakatau, creating 12.09 ha of new land [19]. In October and November 2018, lava flows also occurred more frequently in the SW of the island. Throughout the eruption period, volcanic activity was frequently accompanied by explosions with large ash plumes.

On 22 December 2018, the described eruptive phase (Krakatau I) subsequently resulted in a powerful explosion (Krakatau II), which brought significant morphological changes [42]. The explosion was accompanied by seismic activity that led to the collapse of the SW flank of Anak Krakatau [10]. Hence, a devastating tsunami developed which reached the coastal regions of Java and Sumatra and the surrounding islands of the Krakatau volcanic complex [9]. In total, 437 people were killed and over 31,000 were injured [42]. The eruption also resulted in large-scale tephra deposition on the surrounding islands [10,19].

3. Datasets

The basis of the developed change analyses for the detection of volcanic hazards is PlanetScope data. PlanetScope is a satellite constellation of over 150 CubeSats (Doves) (10 cm × 10 cm × 30 cm) capable of recording the entire Earth's landmass (approx. 1.49×108 km^2) on a daily basis [43,44]. The constellation has been continuously recording data since June 2016 and has been extended by additional Dove generations in subsequent years. In total, three different generations exist, all of which have slightly different sensor systems (1st generation: PS2, 2nd generation: PS2.SD, 3rd generation: PSB.SD). Depending on the generation, PlanetScope satellites are equipped with three (blue, wavelength λ = 455-515 nm; green λ = 500-590 nm; red λ = 590-670 nm) or four bands (additional near infrared (NIR) λ = 780-860 nm) [44]. PlanetScope has a geometric resolution of 3.7 m, which is resampled to 3 m within the provided data products [44]. Depending on the different Dove generations, the swath width thereby varies between 25–30 km $\times$ 8–10 km [18,44]. Furthermore, almost all satellites fly at an altitude of 475 km in a near-polar orbit with inclinations of 8° and 98° [43]. In addition to daily coverage of the entire Earth's landmass, overlap areas occur with successive Doves, which are recorded multiple times with a time offset of a few minutes and can thus be used to observe short-term, dynamic processes at the Earth's surface [45].

Complementary to the PlanetScope imagery, Sentinel-2 Multispectral Instrument (MSI) and Landsat-8 Operational Land Imager (OLI) data were included in the change analysis. First, these datasets were used to create multispectral false-color representations during eruption periods at the analyzed volcanoes. These were used for visual analysis and validation of the lava area derived from the PlanetScope. Second, thermal anomalies (so-called hotspot), detected by the Normalized Hotspot Indices (NHI) [46], were derived from the Sentinel-2 and Landsat-8 imagery during an eruption period and used as an input dataset for object-oriented lava flow mapping (Section 4.3). The Sentinel-2 satellites (2A and 2B) fly in a near-polar sun-synchronous orbit at an altitude of 786 km and have a temporal resolution of 5 days. The MSI sensor on board of the Sentinel-2 satellites has thirteen spectral bands in the wavelength range between visible light (VIS) (λ = 443 nm) and shortwave infrared (SWIR) (λ = 2190 nm) in three different geometric resolutions (10 m, 20 m, 60 m) [47].

Landsat-8 flies in a near-polar sun-synchronous orbit at an altitude of 705 km and has a repetition rate of 16 days. The OLI sensor has nine spectral bands with a geometric resolution of 15 m (panchromatic) or 30 m (VIS-SWIR) in the wavelength range from VIS (λ = 443 nm) to SWIR (λ = 2290 nm) [48]. In addition, Landsat-8 has the Thermal Infrared Sensor (TIRS), which acquires two bands in the thermal wavelength range (λ = 10.6–12.51 μm) with a geometric resolution of 100 m [48].

For the generation of false-color representations for monitoring thermal anomalies (Section 6.1), Sentinel-2 and Landsat-8 data were obtained via Google Earth Engine (GEE) for the Karangetang I, Karangetang II, and Krakatau I eruption events. Furthermore, two additional Sentinel-2 L2A scenes were downloaded for eruption phase Karangetang II. These were used to test the application and functionality of the developed object-oriented change detection method (Section 4.3) with data having a lower geometric resolution than PlanetScope (Section 6.3). For this purpose, a pre- (9 May 2019) and post- (4 December 2020) eruption scene were selected.

Due to the good geometric resolution and the availability of SWIR bands, Sentinel-2 and Landsat-8 are well suited for monitoring thermal activities at active volcanoes [15,49].

Using the Normalized Hotspot Indices (NHI) [46], volcanic hotspots can be mapped in detail based on the well-resolved IR bands of Sentinel-2 (20 m) and Landsat 8 (30 m). Based on the particularly strong emission of hot objects at the Earth's surface in the SWIR wavelength range, as well as the shift of the maximum emission towards shorter SWIR wavelengths as the temperature of the objects increases, volcanic hotspots can be identified and classified according to their intensity. The detailed methodology of the NHI is described in Marchese et al. [46]. The NHI hotspots for all active Holocene volcanoes, listed by the Global Volcanism Program (GVP), can be accessed online using the GEE tool (https://nicogenzano.users.earthengine.app/view/nhi-tool) [50]. For the analyzed events (Karangetang I, Karangetang II, Krakatau I), NHI hotspot pixels were derived and used as input data for the object-oriented change detection methodology for lava flow classification (Section 4.3).

Digital elevation models (DEM) were used to generate runoff models which allowed us to constrain potential lava runoff areas in the object-oriented change detection methodology

(Section 4.3). Depending on the event studied, a TanDEM-X DEM (12 m) [51], Copernicus GLO-30 DEM (30 m) [52], Shuttle Radar Topography Mission (SRTM) DEM (30 m) [53], and an ALOS PALSAR RTC Product DEM (12.5 m) [54] were used.

4. Methods

4.1. Pre-Processing

To ensure the comparability of the pre- and post-scenes, PlanetScope data were downloaded as an "Ortho Scene" product [44] and projected onto a local UTM zone of the volcanoes. Individual PlanetScope scenes (of the same acquisition date) were then mosaiced for the respective study area. In the process, the overlap areas of the scenes were averaged. Subsequently, clouds or vapor and smoke plumes were manually masked out in all scenes. To minimize geometric redundancies between the pre- and post-scene, the PlanetScope scenes were co-registered. Histogram matching was then applied to the PlanetScope image pairs. Finally, the *NDVI* (Equation (1)) [55] was calculated for each scene and appended to the respective dataset.

For monitoring of thermal anomalies using Sentinel-2 and Landsat-8 the RGB band combination SWIR2, SWIR1, NIR was used. Thus, for Sentinel-2 MSI data, bands 12, 11, and 8a were used. For Landsat-8 OLI, bands 7, 6, and 5 were applied. The two Sentinel-2 scenes (pre- and post-scenes Karangetang II) were reduced to the bands with a pixel size of 10 m (B02 (blue), B03 (green), B04 (red), B08 (NIR)). Subsequently, clouds and vapor or smoke plumes were manually masked out and image-to-image co-registration was performed. In the last step, the *NDVI* was calculated for both scenes and added to the dataset.

The NHI hotspot pixels were extracted with a corresponding spatial distance filter, so that hotspots were detected exclusively in the close surroundings of a volcano. The NHI vector data include the total hotspot area per observation from Sentinel-2 and Landsat-8, respectively, for which the three intensity classes (low-mid, high, extreme) [50] were combined. The output NHI pixels were aggregated over the entire eruption period so that a final dataset was available for each eruptive phase. This represents the total area of all hotspot pixels during the eruption and suggests that these areas have been affected by volcanic activity (e.g., lava flows). The total hotspot area aggregated over time was then implemented in the object-oriented change analysis (Section 4.3).

The DEMs (TanDEM-X 12 m, Copernicus GLO-30 m, SRTM 30 m, ALOS PALSAR RTC DEM 12.5 m) were projected onto the UTM zones of Karangetang, Agung, and Krakatau, respectively, using a bilinear resampling method. The DEM scenes were then reduced to the extents of the volcanoes. These data were then used to generate the respective volcanoes' drainage system.

The scheme of O'Callaghan and Mark (1984) was used to derive the runoff networks from DEM data. First, the sinks present in the DEMs were filled to avoid the derivation of discontinuous flows. The subsequently computed flow direction matrix includes a directional coding that channels the runoff flow in the direction of the adjacent pixel with the steepest slope [56,57]. From this, the flow accumulation matrix can be derived. This layer represents the cumulative weighting of all cells located upstream and drain out into a given pixel [57,58]. Flow accumulation was calculated using the D8 algorithm [56], with no weighting of individual cells [59]. Accordingly, cells with high flow accumulation values are areas where runoff is concentrated. These areas can be considered the main outlets of the drainage network [57,58]. Thus, by setting a threshold, the main flows can be identified and extracted. In this process, finding an appropriate threshold can be done manually based on geographic and site-specific factors [56]. Due to different environments at the volcanoes, as well as the different geometric resolutions of the DEMs, the threshold values for extracting the main flows were identified manually. For the Karangetang I and II events the TanDEM-X DEM (12 m) (flow accumulation threshold 50) was used. The Krakatau I event was analyzed using the Copernicus DEM (30 m) (flow accumulation threshold 10). Finally, the extracted flow networks were converted to a binary raster, which was used as input for the object-oriented mapping of lava flows (Section 4.3).

4.2. Volcanic Ash Deposits Mapping

For mapping volcanic hazards in vegetated areas, a *dNDVI* analysis was performed. The *NDVI* describes the normalized ratio of reflectance in the NIR and red (Equation (1)) and provides information about the photosynthetic activity or health of plants [55],

$$NDVI = \frac{\rho_{NIR} - \rho_{RED}}{\rho_{NIR} + \rho_{RED}} \tag{1}$$

where ρ_{NIR} and ρ_{RED} represent the corresponding reflections in the NIR and red wavelength ranges.

In this work, *dNDVI* analyses were performed for the Agung I and Krakatau II eruption events to identify and quantify ash and tephra deposits. In addition, *dNDVI* layers were created for the Karangetang I, Karangetang II, and Krakatau I eruption phases, which were subsequently incorporated into the object-oriented methodology for lava flow mapping (Section 4.3).

For the identification of volcanic induced vegetation loss, the *NDVI* layers of the pre- and post-scenes (exact dates of imagery are reported in Sections 5.1 and 5.2) of the respective eruption phase were differentiated using Equation (2):

$$dNDVI = NDVI_{post} - NDVI_{pre} \tag{2}$$

The direction of change can then be inferred from the resulting *dNDVI* layer [60], whereby areas of negative *dNDVI* values represent volcanic deposits that cover or destroyed vegetation. A threshold classification of the *dNDVI* layer was then used to classify the intensity of change. In doing so, the events used to analyze ash and tephra deposits (Agung I and Krakatau II) were classified with |0.2| intervals. For the analysis of eruption events with the lava flows (Karangetang I, Karangetang II, Krakatau I), the *dNDVI* change areas were classified using a −0.3 threshold and added to the object-oriented change detection method as an input dataset.

4.3. Lava Flow Mapping

The separation of cooled lava flows or other volcanic deposits and the background in unvegetated areas is hardly possible due to spectral similarities in the VIS and NIR [22], which is why a *dNDVI* analysis (Section 4.2) does not yield any valuable results [18]. Especially at volcanoes with high eruptive activity, more overlapping lava effusions take place. Due to these dynamics, the overlapping lava deposits exhibit mineralogical and morphological similarities that lead to identical spectral signatures [22,24,26]. Accordingly, at the end of an eruptive phase, it is difficult to distinguish which individual lava flows drained where and what its age structures are [22,24]. Vegetation overprinting of cooled lava or lahar deposits can also complicate the detection with optical remote sensing data [22,24,61].

To address this issue, a change detection method was developed to map lava deposits in unvegetated areas. The methodology follows an object-based image analysis (OBIA) approach [62]. This allows for the inclusion of spatial, structural, and contextual information into the classification, leading to the exploitation of the full data potential, especially in the case of VHR satellite imagery [62,63].

Input data of the OBIA change detection are the PlanetScope scenes before and after an eruption event (pre- and post-scenes), the classified *dNDVI* layer (threshold <−0.3) (Section 4.2), the temporally aggregated NHI layer, and the extracted drainage system (Section 4.1). The schematic workflow of the OBIA is shown in Figure 2.

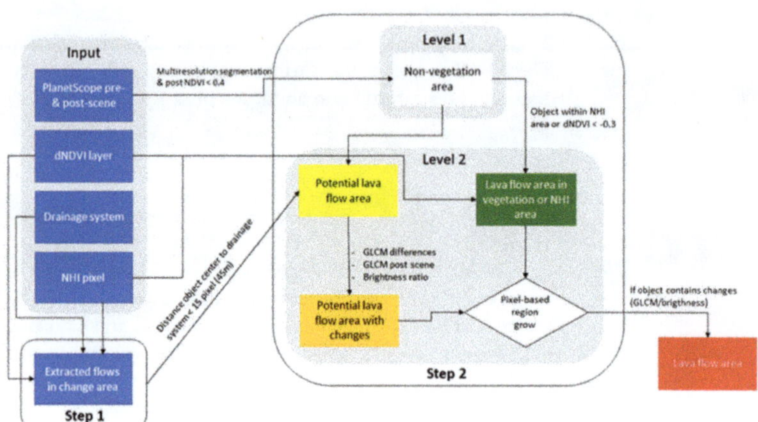

Figure 2. Workflow of OBIA change detection based on NHI pixels and the *dNDVI* area. Input data: pre- and post-PlanetScope scene, temporally aggregated NHI pixels, *dNDVI* layer, and the drainage system. Step 1: extraction of the drainages in the overprint area (NHI or *dNDVI* area); step 2: classification of segments in the unvegetated areas using the combination of distance to drainage and changes in texture (GLCM) and brightness. Starting from the "lava flow area in vegetation or NHI area" objects, a pixel-based region growing into the "potential lava flow area with change" change areas is used to model the lava flow to its point of origin.

In the first step (Figure 2), the *dNDVI* and NHI surfaces are intersected with the drainage system. In this way, the drains can be identified which lie within the *dNDVI* or NHI areas and therefore could potentially have carried a lava flow because of a volcanic or thermal overprint. Then, starting from the identified drainage sub-areas in the *dNDVI* or NHI surfaces, the entire coherent drainage flow is detected using a region-growing algorithm. In this process, starting from the runoff flow segments within the *dNDVI* or NHI areas, the object grows exclusively into areas that are unvegetated in the post-scene (*NDVIpost* layers <0.4). This step ensures that only runoff networks within unvegetated areas are detected in the post-scene, as these are the only areas where lava deposits may be present. If no vegetation change occurred, only the NHI pixels can be used as the initial area for intersection with the drainage system. Similarly, if no NHI pixels are present during the eruption period, only the *dNDVI* layer can also be used as the initial change area. The outflows identified in this step are then used as the basis for OBIA lava flow mapping.

In the second step (Figure 2), the extracted drains and the VHR PlanetScope data are used to identify areas where lava could potentially have flowed. These are inferred from the distance to the drainage system and texture and brightness changes derived from the pre- and post-event PlanetScope imagery. This combination can be used to ensure that differentiation between lava flows and background can be achieved even in the unvegetated area. For this purpose, the unvegetated area in the post-scene is first identified using a threshold value (*NDVIpost* < 0.4). Then, a multi-resolution segmentation [63] is performed within this area to create the segments in the highest hierarchical object level 1 (Figure 2). Lava flows are not bound to a specific shape and can take on irregular shapes, especially for smaller objects that do not cover an entire lava flow. At the same time, despite spectral similarities between lava flows and background, differentiations can be made between objects based on spectral properties [24]. For this reason, the used multi-resolution segmentation placed a stronger emphasis on spectral properties rather than geometric properties.

After segmentation, the distance of the segment center of each object to the nearest point of the flow line is calculated. Objects that are within a distance of ≤15 PlanetScope pixels (45 m) are assigned to the class "potential lava flow area" and form the object level 2. In addition, all objects of the class "non-vegetation area" are intersected with the *dNDVI* and NHI layers to identify objects with vegetation changes or thermal overprinting (class "lava flow area in vegetation or NHI area") (Figure 2). In the next step, texture and brightness changes in the objects of the potential lava area (class "potential lava flow area") are calculated. Since almost no spectral differentiation is possible between objects with lava deposits and the background [22,24,26], information about texture and brightness are included in the classification. Texture was calculated using a grey level co-occurrence matrix (GLCM) based on Haralick et al. [64] for all objects in the potential lava flow area class. In this analysis, the GLCM was calculated based on the individual objects using the red band for the pre- and post-PlanetScope scenes. Following the recommendation of Haralick et al. [64], the average of all four possible viewing diagonals ($0°, 45°, 90°, 135°$) was calculated. The differences of the secondary statistics homogeneity, entropy and contrast were used. For this purpose, each feature was calculated in the pre- and post-scene and then subtracted from each other. In addition, two post-scene GLCM statistics, contrast and spatial standard deviation, were integrated into the classification. Another differentiator is the overall brightness of the objects. Here, a ratio of the overall brightness from the pre- and post-scene was formed to detect changes in the brightness of volcanic deposits. The mentioned texture and brightness parameters were used with the help of a respective threshold value for the classification of the objects of the class "potential lava flow area". The thresholds are strongly dependent on the local factors of the volcanic environment and were therefore defined manually for each individual volcano. The exact threshold values for the individual analyses are presented in Table 2. If an object exceeds or falls below one of the defined limits, it is assigned to the class "potential lava flow area with changes". These objects are located at a plausible distance from the drainage system and contain texture or brightness changes in the PlanetScope scenes.

Table 2. Optimal threshold values of GLCM statistics and overall brightness ratio for the classification of lava flows of the events Karangetang I and II and Krakatau I. If a segment meets one of the defined thresholds, it is assigned to the class "potential lava flow area with changes" (see Figure 2).

Classification Feature	Karangetang I	Karangetang II	Krakatau I
GLCM difference homogeneity	>0.02125	>0.02125	>0.005
GLCM difference contrast	<−1000	<−600	<−250
GLCM difference entropy	<−0.075	<−0.05	<−0.5
GLCM contrast (post-scene)	<710	<900	<1000
GLCM spatial standard deviation (post scene)	<44	<44	<42.5
Overall brightness ratio	<0.8	<1.8	<0.8

In the final process of the OBIA change detection, a pixel-based region-growing procedure is performed, which has as a starting point the objects of the class "lava flow area in vegetation or NHI area" (Figure 2). From these objects, which have a certain volcanic overprint due to vegetation changes or thermal activities, the algorithm grows pixel-based into all neighboring objects belonging to the class "potential lava flow area with changes". A pixel is included in the objects of the initial class if it fulfils one of the defined conditions of GLCM differences, GLCM post-scene, or overall brightness ratio (Table 2). This process is iterated until none of the adjacent pixels of the object class "potential lava flow area with changes" fulfil any of the defined threshold conditions. The resulting class "lava flow area" now contains all objects that represent the area of the lava flow. Finally, all objects of the class "lava flow area" were merged to one object. All objects that are <15 pixels and completely enclosed by the lava area were included in this object. Using a majority filter, the edges and borders of the lava flow object were refined. The final lava area was modelled upstream starting from the volcanically overprinted area ($dNDVI < −0.3$ or NHI hotspot), through the distance to potential runoff streams and changes in the VHR satellite data (texture and brightness), to the potential origin of the lava in the non-vegetated area.

5. Results

5.1. Volcanic Ash Deposits Mapping

Differential *NDVI* analyses (Section 4.2) were used to detect ash and tephra deposits after the Agung I and Krakatau II events.

For the *dNDVI* analysis of ash and tephra deposits after the Agung eruptions, suitable PlanetScope scenes before (18 September 2017) and after (7 December 2017) the event were selected. The post-scene had partially heavy cloud cover, so only the southern region of Agung could be analyzed. Figure 3 visualizes the *NDVI* changes after the eruptions around 21 November 2017 on Agung. The corresponding *dNDVI* layer was classified into several change intervals to visualize the magnitude of change [60]. Here, the lower the *dNDVI* values are, the stronger was the vegetation loss from pre- to post-scene. Areas with low *dNDVI* values have therefore experienced a large vegetation change due to volcanic activity [18]. The total area of ash visible in the imagery (*dNDVI* < −0.4) was quantified to be 25.78 km^2. A striking feature is the spatial distribution of ash deposits on the southern flank of Agung, which become more intense with increasing proximity to the crater. In contrast, the immediate surrounding of Agung again shows higher *dNDVI* values.

Using a *dNDVI* analysis, the detection of ash and tephra deposition on Anak Krakatau and surrounding islands (Krakatau II) was possible. For this purpose, a PlanetScope scene before (14 March 2018) and after (1 July 2019) the eruption on 22 December 2018 was selected. In Figure 4, the effects of ash and tephra deposition as well as vegetation destruction due to the tsunami and explosive force of the eruption are visible at Anak Krakatau and the neighboring island of Krakatau Kecil. Overall, vegetation changes (*dNDVI* < −0.2) are present on a total area of 3.4 km^2. Particularly affected by ash and tephra deposition was Krakatau Kecil, located to the east [42]. Figure 4 shows the decrease in *NDVI* signal over nearly the entire island area, especially on the west coast. On Anak Krakatau, a sharp decline in vegetation signals can be seen in the east of the island.

Figure 3. Ash deposits at Agung (I) after the eruptions around 21 November 2017 based on *dNDVI* changes. Areas with large negative *dNDVI* values represent strong vegetation changes or ash deposition. Background: PlanetScope pre-scene (18 September 2017). The classification could only cover partial areas to the S and SW of Agung due to strong cloud coverage at the post-scene.

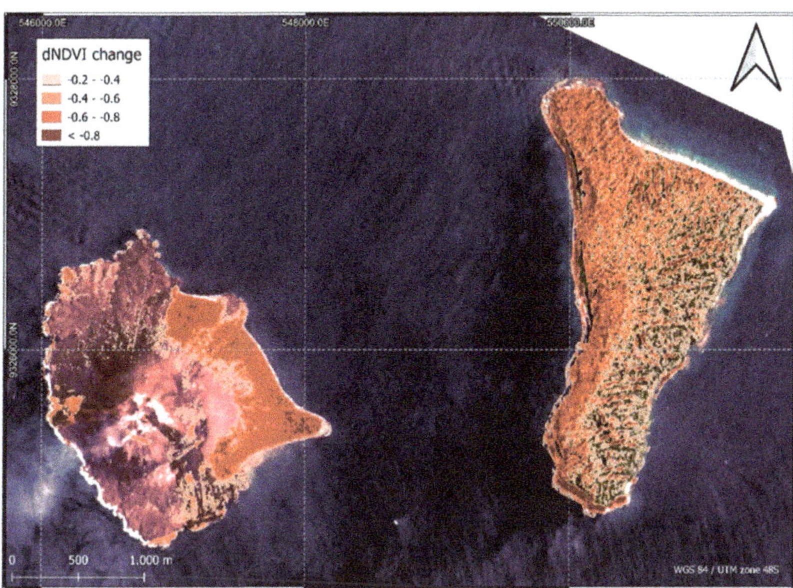

Figure 4. Vegetation changes after the 22 December 2018 eruption at Krakatau (II) based on the *dNDVI*. Former vegetation areas in the E of Anak Krakatau were destroyed by the eruption. On Kecil island E of Anak Krakatau, there were thick ash and tephra deposits, as well as vegetation loss on the W coasts due to the tsunami wave.

5.2. Lava Flow Mapping

Object-oriented change detection methods (Section 4.3) were used to analyze the lava flows of the different eruptive phases Karangetang I, Karangetang II and Krakatau I. Based on GVP reports [31] and NHI hotspot time series, the Karangetang I eruption event was temporally constrained and thus PlanetScope scenes from 19 July 2018 (pre-scene) and 4 May 2019 (post-scene) were selected for change analysis. Lava flow mapping was performed using the OBIA change detection based on vegetation change derived from the classified *dNDVI* layer (Section 4.3, Figure 2). TanDEM-X DEM (12 m) was used for the derivation of the drainage system and subsequent identification of potential lava flows (Section 4.1). The derived lava area and the unaltered pre-scene are shown in Figure 5. Using the OBIA change detection methodology, the total area of the lava flow on the NNW flank of Karangetang was determined to be 0.49 km^2. The total length of the lava flow is 3425 m. At the widest point near the summit area, a width of 250 m was measured. In the region where the road overflow occurred, a width of 140 m was measured. Figure 6 shows the derived classes used to identify and model the potential lava outflow area.

The Karangetang II eruption event was temporally constrained using the GVP reports [32,33] and NHI hotspot time series. OBIA change analysis was performed using PlanetScope scenes from 4 May 2019 (pre-scene) and 3 May 2020 (post-scene). Because the described volcanic activity was mainly confined to the vegetation-free area on the upper W flank of the Karangetang, the temporally aggregated NHI pixels were also included in the OBIA method in addition to the *dNDVI*. The combination of the NHI and *dNDVI* areas formed the initial surface for the pixel-based region growing for mapping the lava surface (Section 4.3, Figure 2).

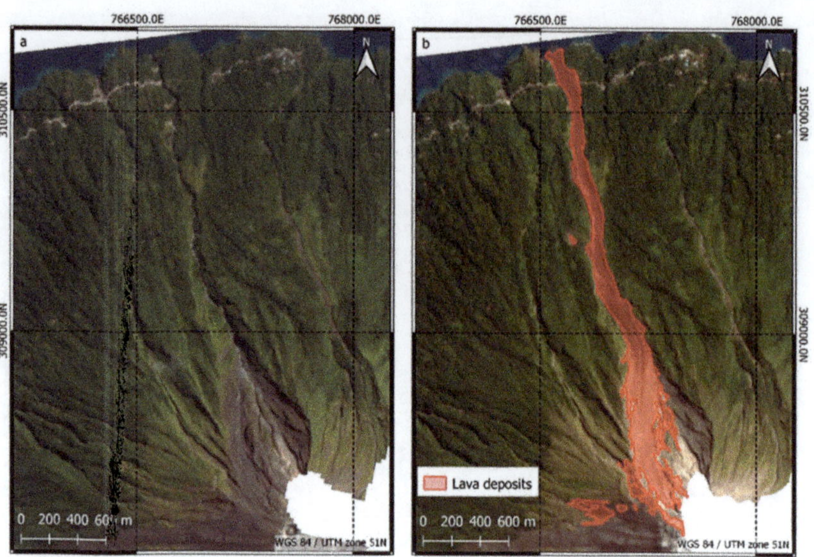

Figure 5. Classification result of the lava flow on the NNW flank of Karangetang (I). (**a**) PlanetScope pre-scene (19 July 2018). (**b**) Classified lava flow area with the post-scene (4 May 2019) in the background.

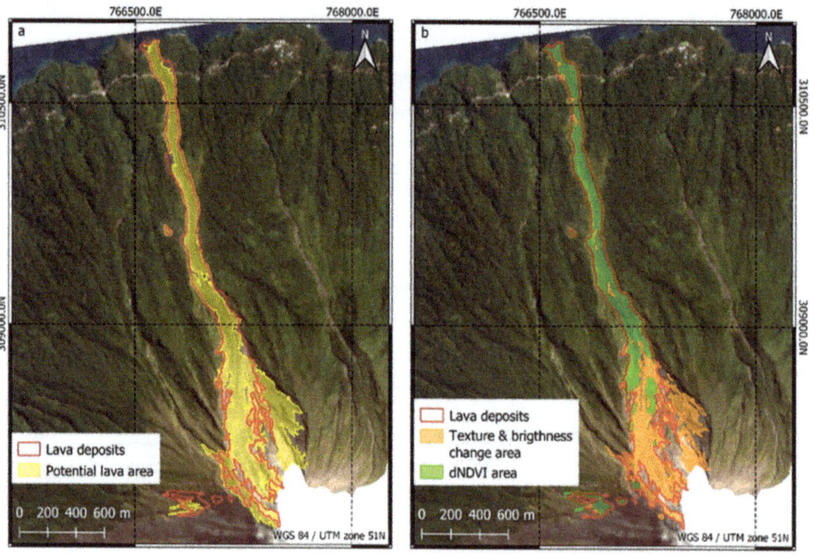

Figure 6. (**a**) Potential lava areas based on distance from the runoff model (derived from TanDEM-X DEM). All areas in the "potential lava area" class are within 45 m of the nearest point on the extracted drainage flow. (**b**) Change areas within vegetation (*dNDVI*) and change areas based on texture (GLCM) and brightness change. Areas in the change area class were classified based on GLCM and brightness thresholds. Based on the *dNDVI* area, region-growing modelling is performed into the "change area". Background: PlanetScope post-scene (4 May 2019).

Figure 7 shows the pre- and post-scene of the W flank together with the derived lava area. During the Karangetang II eruptive phase, 0.66 km^2 was covered by lava. The lava surface occupies almost the entire W flank of Karangetang and measures 880 m at its widest point. The resulting lava field is mainly concentrated in the non-vegetated area of the W flank. In the northern part of the W flank, lava also flowed into the vegetation area. The length of these lava deposits could be determined to be 1650 m. Figure A1 shows the NHI and *dNDVI* areas, the potential lava areas (based on distance from the drainage system) and the change areas (based on texture and brightness changes).

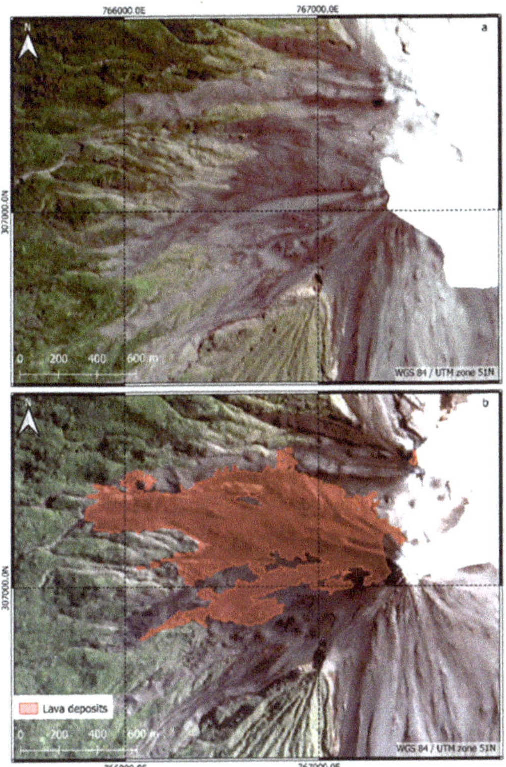

Figure 7. Classification result of lava flows on the W flank of Karangetang (II). (**a**) PlanetScope pre-scene (4 May 2019). (**b**) Classified lava flow area with the post-scene (3 May 2020) in the background.

Based on temporal constraints using GVP reports [41,42] and NHI hotspot time series, PlanetScope scenes on 14 March 2018 (pre) and 17 December 2018 (post) were selected for the Krakatau I eruption event. Since hardly any vegetated areas exist on Anak Krakatau, no corresponding *dNDVI* changes could be included in the methodology. Therefore, only the temporally aggregated NHI pixels were used to classify and model lava flows using OBIA change detection.

Figure 8 shows the pre- and post-scene and the OBIA-derived lava area at Anak Krakatau. A total area of 0.94 km^2 was affected by the lava effusion, which was almost exclusively concentrated on the S flank of Anak Krakatau. The widest part of the lava surface on the S flank has an E-W extension of 1000 m, while the maximum distance from the crater to the newly formed coastline in the SW of the island is 1120 m. Figure A2 includes the derived potential lava areas that lie within 45 m of the extracted drains, as well as the thermally overprinted NHI area and the alteration area based on texture and brightness changes.

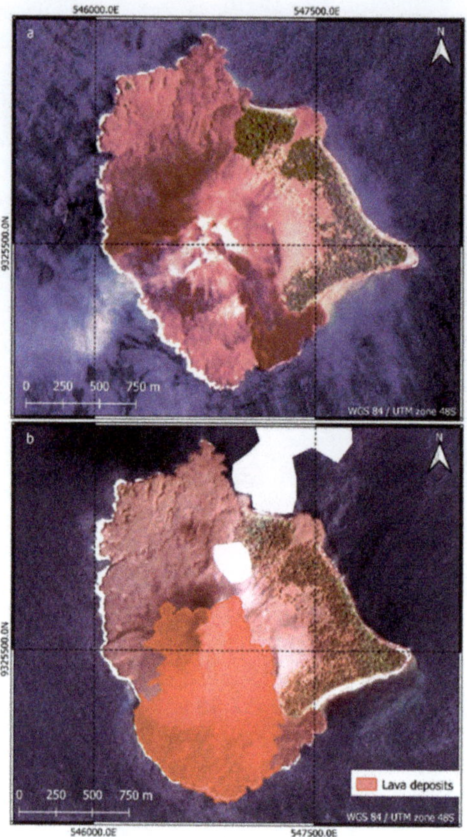

Figure 8. Classification result of lava flows on the S flank of Krakatau (I). (**a**) PlanetScope pre-scene (14 March 2018) of Anak Krakatau (I). (**b**) Classified lava flow area with the post-scene (17 December 2018) in the background.

6. Discussion

6.1. Discussion of the Results

The methodology presented in Section 4.2 for the classification of ash and tephra deposits obtained good results when applied to the case studies of Agung I (Figure 3) and Krakatau II (Figure 4).

Due to cloud cover, only a partial area of the S flank of Agung could be analyzed. Despite this, the PlanetScope scenes used were suitable for analyzing ash and tephra deposition due to the temporal proximity to the eruptions of Agung I. It is reported that the ash deposition happened over a wide area outside of the PlanetScope imagery extent [36,37]. Therefore, the area derived from the *dNDVI* analysis (25.78 km^2) (Section 5.1) massively underestimates the total ash deposit area. However, the spatial distribution of ash deposits in the near surroundings of Agung is comparable to reports from the GVP [36], which shows increased deposition on the SW flank. Here, ash deposition increases with proximity to the crater. The comparatively higher *dNDVI* values in the close vicinity of the crater can be explained with the unvegetated crater environment, which does not reveal any changes in the NDVI signal [18].

Within the analyses of the Krakatau II eruption, much of the negative *dNDVI* values on neighboring Krakatau Kecil Island can be explained in terms of ash and tephra deposits created by the powerful eruption of Anak Krakatau. However, it can be assumed that a large area of change can be attributed to the destruction of vegetation. The surrounding islands of Rakata, Sertung, and Kecil were all massively impacted by the tsunami [9,42]. Modelling of the tsunami shows a wave run-up of 10–23 m on the S, W, and NW coasts of Kecil [9], indicating the destruction of vegetation there and explaining the significantly lower *dNDVI* values in this area. Additionally, in the E of Anak Krakatau, the sharp decrease in the *NDVI* signal can be explained by the complete destruction of vegetation due to explosive eruptive activity [19].

The presented OBIA change detection method for mapping lava flows (Section 4.3) using VHR PlanetScope data also achieved promising results. For the Karangetang I eruption event, based on the *dNDVI* areas, the lava flow could be well mapped in the vegetation area. Despite of the difficulty to spectrally differentiate newly deposited lava from the unvegetated background areas close to the crater, good results could be obtained using the OBIA method. During the Karangetang I eruptive phase, according to the GVP [31], the lava flowed over a length of up to 3.5 km and was about 160 m wide at the road crossing location. These results are in good agreement with the derived lengths (3.425 km) and widths (140 m at the road crossing) of the OBIA classification (Section 5.2). W and E of the lava flow, levees can be recognized (Figure 5b), which could be formed partly due to the channeling of the lava flow and the natural topography of the outflow stream [65]. The color deviation from the dark color, typical for basaltic-andesitic lava, in the upper part of the lava flow can possibly be explained by a renewed overprinting of brighter deposits from a pyroclastic flow or ash [21]. The small lava flow on the WNW flank of Karangetang was recognized almost exclusively based on vegetation changes (*dNDVI*) (Figure 6b). Similarly, no connection between the smaller lava flow in the WNW and the large lava flow in the NNW could be generated. Due to the thresholding used in generating the main flow networks from the DEM, this area was not included in the drainage system modelling. In Figure 9, the classified lava area was compared with a timeseries of multispectral false color imagery (Sentinel-2). Using this methodology, a qualitative assessment about the validity of the classification can be made. The false-color representations clearly show the evolution of the active lava flow on the NNW flank of Karangetang and are consistent with reports from the GVP [31]. Here, the derived lava area includes nearly all of the detected thermal anomalies.

In contrast to reports [32], no lava deposits were identified at the N crater during the Karangetang II eruption phase (Section 5.2). Only a small area within the crater, which can be attributed to a hotspot detection in the NHI layer, was classified as lava. The derived lava area was assigned to the main crater by the region-growing algorithm (Figure 7b). The combination of the NHI and *dNDVI* layers covers mainly the more northerly lava deposits on the W flank (Figure A1). Based on these surfaces, the texture and brightness changes also identified areas located farther south. Brightness was an important factor in detecting these areas of change. Visually, the lava overprinting can be seen in the significantly darker areas in the post-scene, which also make the entire flank area appear much more homogeneous (Figure 7). Furthermore, incandescent debris avalanches are frequently mentioned in the reports of the GVP [32,33]. These deposits could not be identified both visually and by OBIA change detection. Furthermore, the overlap of individual lava flows throughout the eruptive phase makes detection of individual lava flows nearly impossible [24].

The analysis of the multispectral false-color imagery (Sentinel-2 and Landsat-8) allows the identification of individual lava flows during the eruption (Figure A3). There, the development of the lava flow in WNW direction at the beginning of the eruption period is visible. In the following months, sporadic thermal anomalies occurred in the W and WSW direction, but the main activity remained in the WNW. No lava effusion from N crater can be observed in the false-color imagery, confirming the derived lava area. Interrupted and fragmented thermal anomalies with a tongue-like structure are common (Figure A3c,f,h,j,m–p,u). These anomalies may represent the block debris avalanches described in the GVP reports [32,33], which contain incandescent pyroclastic material and usually form at the front of lava flows [66]. On the other hand, fragmented thermal anomalies can also represent differentially cooled regions of a coherent lava flow [11,65,66]. Because the derived lava area includes nearly all of the thermal anomalies in the multispectral false-color time series, a high-quality detection of lava areas can be assumed.

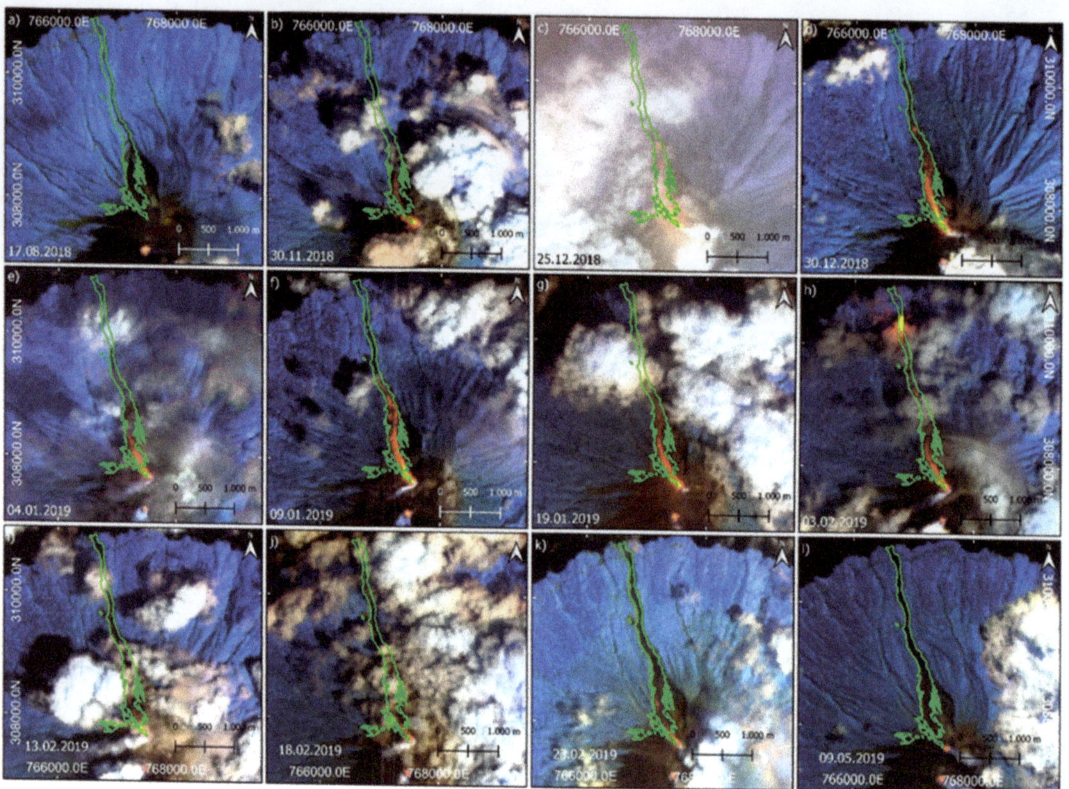

Figure 9. Multispectral false-color time series (Sentinel-2) in the SWIR2/SWIR1/NIR band combination during the Karangetang I eruption period. The classified lava area (in green) outlines the thermal anomalies that occurred. (**a–l**) Volcanic activity and evolution of lava flow towards the NNW flank. Thermal anomalies of the lava flow are displayed in red to yellow, with yellower pixel indicating higher temperatures.

The analyzed eruption phase Krakatau I (Section 5.2) can be seen as a precursor of the catastrophic event on 22 December 2018 (Krakatau II). Walter et al. [10] were able to detect movement of the SW and S flanks within this phase using interferometric SAR (InSAR) time series. This ranged from 4 to 10 mm per month, with higher rates of movement correlating with increased volcanic activity, especially in September and October 2018 [10]. The location of the lava deposits in the S-SW region of the island (Figure 8b) is consistent with the observations of Walter et al. [10]. The results of the Krakatau I event prove that the OBIA method also works using only the NHI pixels as the input (Figure A2b). In addition, the newly formed coastline in the SW and S of Anak Krakatau due to the lava effusion was delineated and confirms the increment of the island quantified by Ginting et al. [19]. In the SE of the island, there is a section of a recent lava deposit, which was formed shortly before the pre-scene (Figure 8). This was partially overprinted by the lava flows of the eruption phase. Here, the part bordering the sea, which was not overprinted during the 2018 activities, could be well demarcated from the lava deposits.

Figure A4 can be used for the qualitative validation of the lava area. The lava area contains nearly all detected thermal anomalies. Only on 14 November 2018 (Figure A4o) thermal anomalies occur outside the detected lava area. Due to the proximity to the crater and the circular distribution of the anomalies, ejected incandescent tephra (e.g., blocks, bombs) can be assumed here [2,12]. Additionally, visible is the increase in thermal activity in September 2018 [10,41]. As the month begins, increasingly large thermal anomalies are detected (Figure A4e–k), which exert visible smearing effects of the thermal signal on their neighboring pixels due to their high emission (Figure A4i,j) [27]. It was during this phase of eruptive activity that there was increased movement of the SW flank, which subsequently collapsed due to seismic activity [10], triggering the catastrophic tsunami [9].

6.2. Advantages and Limitations of the Presented Lava Flow Mapping Approach

Using the *dNDVI* analysis (Section 4.2), an effective methodology was applied to detect vegetation changes with PlanetScope data. Similar to the results of Aldeghi et al. [18], ash and tephra deposits could be detected very well, which in turn is an important parameter for further analysis of possible secondary natural hazards, e.g., secondary lahars [25]. However, this methodology is applicable only in vegetated areas. The *dNDVI* methodology is not suitable for detecting volcanic hazards in unvegetated areas as there are no temporal changes in the *NDVI* signal [18,23]. In addition, the detected vegetation change cannot be automatically assigned to a volcanic hazard. Natural vegetation changes triggered by fire, landslides, or seasonal changes also cause a reduction in the *NDVI* signal [20,23]. Explosive eruptions also often result in complete destruction of vegetation in the impact zone (Section 5.1, Figure 4). To identify these areas and not confuse them with ash or tephra deposition, a preliminary visual analysis with expert knowledge of the pre- and post-scene is necessary [18]. Ash deposits can be very easily eroded by wind or precipitation, which is why the selection of pre- and post-scenes that are close in time to the eruption is particularly important [18]. After deriving the *dNDVI* layer, a threshold must be used to extract the area of change [60]. The definition of this threshold value proves to be difficult. The method used for manually finding a threshold based on local-specific factors and expert knowledge achieved good results in this regard [18,21]. However, this introduces a degree of subjectivity into the classification. Alternatively, automatable adaptive methods of threshold finding exist (e.g., Otsu method [67]), which can speed up and potentially optimize the process [60].

The developed OBIA change analysis (Section 4.3) also allows, in contrast to the *dNDVI* analysis, the classification of volcanic hazards in areas without vegetation. Due to the use of PlanetScope data in combination with modelling based on topographic data (DEM), the methodology offers decisive advantages over classical numerical modelling of volcanic hazards (e.g., MAGFLOW [68], LAHARZ [69]). These modelling efforts rely on strong parameterization of input variables and are used only to predict volcanic natural hazards [58,70]. By linking such modelling with information from satellite imagery, post-eruption or near-real-time analyses can also be performed [14]. The developed OBIA methodology models the potential area of lava flows via a simple distance analysis to the drainage system. Using the PlanetScope data, spectral and textual information from the real world can then in turn be transferred to the model, which is not possible in the aforementioned numerical modelling. In addition, the full potential of the VHR PlanetScope data is exploited via the object-oriented approach. Thus, on the one hand, the reduction in spectral variability in the scene can contribute to an improved classification of lava flows [21]. On the other hand, contextual information about individual image objects can be obtained [63]. Regarding the scarcely existing spectral differences between lava flows and background [22,24], this is crucial for the function of the methodology. In addition, the analysis of texture and brightness changes leads to homogeneous and continuous lava flow segments within the object-based approach. This would not be possible in a pixel-based classification [24].

However, there are some limitations and potential sources of error in the OBIA methodology. For the methodology to function, alteration or overprinting of pixels due to *dNDVI* changes or NHI hotspots is essential. If this is not the case, potential outflows, via intersection with these areas of change, cannot be identified and lava flow modelling fails. In addition, there is always the possibility that areas of volcanic overprinting were not detected by the *dNDVI* analysis or NHI detection.

Another source of error for the modelling of the lava flows is the creation of a runoff model based on the DEMs (Section 4.1). Here, a high degree of generalization is assumed, which has an influence on the result. The resolution and quality of the DEM is elementary, because only in this way morphological structures can be recognized. In Figure 10, the OBIA methodology for the event Karangetang I was tested using different DEMs. All DEMs achieved satisfactory results, with the TanDEM-X (12 m) drainage model representing the lava flow surface most accurately. The comparison confirms that for modelling runoff, coarser resolution does not necessarily lead to worse results [70]. Despite this, higher resolution DEMs achieve smaller-scale runoff models, which in turn allow for a more detailed modelling of lava flows [70]. The DEMs used for the methodology were all generated before the analyzed eruptive phases. Thus, they may not represent the current prevailing morphology of the volcanoes. In particular, explosive eruptions can result in drastic morphological changes [10,71] that are disregarded in this modelling. For applying this classification approach after an eruption with severe morphological changes, an updated DEM must be available. One possible solution for this is to derive DEMs from (tri-) stereoscopic images of VHR optical satellites. These not only provide an actual image of the relief of a volcano after an eruption, but also have high geometric resolution [43]. However, the analyzed events mainly resulted in the discharge of lava with no major changes in the volcanoes morphology (except the lava deposits), which is why the used DEMs provide sufficient quality for modelling the lava flow path.

Another way to optimize drainage network modelling, is to calculate flow accumulation with a multi-flow algorithm. In contrast to single-flow algorithms, multi-flow algorithms achieve better results when modelling diverging drainage networks and parallel flows [72]. The single-flow D8 algorithm used in this study [56] (Section 4.1), models runoff into one of the adjacent cells based on the slope gradient. In contrast, multi-flow algorithms, e.g., FD8 [73], DINF [74], MFD-md [72], assume that runoff can flow into more than one adjacent cell, which gives a more realistic representation of the flows [72].

A strong generalization, and thus potential source of error, arises when defining a threshold for distance analysis of objects to the drainage network. If a drain is identified as a potential lava flow, all objects that lie within a defined distance are assigned to the class "potential lava flow area" (Section 4.3, Figure 2). Thereby a spatial generalization takes place, which does not correctly reflect the morphological structures of a complex runoff system. A possible optimization to derive the threshold value in the classification of potential lava areas, can be achieved by using the HAND (height above the nearest drainage) layer [75]. The HAND layer is calculated using a topographic algorithm that identifies potential floodplains based on the relative vertical distance to the drainage network [75].

Using the NHI (Section 3), thermal anomalies during the eruptive phase can be detected and used as an initial area for modelling a lava flow. However, very weak thermal anomalies can often be difficult to detect [50]. If a thermal anomaly is detected by the NHI, the entire pixel is classified as a hotspot. There is no sub-pixel analysis to identify the hot object within a pixel, so the total hotspot area is overestimated [50]. In addition, cloud cover and the temporal resolution of Sentinel-2 and Landsat-8 are limiting factors for hotspot detection [46]. Furthermore, no atmospheric correction is applied to the data used in the NHI tool, which can sometimes lead to noise in the SWIR signal [50]. Figure 11 shows the results of the OBIA methodology for the Karangetang II event in a version with and without the inclusion of NHI hotspots. This comparison shows that for eruptions that produce volcanic deposits in unvegetated areas, significantly more accurate results can be obtained by incorporating the NHI pixels. Due to minor changes within the vegetation, the *dNDVI* area only intersects a small portion of the drainage network. Accordingly, based on the distance analysis, the "potential lava flow area" class (Figure 2) does not represent the entire lava flow area. Thus, the NHI hotspot area contributes significantly to the improvement of the results, especially in the vegetation-free space.

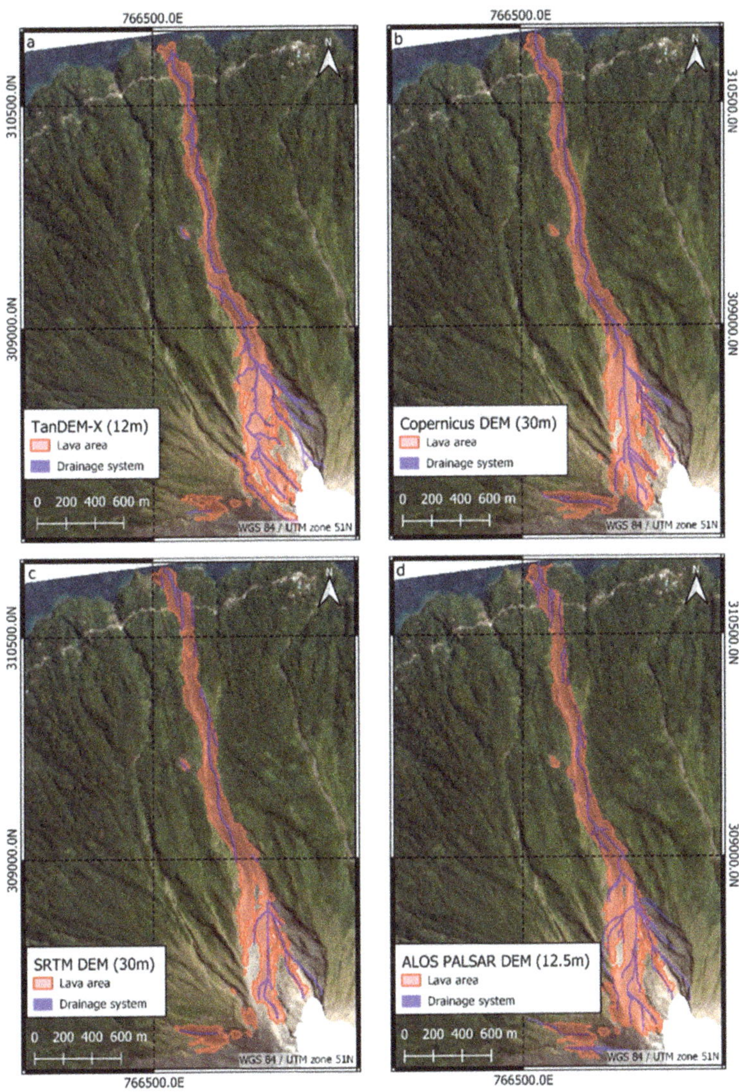

Figure 10. Results of the OBIA method with different drainage system models from TanDEM-X (**a**), Copernicus (**b**), SRTM (**c**), and ALOS PALSAR (**d**) DEMs for the Karangetang I eruptive phase. A threshold value for the extraction of the main outflows was set manually according to the resolution of the DEMs. The TanDEM-X discharge model most accurately derives the lava flow area in the unvegetated area. In the E of the unvegetated area, misclassifications are found in all results. Additionally, noticeable are the partially interrupted outflows in TanDEM-X, SRTM, and ALOS PALSAR DEMs, which are probably due to data errors.

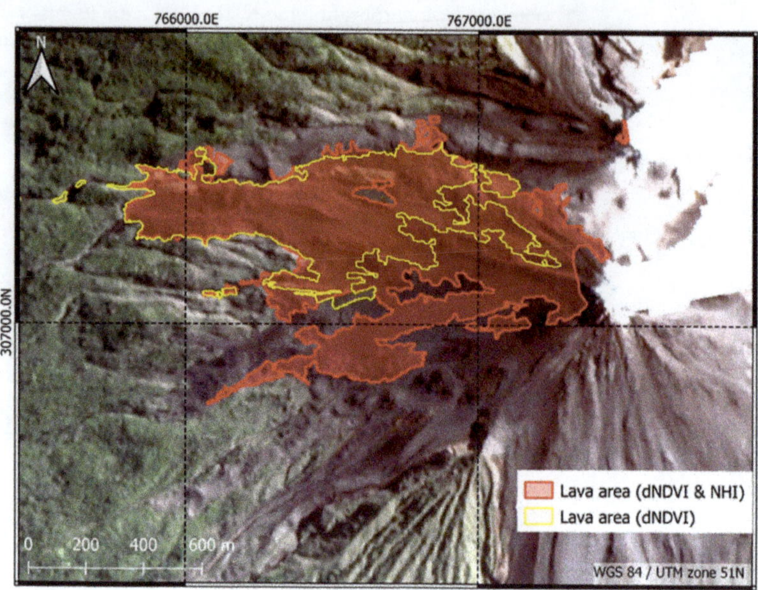

Figure 11. Results of OBIA change detection methodology with and without NHI hotspots. The lava area derived without the incorporation of NHI pixels (in yellow) is significantly smaller and concentrated on the northern part of the W flank of the Karangetang. This is because in step 1 of the OBIA methodology, the areas of change based on the *dNDVI* were intersected with the drainage network. Due to the only minor changes in vegetation, which were limited to the WNW outcropping lava tongue, only the area to the N could be detected as a potential lava area. Subsequent analysis of texture and brightness changes was limited to this area. The lava area with NHI hotspots (in red) is located further south on the W flank. This is due to the detection of hotspots in this area by the NHI, which were subsequently intersected with the drainage network.

Based on the texture and brightness changes, objects that have undergone a change, but are located outside the volcanically overprinted *dNDVI* or NHI areas can be identified. However, for the used classification variables (Table 2), the threshold values for class separation have to be manually adjusted for each application. A common problem with this is that the GLCM changes between pre- and post-scene are very small and do not robustly represent volcanic deposits in different study areas. Therefore, the optical derivation and subsequent definition of the respective thresholds proves to be difficult. By adding SAR amplitude information, the object-oriented classification of the alteration surfaces could be optimized. The deposits of volcanic hazards (e.g., lava flows, pyroclastic flows) change the surface roughness, which in turn leads to a change of the backscatter intensity of the radar signal [27,76]. Changes at the surface due to volcanic activity also cause decorrelation of the interferometric coherence when comparing coherence data derived from pre-event and from co-event SAR image pairs [77]. In addition, the use of SAR data is highly recommended, especially in tropical regions which have a high probability of a cloudy sky [17].

A quantitative validation of the presented results is not an easy task. Reference data are scarce due to costly and sometimes hazardous surveys [20,21,24]. Accordingly, a classical accuracy assessment cannot be performed. One possibility is validation via visual derivation of the deposits from VHR satellite or aerial imagery. Usually, the deposits of volcanic hazards are very complex and difficult to interpret visually [21,25], especially in unvegetated areas where there is little contrast between deposits and background. Accordingly, expert knowledge has to be available for this type of validation. Despite all this, subjective assessments may occur [21,24,66]. In Figure 12, the lava flow of the Karangetang I eruption phase was visually interpreted and manually digitized. This was compared with the semi-automatically generated results of the OBIA method. The agreement between the manually and semi-automatically derived areas is 85.17%, whereby the OBIA-derived lava area showed an overestimation by 14.83% compared to the manually derived reference. Another way to qualitatively validate the lava areas is to produce multispectral false color time series (Sentinel-2 and Landsat-8) as performed in Section 6.1 (Figures 9, A3 and A4). Due to the sensitivity of the SWIR bands to thermal anomalies, greater contrast can thus be produced [66] and lava flows can be tracked during an eruption period. Other validation opportunities include comparison with current geologic maps, which are scarce at remote volcanoes [21,71], or comparison with numerical modelling results [50,58].

The OBIA methodology can be applied to other tropical volcanoes. It is important that the study area around the volcano has vegetation, as this allows potential *dNDVI* changes to be measured [23]. Application of the methodology in areas with little vegetation depends heavily on the availability of NHI pixels and is unlikely to provide satisfactory results (Figure 11). In addition, the OBIA change detection methodology can theoretically be applied to map other volcanic hazards that exhibit flow behavior, e.g., pyroclastic flows and lahars, with the latter requiring vegetated volcano flanks, as lahars cause no thermal anomaly that could be detected by the NHI algorithm.

6.3. How Volcano Monitoring Benefits from PlanetScope Imagery

In this work, the potentials of PlanetScope data for monitoring volcanic hazards were investigated. Based on the results presented in Section 5, it can be confirmed that PlanetScope is a useful data source in volcano monitoring when using appropriate methods, especially the high geometric resolution (3 m), which brings great advantages over other sensor systems. With this, it is possible to identify small-scale volcanic deposits and activities. Harris et al. [23] encountered a significant underestimation of marginal areas due to partially covered pixels when mapping lahar deposits using Landsat data (30 m). This problem is resolved with the use of higher spatial resolution PlanetScope data. When mapping ash and tephra deposits, satisfactory results can also be obtained even with lower-resolution sensors (e.g., Sentinel-2 MSI (10 m)) [18]. Figure 13 compares the use of the OBIA change detection methodology using Sentinel-2 with the PlanetScope results. For the mapping of the lava area after the Karangetang II eruption phase, only the 10 m bands (B02 (blue), B03 (green), B04 (red), B08 (NIR)) of Sentinel-2 MSI were used to ensure comparability with PlanetScope. The methodology provides satisfactory results even with lower resolution Sentinel-2 (10 m) data. The agreement of the Sentinel-2 lava area with the PlanetScope lava area is 79.65%. At the same time, 13.4% of the Sentinel-2 lava area lies outside the boundaries of the PlanetScope lava area. This suggests that the methodology is robust to the geometric resolution of the input data. However, with lower geometric resolution, the detection of the boundary areas deteriorates, which was also observed in Harris et al. [23]. In addition, the advantages of the object-based approach become blurred [63]. Accordingly, high geometric resolution is an important factor for the application of object-based analysis methods in volcano monitoring [21,24]. Furthermore, coarser resolution makes texture analysis via GLCM more difficult because there is less contrast in the image [24]. However, when classifying small-scale volcanic hazards, e.g., fine lahar deposits within watercourses, even PlanetScope reaches its limits. Here, even better resolution systems (e.g., WorldView (0.30 m), Pléiades (0.50 m)) are useful.

Figure 12. Validation of lava area derived from OBIA methodology based on TanDEM-X drainage model (red) and manually digitized lava area (light blue). The manual digitization took place based on the PlanetScope post-scene (4 May 2019). In the lower area, the edge areas were manually captured in a slightly coarser way. In contrast, the lava delta in the sea was mapped better by manual digitization. In the upper area, the potential misclassifications in the E of the lava flow were omitted. In addition, there is the connection of the small lava flow in the WNW to the North Crater. In the sub-areas of the unvegetated area, manual digitization was difficult to perform because there was little contrast with the background. Accordingly, the optical validation of the lava flow from the OBIA method is given only qualitatively.

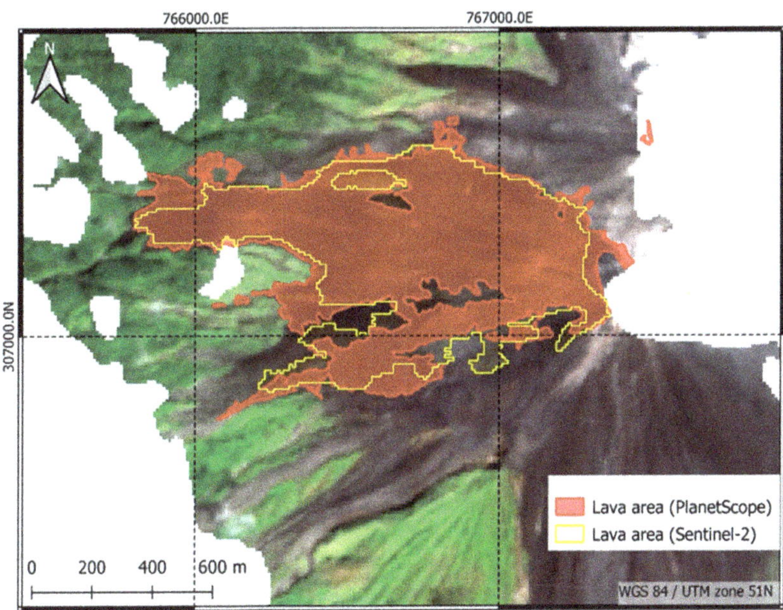

Figure 13. Lava areas derived from PlanetScope and Sentinel-2 data for the Karangetang II erup-tion event. Both lava flows were classified under the same conditions using the OBIA change detection methodology based on *dNDVI* and NHI. A pre- (9 May 2019) and post-scene (4 December 2020) were used for classification and modelling using Sentinel-2. The results with the 10 m Sentinel-2 imagery are also satisfactory. However, marginal areas to the N of the surface are recorded much more coarsely. In the S of the lava field larger differences to the PlanetScope derived lava area were detected. Due to the lower resolution, the GLCM differences could possibly not detect detailed differences between the objects, which is why the lava areas appears somewhat coarser here.

The spectral resolution of PlanetScope is a limitation for monitoring active volcanoes. Due to the missing SWIR or TIR bands hotspots cannot be detected. Work is currently underway to improve the spectral resolution of PlanetScope. In the future, the Doves will be equipped with an 8-band sensor system, with the addition of a Coastal Blue, Green II, Yellow, and Red Edge band to the existing bands, but unfortunately no SWIR channels [44]. For other sensors with higher spectral resolution (e.g., Moderate Resolution Imaging Spectroradiometer (MODIS), Visible Infrared Imaging Radiometer Suite (VIIRS)), spatial resolution limits the exact localization of thermal anomalies [17]. Thus, via the combination of geometrically high-resolution PlanetScope data and spectrally high-resolution IR data, the full potential of both data sources can be combined. Sentinel-2 and Landsat-8 provide a hybrid solution here, combining IR bands with good geometric resolution, allowing hotspots to be located spatially (e.g., NHI) [16]. By integrating these data with PlanetScope, the spectral limitations of PlanetScope can be circumvented.

Another major advantage of PlanetScope data is the daily coverage of the entire earth's landmass. Due to the constellation with more than 150 Doves, this high temporal resolution can be achieved [44]. Especially in volcano monitoring, this is a drastic increase compared to the otherwise used optical sensor systems such as Sentinel-2 and Landsat. Due to the highly dynamic nature of volcanic hazards, a repetition rate of several days prevents the detection of fast-moving changes [26,50]. Individual lava flows of different eruptive phases separated in time can be distinguished [22,24,61,66]. However, differentiating lava flows within a single eruptive phase is difficult. Based on the high temporal resolution of PlanetScope data, it is possible to analyze individual lava flows and their dynamics during an eruptive phase, especially in conjunction with the developed OBIA change detection methodology (Section 4.3). The daily coverage of PlanetScope allows an almost daily (depending on cloud cover) mapping of the lava flows, which in turn provides information about the eruption activity. Hence, especially in near-real-time monitoring of volcanic activity, useful information can be generated [13].

7. Conclusions

In this work, a new change detection methodology for object-based mapping of lava flows was developed using PlanetScope data. In addition, ash and tephra deposits could be detected using difference analysis of the Normalized Difference Vegetation Index (*dNDVI*).

The developed change detection method combines the advantages of the very high resolution (VHR) PlanetScope data in an object-oriented approach. By combining the data with a drainage network model, Normalized Hotspot Indices (NHI) hotspot pixels (derived from Sentinel-2 and Landsat-8 data), and *dNDVI* analyses, lava flows could be mapped even in unvegetated areas where volcanic deposits and the background have high spectral similarities. This is a major added value compared to methods described in the literature for detecting volcanic deposits (e.g., lava flows) using PlanetScope data based solely on *NDVI* changes. The object-oriented change analysis obtained promising results, despite some generalizations in the modelling of potential lava areas.

The PlanetScope data turned out to be highly useful in the analysis of volcanic hazards. Due to the high geometric resolution, even small-scale change processes can be detected. In addition, the high temporal resolution allows for the monitoring of highly dynamic processes in volcanic environments with high eruptive activity. This represents a significant advantage over comparable remote sensing systems that are commonly used for volcano monitoring. Due to the lack of bands in the shortwave or thermal infrared, no thermal information about volcanic activity can be collected with PlanetScope data. Using a multi-sensor analysis, e.g., combining PlanetScope with infrared data, these spectral limitations of PlanetScope can be circumvented. The data sources can thus complement each other and exploit the full potential of Earth observation in volcano monitoring. The results of this work and the aforementioned advantages of the data show that PlanetScope is very well suited for monitoring volcanic hazards and active volcanoes. The derived lava areas or ash and tephra deposits can serve as a basis or additional data for further work and thus support natural disaster management in the risk analysis of volcanic hazards. In addition, the developed methodology based on PlanetScope data can make an important contribution to rapid damage mapping in disaster areas in the future, e.g., in the context of the International Charter 'Space and Major Disasters'.

Author Contributions: Conceptualization, M.R. and S.P.; methodology, M.R.; validation, M.R.; writing—original draft preparation, M.R.; writing—review and editing, S.P.; visualization, M.R.; supervision, S.P. All authors have read and agreed to the published version of the manuscript.

Funding: Financial support by the Client II Project Tsunami_Risk (BMBF Funding reference number 03G0906B) is acknowledged.

Data Availability Statement: The PlanetScope imagery used in this work was provided by Planet Labs through DLR's RESA program (RESA—RESA, BMWi grant 50 EE 1612). The datasets generated during the study are available from the corresponding author on reasonable request.

Acknowledgments: The NHI hotspots were kindly provided by Nicola Genzano (University of Basilicata).

Conflicts of Interest: No potential conflict of interest was reported by the authors.

Appendix A

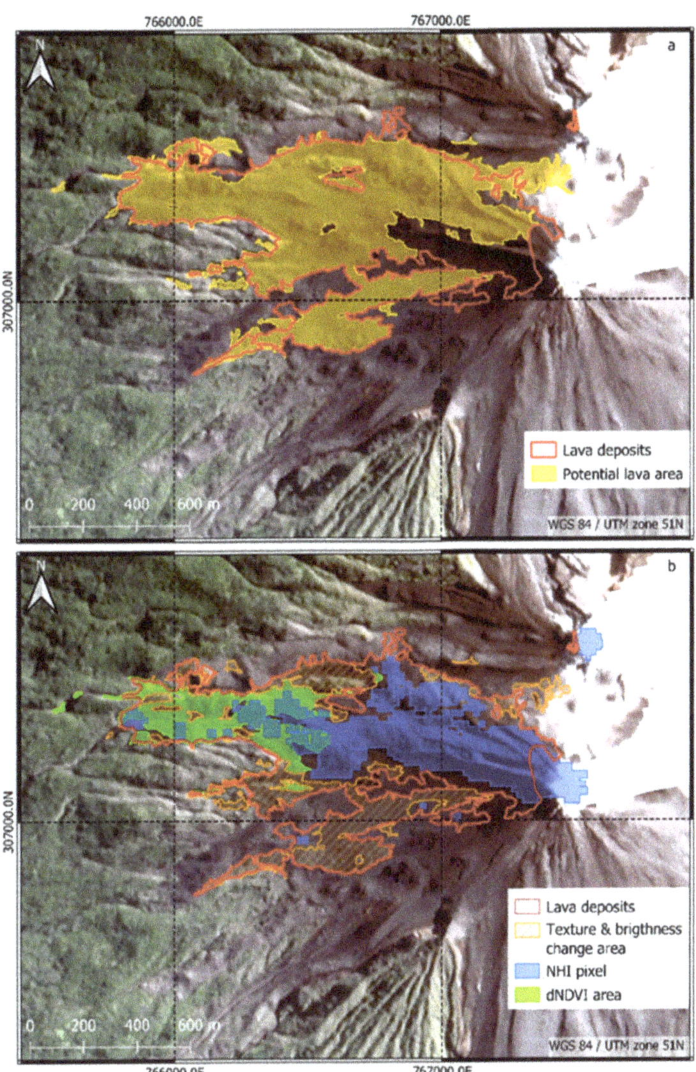

Figure A1. (**a**) Potential lava areas based on distance to discharge model (TanDEM-X). All areas in the "potential lava area" class are within 45 m of the nearest point to the extracted drainage flow. (**b**) Areas based on thermal overprinting (NHI area) and vegetation changes (*dNDVI* area). These areas are considered to be volcanically overprinted and thus are certainly covered with lava deposits. They form the basis for modelling lava flows in the areas with texture and brightness changes. Areas in the "change area" class were classified based on GLCM and brightness thresholds and are within the "potential lava area" class. Based on the NHI and *dNDVI* area, region-growing modelling is performed into the "change area".

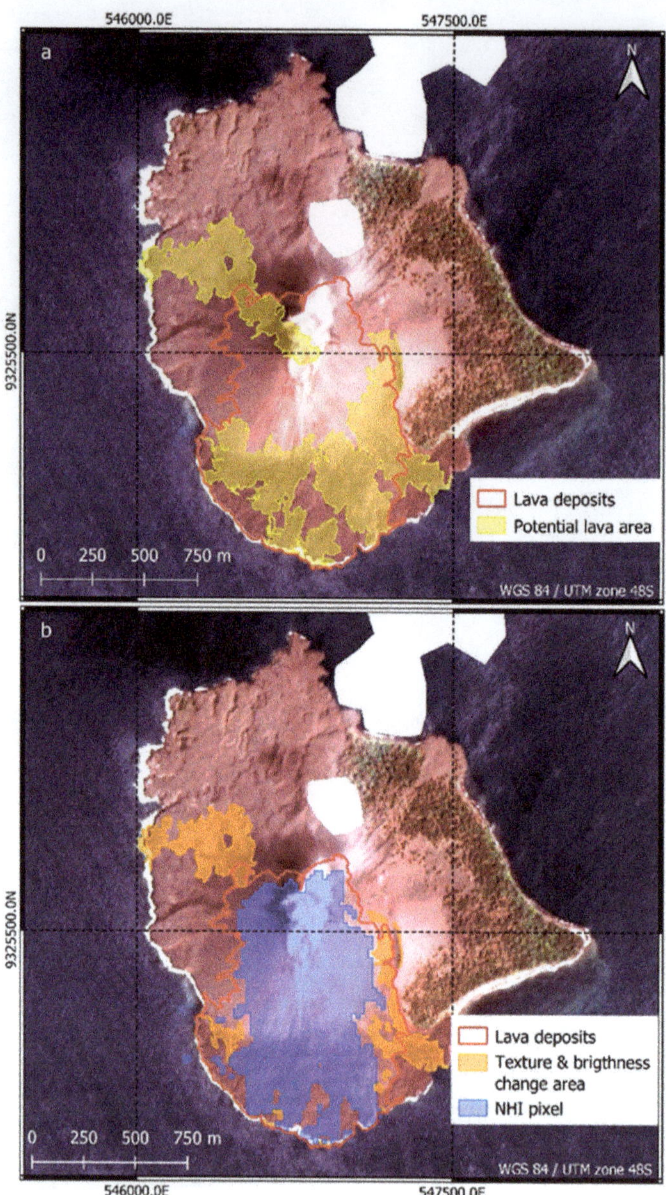

Figure A2. (**a**) Potential lava areas based on distance to the outflow model (Copernicus DEM). All areas in the "potential lava area" class are within 45 m of the nearest point to the extracted drainage flows. (**b**) Areas based on thermal overprinting (NHI area). This area was considered to be volcanically overprinted and thus is certainly covered with lava deposits. It forms the basis for modelling the lava flows in the areas with texture and brightness changes. Areas in the "change area" class were classified using the GLCM and brightness thresholds and are within the "potential lava area" class. Based on the NHI surface, region-growing modelling is performed into the "change area".

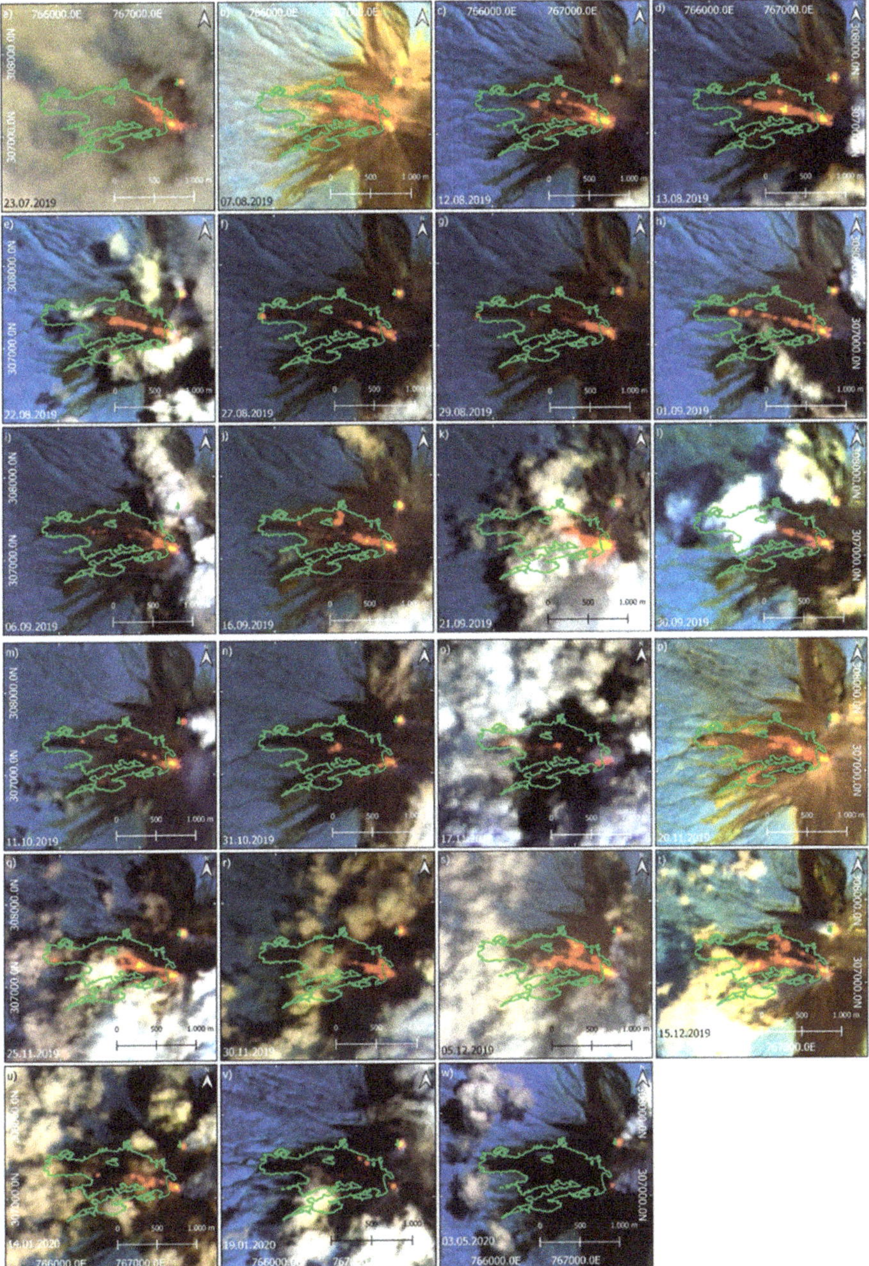

Figure A3. Multispectral false-color time series (Sentinel-2 and Landsat-8) in the SWIR2/SWIR1/NIR band combination during the Karangetang II eruption period. The classified lava area (in green) outlines the thermal anomalies that occurred. (**a–w**) Volcanic activity and evolution of lava flow towards the W flank. Thermal anomalies of the lava flow are displayed in red to yellow, with yellower pixel indicating higher temperatures.

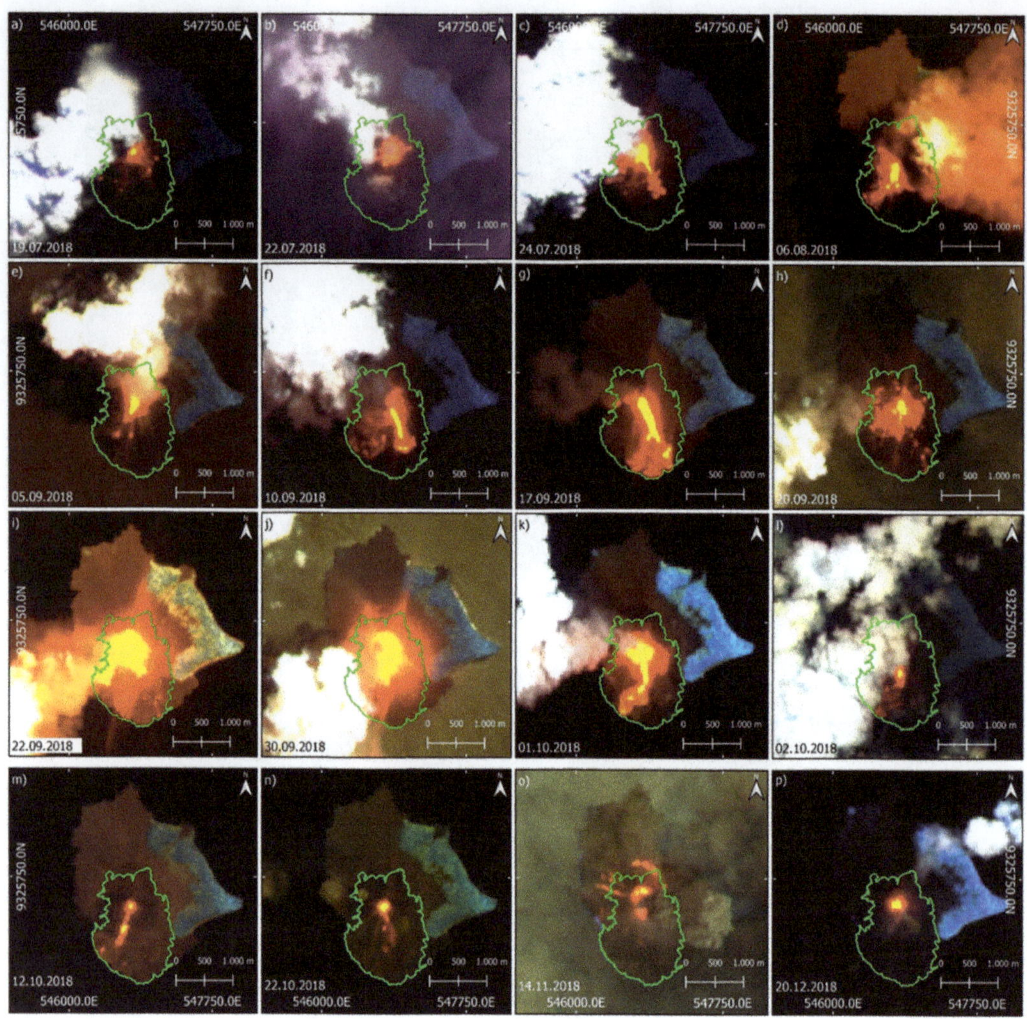

Figure A4. Multispectral false-color time series (Sentinel-2 and Landsat-8) in the SWIR2/SWIR1/NIR band combination during the Krakatau I eruption period. The classified lava area (in green) outlines the thermal anomalies that occurred. (**a–p**) Volcanic activity and evolution of lava flow towards the S flank. Thermal anomalies of the lava flow are displayed in red to yellow, with yellower pixel indicating higher temperatures.

References

1. Auker, M.R.; Sparks, R.S.J.; Siebert, L.; Crosweller, H.S.; Ewert, J. A statistical analysis of the global historical volcanic fatalities record. *J. Appl. Volcanol.* **2013**, *2*, 1–24. [CrossRef]
2. Brown, S.K.; Loughlin, S.C.; Sparks, R.; Vye-Brown, C.; Barclay, J.; Calder, E.; Cottrell, E.; Jolly, G.; Komorowski, J.-C.; Mandeville, C.; et al. Global volcanic hazard and risk. In *Global Volcanic Hazards and Risk*; Loughlin, S., Sparks, S., Brown, S.K., Jenkins, S.F., Vye-Brown, C., Eds.; Cambridge University Press: Cambridge, UK; New York, NY, USA; Port Melbourne, Australia; New Delhi, India; Singapore, 2015; pp. 81–172, ISBN 9781316276273.
3. Patrick, M.R.; Dietterich, H.R.; Lyons, J.J.; Diefenbach, A.K.; Parcheta, C.; Anderson, K.R.; Namiki, A.; Sumita, I.; Shiro, B.; Kauahikaua, J.P. Cyclic lava effusion during the 2018 eruption of Kīlauea Volcano. *Science* **2019**, *366*, eaay9070. [CrossRef] [PubMed]

4.   Wilson, T.M.; Stewart, C.; Sword-Daniels, V.; Leonard, G.S.; Johnston, D.M.; Cole, J.W.; Wardman, J.; Wilson, G.; Barnard, S.T. Volcanic ash impacts on critical infrastructure. *Physics Chem. Earth Parts A/B/C* **2012**, *45–46*, 5–23. [CrossRef]

5.   Horwell, C.J.; Baxter, P.J. The respiratory health hazards of volcanic ash: A review for volcanic risk mitigation. *Bull. Volcanol.* **2006**, *69*, 1–24. [CrossRef]

6.   Baxter, P.J.; Jenkins, S.; Seswandhana, R.; Komorowski, J.-C.; Dunn, K.; Purser, D.; Voight, B.; Shelley, I. Human survival in volcanic eruptions: Thermal injuries in pyroclastic surges, their causes, prognosis and emergency management. *Burns* **2017**, *43*, 1051–1069. [CrossRef]

7.   Charbonnier, S.J.; Germa, A.; Connor, C.B.; Gertisser, R.; Preece, K.; Komorowski, J.-C.; Lavigne, F.; Dixon, T.; Connor, L. Evaluation of the impact of the 2010 pyroclastic density currents at Merapi volcano from high-resolution satellite imagery, field investigations and numerical simulations. *J. Volcanol. Geotherm. Res.* **2013**, *261*, 295–315. [CrossRef]

8.   Vallance, J.W.; Iverson, R.M. Lahars and Their Deposits. In *Encyclopedia of Volcanoes*; Sigurdsson, H., Ed.; Academic Press: San Diego, CA, USA, 2015; pp. 649–665.

9.   Grilli, S.T.; Tappin, D.R.; Carey, S.; Watt, S.F.L.; Ward, S.N.; Grilli, A.R.; Engwell, S.L.; Zhang, C.; Kirby, J.T.; Schambach, L.; et al. Modelling of the tsunami from the December 22, 2018 lateral collapse of Anak Krakatau volcano in the Sunda Straits, Indonesia. *Sci. Rep.* **2019**, *9*, 11946. [CrossRef]

10.   Walter, T.R.; Haghshenas Haghighi, M.; Schneider, F.M.; Coppola, D.; Motagh, M.; Saul, J.; Babeyko, A.; Dahm, T.; Troll, V.R.; Tilmann, F.; et al. Complex hazard cascade culminating in the Anak Krakatau sector collapse. *Nat. Commun.* **2019**, *10*, 4339. [CrossRef]

11.   Francis, P.; Rothery, D. Remote Sensing of Active Volcanoes. *Annu. Rev. Earth Planet. Sci.* **2000**, *28*, 81–106. [CrossRef]

12.   Hidayat, A.; Marfai, M.A.; Hadmoko, D.S. Eruption on Indonesia's volcanic islands: A review of potential hazards, fatalities, and management. *IOP Conf. Ser. Earth Environ. Sci.* **2020**, *485*, 12061. [CrossRef]

13.   Pallister, J.S.; Schneider, D.J.; Griswold, J.P.; Keeler, R.H.; Burton, W.C.; Noyles, C.; Newhall, C.G.; Ratdomopurbo, A. Merapi 2010 eruption—Chronology and extrusion rates monitored with satellite radar and used in eruption forecasting. *J. Volcanol. Geotherm. Res.* **2013**, *261*, 144–152. [CrossRef]

14.   Ganci, G.; Cappello, A.; Bilotta, G.; Del Negro, C. How the variety of satellite remote sensing data over volcanoes can assist hazard monitoring efforts: The 2011 eruption of Nabro volcano. *Remote Sens. Environ.* **2020**, *236*, 111426. [CrossRef]

15.   Marchese, F.; Neri, M.; Falconieri, A.; Lacava, T.; Mazzeo, G.; Pergola, N.; Tramutoli, V. The Contribution of Multi-Sensor Infrared Satellite Observations to Monitor Mt. Etna (Italy) Activity during May to August 2016. *Remote Sens.* **2018**, *10*, 1948. [CrossRef]

16.   Plank, S.; Marchese, F.; Genzano, N.; Nolde, M.; Martinis, S. The short life of the volcanic island New Late'iki (Tonga) analyzed by multi-sensor remote sensing data. *Sci. Rep.* **2020**, *10*, 22293. [CrossRef]

17.   Plank, S.; Walter, T.R.; Martinis, S.; Cesca, S. Growth and collapse of a littoral lava dome during the 2018/19 eruption of Kadovar Volcano, Papua New Guinea, analyzed by multi-sensor satellite imagery. *J. Volcanol. Geotherm. Res.* **2019**, *388*, 106704. [CrossRef]

18.   Aldeghi, A.; Carn, S.; Escobar-Wolf, R.; Groppelli, G. Volcano Monitoring from Space Using High-Cadence Planet CubeSat Images Applied to Fuego Volcano, Guatemala. *Remote Sens.* **2019**, *11*, 2151. [CrossRef]

19.   Ginting, F.I.; Nelson, M.; Minasny, B.; Fiantis, D. Changes in Anak Krakatau landscape after December 2018 eruption. *IOP Conf. Ser. Earth Environ. Sci.* **2021**, *708*, 12088. [CrossRef]

20.   Solikhin, A.; Thouret, J.-C.; Liew, S.C.; Gupta, A.; Sayudi, D.S.; Oehler, J.-F.; Kassouk, Z. High-spatial-resolution imagery helps map deposits of the large (VEI 4) 2010 Merapi Volcano eruption and their impact. *Bull. Volcanol.* **2015**, *77*, 1–23. [CrossRef]

21.   Thouret, J.-C.; Kassouk, Z.; Gupta, A.; Liew, S.C.; Solikhin, A. Tracing the evolution of 2010 Merapi volcanic deposits (Indonesia) based on object-oriented classification and analysis of multi-temporal, very high resolution images. *Remote Sens. Environ.* **2015**, *170*, 350–371. [CrossRef]

22.   Corradino, C.; Ganci, G.; Cappello, A.; Bilotta, G.; Hérault, A.; Del Negro, C. Mapping Recent Lava Flows at Mount Etna Using Multispectral Sentinel-2 Images and Machine Learning Techniques. *Remote Sens.* **2019**, *11*, 1916. [CrossRef]

23.   Harris, A.J.L.; Vallance, J.W.; Kimberly, P.; Rose, W.I.; Matías, O.; Bunzendahl, E.; Flynn, L.P.; Garbeil, H. Downstream aggradation owing to lava dome extrusion and rainfall runoff at Volcán Santiaguito, Guatemala. In *Volcanic hazards in Central America*; Rose, W.I., Bluth, G.J.S., Carr, M.J., Ewert, J.W., Patino, L.C., Vallance, J.W., Eds.; Geological Society of America: Boulder, CO, USA, 2006; pp. 85–104, ISBN 9780813724126.

24.   Li, L.; Solana, C.; Canters, F.; Kervyn, M. Testing random forest classification for identifying lava flows and mapping age groups on a single Landsat 8 image. *J. Volcanol. Geotherm. Res.* **2017**, *345*, 109–124. [CrossRef]

25.   Lavigne, F.; Thouret, J.; Voight, B.; Suwa, H.; Sumaryono, A. Lahars at Merapi volcano, Central Java: An overview. *J. Volcanol. Geotherm. Res.* **2000**, *100*, 423–456. [CrossRef]

26.   Aufaristama, M.; Hoskuldsson, A.; Jonsdottir, I.; Ulfarsson, M.; Thordarson, T. New Insights for Detecting and Deriving Thermal Properties of Lava Flow Using Infrared Satellite during 2014–2015 Effusive Eruption at Holuhraun, Iceland. *Remote Sens.* **2018**, *10*, 151. [CrossRef]

27.   Bignami, C.; Chini, M.; Amici, S.; Trasatti, E. Synergic Use of Multi-Sensor Satellite Data for Volcanic Hazards Monitoring: The Fogo (Cape Verde) 2014–2015 Effusive Eruption. *Front. Earth Sci.* **2020**, *8*, 22. [CrossRef]

28.   Plank, S.; Marchese, F.; Filizzola, C.; Pergola, N.; Neri, M.; Nolde, M.; Martinis, S. The July/August 2019 Lava Flows at the Sciara del Fuoco, Stromboli–Analysis from Multi-Sensor Infrared Satellite Imagery. *Remote Sens.* **2019**, *11*, 2879. [CrossRef]

29. Morrice, M.G.; Jezek, P.A.; Gill, J.B.; Whitford, D.J.; Monoarfa, M. An introduction to the Sangihe arc: Volcanism accompanying arc—Arc collision in the Molucca Sea, Indonesia. *J. Volcanol. Geotherm. Res.* **1983**, *19*, 135–165. [CrossRef]

30. Global Volcanism Program. Karangetang (267020). In *Volcanoes of the World, v. 4.10.1 (29 Jun 2021)*; Venzke, E., Ed.; Smithsonian Institution: Washington, DC, USA, 2013. Available online: https://volcano.si.edu/volcano.cfm?vn=267020 (accessed on 29 June 2021).

31. Global Volcanism Program. Report on Karangetang (Indonesia)—May 2019. *Bull. Glob. Volcanism Netw.* **2019**, *44*, 5. [CrossRef]

32. Global Volcanism Program. Report on Karangetang (Indonesia)—December 2019. *Bull. Glob. Volcanism Netw.* **2019**, *44*, 12. [CrossRef]

33. Global Volcanism Program. Report on Karangetang (Indonesia)—June 2020. *Bull. Glob. Volcanism Netw.* **2020**, *4*, 6. [CrossRef]

34. Fontijn, K.; Costa, F.; Sutawidjaja, I.; Newhall, C.G.; Herrin, J.S. A 5000-year record of multiple highly explosive mafic eruptions from Gunung Agung (Bali, Indonesia): Implications for eruption frequency and volcanic hazards. *Bull. Volcanol.* **2015**, *77*, 1–15. [CrossRef]

35. Global Volcanism Program. Agung (264020). In *Volcanoes of the World, v. 4.10.1 (29 Jun 2021)*; Venzke, E., Ed.; Smithsonian Institution: Washington, DC, USA, 2013. Available online: https://volcano.si.edu/volcano.cfm?vn=264020 (accessed on 29 June 2021).

36. Global Volcanism Program. Report on Agung (Indonesia)—January 2018. *Bull. Global Volcanism Netw.* **2018**, *43*, 1. [CrossRef]

37. Marchese, F.; Falconieri, A.; Pergola, N.; Tramutoli, V. Monitoring the Agung (Indonesia) Ash Plume of November 2017 by Means of Infrared Himawari 8 Data. *Remote Sens.* **2018**, *10*, 919. [CrossRef]

38. Global Volcanism Program. Krakatau (262000). In *Volcanoes of the World, v. 4.10.1 (29 Jun 2021)*; Venzke, E., Ed.; Smithsonian Institution: Washington, DC, USA, 2013. Available online: https://volcano.si.edu/volcano.cfm?vn=262000 (accessed on 29 June 2021).

39. Self, S.; Rampino, M.R. The 1883 eruption of Krakatau. *Nature* **1981**, *294*, 699–704. [CrossRef]

40. Mutaqin, B.W.; Lavigne, F.; Hadmoko, D.S.; Ngalawani, M.N. Volcanic Eruption-Induced Tsunami in Indonesia: A Review. *IOP Conf. Ser. Earth Environ. Sci.* **2019**, *256*, 12023. [CrossRef]

41. Global Volcanism Program. Report on Krakatau (Indonesia)—October 2018. *Bull. Glob. Volcanism Netw.* **2018**, *43*, 10. [CrossRef]

42. Global Volcanism Program. Report on Krakatau (Indonesia)—March 2019. *Bull. Glob. Volcanism Netw.* **2019**, *44*, 3. [CrossRef]

43. Ghuffar, S. DEM Generation from Multi Satellite PlanetScope Imagery. *Remote Sens.* **2018**, *10*, 1462. [CrossRef]

44. Planet Labs. PlanetScope Product Specifications. 2021. Available online: https://assets.planet.com/docs/Planet_PSScene_Imagery_Product_Spec_June_2021.pdf (accessed on 8 July 2021).

45. Kääb, A.; Altena, B.; Mascaro, J. River-ice and water velocities using the Planet optical cubesat constellation. *Hydrol. Earth Syst. Sci.* **2019**, *23*, 4233–4247. [CrossRef]

46. Marchese, F.; Genzano, N.; Neri, M.; Falconieri, A.; Mazzeo, G.; Pergola, N. A Multi-Channel Algorithm for Mapping Volcanic Thermal Anomalies by Means of Sentinel-2 MSI and Landsat-8 OLI Data. *Remote Sens.* **2019**, *11*, 2876. [CrossRef]

47. European Space Agency. Sentinel-2 User Handbook. Available online: https://sentinel.esa.int/documents/247904/0/Sentinel-2_User_Handbook/8869acdf-fd84-43ec-ae8c-3e80a436a16c (accessed on 20 July 2021).

48. U.S. Geological Survey. Landsat 8 (L8) Data Users Handbook. Available online: https://prd-wret.s3.us-west-2.amazonaws.com/assets/palladium/production/atoms/files/LSDS-1574_L8_Data_Users_Handbook-v5.0.pdf (accessed on 15 July 2021).

49. Massimetti, F.; Coppola, D.; Laiolo, M.; Valade, S.; Cigolini, C.; Ripepe, M. Volcanic Hot-Spot Detection Using SENTINEL-2: A Comparison with MODIS–MIROVA Thermal Data Series. *Remote Sens.* **2020**, *12*, 820. [CrossRef]

50. Genzano, N.; Pergola, N.; Marchese, F. A Google Earth Engine Tool to Investigate, Map and Monitor Volcanic Thermal Anomalies at Global Scale by Means of Mid-High Spatial Resolution Satellite Data. *Remote Sens.* **2020**, *12*, 3232. [CrossRef]

51. Krieger, G.; Moreira, A.; Fiedler, H.; Hajnsek, I.; Werner, M.; Younis, M.; Zink, M. TanDEM-X: A Satellite Formation for High-Resolution SAR Interferometry. *IEEE Trans. Geosci. Remote Sens.* **2007**, *45*, 3317–3341. [CrossRef]

52. Fahrland, E. Copernicus Digital Elevation Model (DEM): Product Handbook. 2020. Available online: https://spacedata.copernicus.eu/documents/20126/0/GEO1988-CopernicusDEM-SPE-002_ProductHandbook_I1.00.pdf/082dd479-f908-bf42-51bf-4c0053129f7c?t=1586526993604 (accessed on 20 July 2021).

53. U.S. Geological Survey. USGS EROS Archive—Digital Elevation: Shuttle Radar Topography Mission (SRTM) 1 Arc-Second Global. Available online: https://www.usgs.gov/centers/eros/science/usgs-eros-archive-digital-elevation-shuttle-radar-topography-mission-srtm-1-arc?qt-science_center_objects=0#qt-science_center_objects (accessed on 20 July 2021).

54. Alaska Satellite Facility. ALOS PALSAR—Radiometric Terrain Correction. Available online: https://asf.alaska.edu/data-sets/derived-data-sets/alos-palsar-rtc/alos-palsar-radiometric-terrain-correction/ (accessed on 20 July 2021).

55. Tucker, C.J.; Sellers, P.J. Satellite remote sensing of primary production. *Int. J. Remote Sens.* **1986**, *7*, 1395–1416. [CrossRef]

56. O'Callaghan, J.F.; Mark, D.M. The Extraction of Drainage Networks from Digital Elevation Data. *Comput. Vis. Graph. Image Processing* **1984**, *28*, 323–344. [CrossRef]

57. Jenson, S.K.; Domingue, J.O. Extracting Topographic Structure from Digital Elevation Data for Geographic Information System Analysis. *Photogramm. Eng. Remote Sens.* **1988**, *54*, 1593–1600.

58. Cando-Jácome, M.; Martínez-Graña, A. Determination of Primary and Secondary Lahar Flow Paths of the Fuego Volcano (Guatemala) Using Morphometric Parameters. *Remote Sens.* **2019**, *11*, 727. [CrossRef]

59. Tarboton, D.G.; Bras, R.L.; Rodriguez-Iturbe, I. On the extraction of channel networks from digital elevation data. *Hydrol. Process.* **1991**, *5*, 81–100. [CrossRef]

60. Panuju, D.R.; Paull, D.J.; Griffin, A.L. Change Detection Techniques Based on Multispectral Images for Investigating Land Cover Dynamics. *Remote Sens.* **2020**, *12*, 1781. [CrossRef]

61. Li, L.; Bakelants, L.; Solana, C.; Canters, F.; Kervyn, M. Dating lava flows of tropical volcanoes by means of spatial modeling of vegetation recovery. *Earth Surf. Process. Landf.* **2018**, *43*, 840–856. [CrossRef]

62. Blaschke, T. Object based image analysis for remote sensing. *ISPRS J. Photogramm. Remote Sens.* **2010**, *65*, 2–16. [CrossRef]

63. Benz, U.C.; Hofmann, P.; Willhauck, G.; Lingenfelder, I.; Heynen, M. Multi-resolution, object-oriented fuzzy analysis of remote sensing data for GIS-ready information. *ISPRS J. Photogramm. Remote Sens.* **2004**, *58*, 239–258. [CrossRef]

64. Haralick, R.M.; Shanmugam, K.; Dinstein, I. Textural Features for Image Classification. *IEEE Trans. Syst. Man, Cybern.* **1973**, *SMC-3*, 610–621. [CrossRef]

65. Harris, A.J.L.; Rowland, S.K. Lava Flows and Rheology. In *Encyclopedia of Volcanoes*; Sigurdsson, H., Ed.; Academic Press: San Diego, CA, USA, 2015; pp. 321–343.

66. Calvari, S.; Di Traglia, F.; Ganci, G.; Giudicepietro, F.; Macedonio, G.; Cappello, A.; Nolesini, T.; Pecora, E.; Bilotta, G.; Centorrino, V.; et al. Overflows and Pyroclastic Density Currents in March-April 2020 at Stromboli Volcano Detected by Remote Sensing and Seismic Monitoring Data. *Remote Sens.* **2020**, *12*, 3010. [CrossRef]

67. Otsu, N.A. Threshold Selection Method from Gray-Level Histograms. *IEEE Trans. Syst. Man Cybern.* **1979**, *9*, 62–66. [CrossRef]

68. Del Negro, C.; Fortuna, L.; Herault, A.; Vicari, A. Simulations of the 2004 lava flow at Etna volcano using the magflow cellular automata model. *Bull. Volcanol.* **2008**, *70*, 805–812. [CrossRef]

69. Iverson, R.M.; Schilling, S.P.; Vallance, J.W. Objective delineation of lahar-inundation hazard zones. *Geol. Society Am. Bull.* **1998**, *110*, 972–984. [CrossRef]

70. Huggel, C.; Schneider, D.; Miranda, P.J.; Delgado Granados, H.; Kääb, A. Evaluation of ASTER and SRTM DEM data for lahar modeling: A case study on lahars from Popocatépetl Volcano, Mexico. *J. Volcanol. Geotherm. Res.* **2008**, *170*, 99–110. [CrossRef]

71. Kassouk, Z.; Thouret, J.-C.; Gupta, A.; Solikhin, A.; Liew, S.C. Object-oriented classification of a high-spatial resolution SPOT5 image for mapping geology and landforms of active volcanoes: Semeru case study, Indonesia. *Geomorphology* **2014**, *221*, 18–33. [CrossRef]

72. Qin, C.; Zhu, A.-X.; Pei, T.; Li, B.; Zhou, C.; Yang, L. An adaptive approach to selecting a flow-partition exponent for a multiple-flow-direction algorithm. *Int. J. Geogr. Inf. Sci.* **2007**, *21*, 443–458. [CrossRef]

73. Quinn, P.; Beven, K.; Chevallier, P.; Planchon, O. The prediction of hillslope flow paths for distributed hydrological modelling using digital terrain models. *Hydrol. Process.* **1991**, *5*, 59–79. [CrossRef]

74. Tarboton, D.G. A new method for the determination of flow directions and upslope areas in grid digital elevation models. *Water Resour. Res.* **1997**, *33*, 309–319. [CrossRef]

75. Rennó, C.D.; Nobre, A.D.; Cuartas, L.A.; Soares, J.V.; Hodnett, M.G.; Tomasella, J.; Waterloo, M.J. HAND, a new terrain descriptor using SRTM-DEM: Mapping terra-firme rainforest environments in Amazonia. *Remote Sens. Environ.* **2008**, *112*, 3469–3481. [CrossRef]

76. Wadge, G.; Cole, P.; Stinton, A.; Komorowski, J.-C.; Stewart, R.; Toombs, A.C.; Legendre, Y. Rapid topographic change measured by high-resolution satellite radar at Soufriere Hills Volcano, Montserrat, 2008–2010. *J. Volcanol. Geotherm. Res.* **2011**, *199*, 142–152. [CrossRef]

77. Dietterich, H.R.; Poland, M.P.; Schmidt, D.A.; Cashman, K.V.; Sherrod, D.R.; Espinosa, A.T. Tracking lava flow emplacement on the east rift zone of Kīlauea, Hawai'i, with synthetic aperture radar coherence. *Geochem. Geophys. Geosyst.* **2012**, *13*, 1–17. [CrossRef]

Article

Comparison of Digital Image Correlation Methods and the Impact of Noise in Geoscience Applications

Niccolò Dematteis and Daniele Giordan *

Research Institute for Geo-Hydrological Protection, National Research Council of Italy, Strada delle Cacce, 73, 10135 Turin, Italy; niccolo.dematteis@irpi.cnr.it
* Correspondence: daniele.giordan@irpi.cnr.it

Abstract: Digital image correlation (DIC) is a commonly-adopted technique in geoscience and natural hazard studies to measure the surface deformation of various geophysical phenomena. In the last decades, several different correlation functions have been developed. Additionally, some authors have proposed applying DIC to other image representations, such as image gradients or orientation. Many works have shown the reliability of specific methods, but they have been rarely compared. In particular, a formal analysis of the impact of different sources of noise is missing. Using synthetic images, we analysed 15 different combinations of correlation functions and image representations and we investigated their performances with respect to the presence of 13 noise sources. Besides, we evaluated the influence of the size of the correlation template. We conducted the analysis also on terrestrial photographs of the Planpincieux Glacier (Italy) and Sentinel 2B images of the Bodélé Depression (Chad). We observed that frequency-based methods are in general less robust against noise, in particular against blurring and speckling, and they tend to underestimate the displacement value. Zero-mean normalised cross-correlation applied to image intensity showed high-quality results. However, it suffers variations of the shadow pattern. Finally, we developed an original similarity function (DOT) that proved to be quite resistant to every noise source.

Keywords: digital image correlation; template matching; natural hazards; surface deformations; optical remote sensing; time-lapse camera

Citation: Dematteis, N.; Giordan, D. Comparison of Digital Image Correlation Methods and the Impact of Noise in Geoscience Applications. *Remote Sens.* **2021**, *13*, 327. https://doi.org/10.3390/rs13020327

Received: 30 November 2020
Accepted: 14 January 2021
Published: 19 January 2021

1. Introduction

Since its appearance in the 80s, digital image correlation (DIC) has been widely used in different fields of research, such as medical imagery [1], fluid dynamics (PIV) [2,3], experimental mechanics [4], glaciology [5–7], potamology (LSPIV) [8,9] and landslide monitoring [10,11]. In laboratory experiments, the environmental conditions (e.g., lighting, camera settings, orientation, timing) are controlled and kept (almost) ideal. On the contrary, in real geoscience applications, the images typically capture natural environments. Consequently, they are affected by disturbances related to non-uniform illumination, shadows, blurring and environmental noise.

To the aim of minimising such effects, various approaches have been developed. Typically, they rely on: (i) image morphological operations [12] and colour manipulation [13], or (ii) they apply post-processing algorithms to correct outliers [14]. (iii) Probability analysis [15] and redundancy networks [16] have been proposed too.

Even though many efforts have been spent to the aim of minimising such disturbances, their quantitative influence has been rarely analysed. Travelletti et al. [11] and Dematteis et al. [17] conducted experiments to evaluate the effects of the shadow pattern change due to different positions of the lighting source. Similarly, only a few studies have been published recently where multiple DIC methods were compared, using aerospace photographs of glaciers [18] and landslides [19], obtaining partially-coherent results.

This paper aims to compare the performances of various DIC methods with respect to different types and levels of noise and the incidence of the template size. We considered several similarity functions (whose one is original) in combination with three common image representations. In order to make the comparison as exhaustive as possible, we conducted the comparison on synthetic and real images acquired by ground- and satellite-borne sensors.

2. Previous Works

The literature that compares different DIC methods is quite limited, as well as studies which investigate the impact of possible noise sources on the DIC performances. Several works have been dedicated to this aim in the field of particle image velocimetry (PIV). PIV concerns laboratory experiments, thus it is an area quite different from the geosciences, on which we focus in our study. However, since the number of publications dedicated to comparing DIC methods is small, we shall include those studies in this brief review. Besides, we will report the results of the few works available in the literature that analysed different techniques and the impact of noise on DIC applications in the field of the geosciences. A synthetic description of the below-mentioned correlation functions can be found in Section 3.1.

Martin and Crowley [20] conducted an experimental comparison of correlation techniques. They considered the sum of squared differences (SSD), the normalised cross-correlation (NCC) and the zero-mean normalised cross-correlation (ZNCC). They applied these methods to different image representations: the image intensity, the intensity of the gradient and the Laplacian. They investigated the stability of the correlation indexes with respect to varying levels of illumination intensity, gaussian and salt and pepper noise and the interrogation template size. To this aim, they calculated the correlation between two identical templates whose one was corrupted with some kind of noise. They found that, in general, SSD provided the most stable similarity index, while the gradient was the best image representation.

Merzkirch and Gui [21] analysed the performances of minimum quadratic difference (MQD), correlation interrogation (i.e., spatial cross-correlation) and correlation tracking, using synthetic particle image velocimetry (PIV) images. It is worth noting that MQD and SSD are the same similarity function. However, we decided to use the same authors' convention for coherence with their publications. They examined the dependence of the results to the displacement amount, the interrogation template size, the particle image density and particle size. However, they considered only rigid displacement of a single template and as a matter of fact, the observed errors were a small fraction of pixels. The study showed that MQD was the most performing method in all the considered circumstances.

Pust [22] compared the performances of several frequency-based correlation methods, ZNCC and MQD using real PIV data. He proved that ZNCC outperformed the other methods. He discussed the lower performances of MQD with respect to the findings of Merzkirch and Gui [21], ascribing such an underperformance to a larger sensitivity of MQD to noise.

Heid and Kääb [18] applied six DIC methods to Landsat images of five glacierised areas. They compared the techniques by analysing the percentage of correct matches in the moving part and the root mean squared error in the stable region, where they assumed null displacement. They founded that the COSI-Corr algorithm [23] and the cross-correlation calculated in the Fourier domain applied to orientation images (FFT-OR) generally performed better. However, NCC gave better results using smaller interrogation templates and in areas that largely changed between two acquisitions. They showed that FFT applied to image intensity (FFT-IN) outperformed normalised FFT (phase correlation, PC) in homogeneous areas. However, FFT was dominated by non-moving features and sometimes it returned underestimated values.

Travelletti et al. [11] analysed the impact of the light source position on the correlation coefficient. They applied NCC to a series of shaded relief produced with different lighting directions. They observed a decrease in the correlation index at the increasing of the difference in the light source position. Furthermore, they analysed the influence of the interrogation template size and Gaussian random noise intensity. To this aim, they used synthetic images to which they applied rigid displacement. They obtained higher precision with larger template sizes, at the cost of a reduced spatial resolution. The impact of the Gaussian noise decreased with larger templates.

Bickel et al. [19] conducted DIC measurements on a pair of aerial images of a landslide. They compared the results obtained with ZNCC, a variant of FFT [24] and COSI-Corr [23] with GNSS measurements. They investigated the impact of different template sizes, pre- and post-processing filters and resampling techniques. They observed that COSI-Corr and ZNCC performed worse with small and large templates respectively, while FFT returned similar results with every considered size and it showed the best spatial resolution. They found a general underestimation of the outcomes of every DIC method compared to GNSS measurements. They did not observe any advantages of using pre-processing filters or resampling methods. Instead, image downsampling worsened the results because it introduced blurring. On the other hand, they showed that the use of post-processing filters allowed to identify and remove most of the outliers.

Sjödahl [25] used synthetic and real laser speckle images with different speckle size to compare the performances of seven correlation functions. He considered NCC applied to image intensity (CI), gradient (G) and Hessian (H). Since NCC is normalised, he defined four more image representations as the product of every possible combination: i.e., CIG, CIH, GH, CIGH. He showed that G and H had a sharper autocorrelation peak but were more prone to noise. He obtained the best performances with CIGH (i.e., the product of the three image representations).

Dematteis et al. [17] analysed the impact of the shadow pattern change due to direct illumination using FFT-IN. They produced various shaded reliefs by changing the position of the light source. They observed that the more different were the lighting positions of the two images, the larger was the bias and the standard deviation of the outcomes. They conducted an experiment on real photographs of a stable surface. They calculated the DIC between images acquired in diffuse and direct illumination conditions, which they distinguished between similar and different lighting position. They obtained the best performances with diffuse illumination, while direct illumination with different lighting position provided the lowest results.

3. Implementation

In this section, we describe the DIC methods used in the comparison and an original correlation function which can be applied to complex matrixes. In the second and third sections, we present the dataset to conduct the comparison of methods. We adopted synthetic and real images: we produced the synthetic images using a shaded relief of a real digital surface model (DSM). The adoption of synthetic images permitted to add artificial noise and to vary its type and magnitude. Also, the true displacement value was exactly known and thus, it was possible to assess the correctness of the DIC outcomes precisely. Furthermore, we used two sets of real images acquired by different sensors. The real images concern diverse environments in order to make the DIC comparison as much general as possible. Like in many practical surveys, we did not have available ground-truth measurements to validate the real images' outcomes. Rather, the scope of using real images was to evaluate whether the influence of the noise observed in the synthetic cases came up also in the real images in a similar fashion. The final part of the section illustrates the metrics that we adopted as criteria for comparing the DIC method outcomes. The rationale of the study is shown in Figure 1.

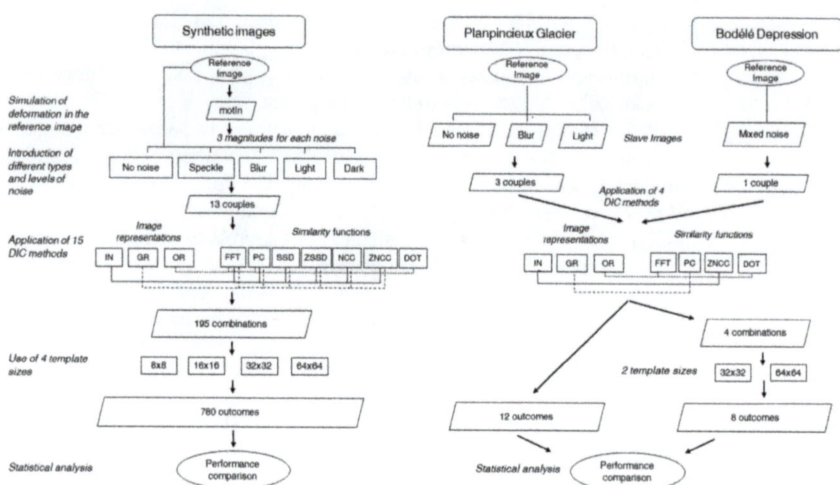

Figure 1. The rationale of the study. On the left, the analysis process using synthetic images is shown, while one right, we present that concerning the real images. In the case of the Planpincieux Glacier images (PG), four DIC methods have been applied between three couples of images composed of a fixed reference one and three images that present (i) no noise, (ii) light noise and (iii) a composition of blur and dark noise. In the case of the Bodélé Depression, both the reference and the slave images present some form of noise, especially shadows and blurring.

3.1. Digital Image Correlation Methods

In our study, we considered three possible image representations: (1) image intensity (IN), (2) intensity of the image gradient (GR) and (3) image orientation (OR), to which we applied various similarity functions. Given an image intensity IN, the intensity of the gradient is defined as:

$$GR = \sqrt{IN_x^2 + IN_y^2} \tag{1}$$

where IN_x, IN_y are the first derivatives in the two directions. A GR image should be insensitive to changes of image intensity and to some extent, to changes of illuminations [20], but it could be more affected by noise [25].

The image orientation [26] is defined as:

$$OR = sgn(IN_x + i \cdot IN_y) \text{ where } sgn(x) = \begin{cases} 0 \text{ if } x = 0 \\ \frac{x}{|x|} \text{ otherwise} \end{cases} \tag{2}$$

An orientation image (OR) is complex and it is intrinsically normalised. As such, it is insensitive to changes of intensity and is robust against changes of colour and local illumination inhomogeneity. Fitch et al. (2002) [26] originally proposed their orientation correlation, which they calculated using the cross-correlation in the Fourier domain.

In its basic form, DIC provides the measurement of the bi-dimensional rigid translation of an image template. The rationale is to search for the position (m, n) where a given similarity function $f(m, n)$ reaches its maximum. $f(m, n)$ is evaluated between a given reference template r and multiple candidates of search templates s within a search band. The search band is a wider area centred in the position of the reference template. For its characteristics, the process is often referred as to template matching, pattern tracking or similar.

Multiple correlation functions have been proposed in the literature. Among them, we considered the most commonly used in geoscience studies. One of the most popular is the (1) normalised cross-correlation (NCC) calculated in the spatial domain. We also examined the (2) zero-mean normalised cross-correlation (ZNCC). Two other methods are the (3) sum of squared differences (SSD), also known as the minimum quadratic difference [27] and the variant (4) zero-mean sum of squared differences (ZSSD). We considered two DIC techniques that operate in the frequency domain: the (5) cross-correlation calculated with Fourier transform (FFT) and the (6) phase correlation (PC). Finally, we included in the comparison an original method which is suitable to apply to complex matrixes, we named it (7) dot multiplication (DOT).

Concerning the correlation functions, NCC and ZNCC are defined as:

$$\text{NCC}(m,n) = \frac{\sum_{m,n} r(i,j)s(i+m,j+n)}{\sqrt{\sum_{m,n} r(i,j)^2 \sum_{m,n} s(i+m,j+n)^2}} \tag{3a}$$

$$\text{ZNCC}(m,n) = \frac{\sum_{i,j}(r(i,j)-\bar{r})(s(i+m,j+n)-\bar{s})}{\sqrt{\sum_{m,n}(r(i,j)-\bar{r})^2 \sum_{m,n}(s(i+m,j+n)-\bar{s})^2}} \tag{3b}$$

where (i,j) is the position of the reference template in the image, (m,n) is the position in the search area of the search template and $\bar{r}$, $\bar{s}$ are the mean values of the reference and search templates, respectively.

Both NCC and ZNCC are simple to evaluate, as they vary between -1 and 1 (respectively the worst and best matching); as such, they can be compared between different correlation attempts. However, large pixel differences can dominate the calculation. ZNCC is expected to perform better than NCC with respect to illumination changes.

An alternative to spatial cross-correlation is SSD [20,27] (and its variant ZSSD), which are simply defined as the Euclidean distance between two templates:

$$\text{SSD}(m,n) = \sum_{m,n}(r(i,j)-s(i+m,j+n))^2 \tag{4a}$$

$$\text{ZSSD}(m,n) = \sum_{m,n}((r(i,j)-\bar{r})-(s(i+m,j+n)-\bar{s}))^2 \tag{4b}$$

The best matching between the templates is found where $\text{SSD}(m,n)$ and $\text{ZSSD}(m,n)$ are minimum. SSD is not normalised and thus, it can suffer changes in image intensity. SSD and ZSSD are easily dominated by large pixel differences for their quadratic form. They both are rarely used in geosciences and are not expected to show great performances. However, we included them in our research for completeness.

We defined a new similarity function, DOT, which applies to complex matrixes, such as OR images. DOT is defined as follows:

$$\text{DOT}(m,n) = \frac{1}{MN} \sum_{m,n} Re(r_{\text{OR}}^*(i,j) \cdot s_{\text{OR}}(i+m,j+n)) \tag{5}$$

where M, N are the dimensions of the search area, while * denotes the complex conjugate. r_{OR} and s_{OR} are respectively the reference and search template of the OR image. The argument of the summation is the real part of the dot product of two complex numbers, which is equivalent to the cosine of the angle amid the two numbers. Therefore, the more the angle is small (i.e., the two numbers are similar), the more the cosine is close to one. DOT can assume values between -1 and 1 for the minimum and maximum similarity, respectively. Thus, it can be compared between multiple correlation attempts.

The computation of DIC in the spatial domain can be quite demanding. A much faster alternative is to operate in the frequency domain. We considered two frequency-based methods:

$$\text{FFT}(i,j) = \mathcal{F}^{-1}\{\mathcal{F}\{r(i,j)\}\mathcal{F}^*\{s(i,j)\}\} \tag{6a}$$

$$\text{PC}(i,j) = \mathcal{F}^{-1}\left\{\frac{\mathcal{F}\{r(i,j)\}\mathcal{F}^*\{s(i,j)\}}{|\mathcal{F}\{r(i,j)\}\mathcal{F}^*\{s(i,j)\}|}\right\} \tag{6b}$$

where $\mathcal{F}$ and $\mathcal{F}^{-1}$ denote the Fourier transform and its inverse. FFT is equivalent to the spatial cross-correlation for the convolution theorem. FFT is not normalised and it can be sensitive to illumination changes. Alternatively, it is possible to ignore the signal amplitude and to consider only the phase correlation, PC [28]. The drawback of PC is that all the phase differences are weighted equally, while one would expect that the negligible components should have less weight [29].

Considering all the possible combinations of image representations and similarity functions, we examined the performances of 21 distinct DIC methods in total.

It is worth noting that the choice between the calculation of DIC in the spatial or frequency domain is not limited to the computational costs. In the spatial domain, the calculus is in the Lagrangian specification, i.e., the observer follows the reference template that slides in different positions within the search area and searches for the best matching between all the possible templates. On the contrary, the frequency domain implies an Eulerian approach. In the Eulerian approach, the observer is static and analyses the same area in two different times. This has two effects: first, since the considered templates are not exactly the same, but they are rather a portion of each other, decorrelation might occur for relatively large displacement [30]. Second, the maximum displacement that can be detected is equal to half the size of the search template for the Nyquist criterion.

To achieve subpixel accuracy, we used the approach proposed by Travelletti et al. (2012) [11] for every DIC method. We up-sampled of a given factor the similarity surfaces with cubic interpolation, using the Matlab function *interp2*. Then we search for the maximum of the interpolated surface. In our study, we used an up-sampling factor of 25.

Where not expressed differently, we calculated the DIC with four template sizes: 8×8, 16×16, 32×32 and 64×64 pixels. When computing DIC in the spatial domain, we used search bands of half the template side's size in every direction (i.e., 4, 8, 16 and 32 pixels). Thereby, the range of possible detected displacement was the same as for the DIC calculated in the frequency domain.

3.2. Synthetic Images and Noise

To conduct a formal comparison of DIC methods, we used synthetic images that we could freely manipulate. We created a series of images where we introduced simulated deformation and noise. Thereby, the real displacement was exactly known and we were able to evaluate the outcomes' correctness. We produced a 540×434 px shaded relief based on a DSM of a natural slope. Characteristics and geographical context of the DSM are shown in Figure S1 from the Supplementary Materials. The lighting position of the shaded relief has been set to $(0°, 90°)$ (azimuth, elevation). The intensity range of the reference image (refIm) varied between $[0, 255]$. Then, we moved the pixels within a region of interest (ROI) using non-linear transformations, in the form:

$$\text{motIm}(i + f(i), j + g(j)) = \text{refIm}(i, j) \tag{7}$$

where motIm is the image with simulated motion and $f(i) = g(j) = 0$ outside the ROI. According to Equation (7), the produced displacement is not a rigid translation, but it is a non-linear surface deformation that simulates real cases. Therefore, within every template couple in the DIC calculus, displacement gradients are present. Before to apply Equation (7), we upsampled the images of a factor of 10. Subsequently, we resampled to the original size. Thereby, we were able to simulate a 1/10th pixel displacement. We conducted the resampling operation using the *imresize* Matlab function using bicubic interpolation.

Then, we introduced various types and magnitudes of noise into motIm (Table 1): (i) speckling, (ii) blurring, (iii) different positions of light source and iv) illumination intensity changes. The Speckle noise is defined in the form $\text{IN} = \text{IN} + n \cdot \text{IN}$, where n varies uniformly with mean 0 and a given variance. We considered speckle variances of 0.03, 0.05 and 0.07. The Blur noise has been produced with a low-pass mean filter of different sizes: 3×3, 5×5 and 7×7 pixels. We simulated Light noise by using three positions of the light source to create the shaded relief: $(10°, 80°)$, $(30°, 60°)$ and $(50°, 40°)$.

Table 1. List of the synthetic images used in this study.

Noise Type	Noise Type and Magnitude	Acronym
None		motIm
Blurring	3 × 3 low-pass mean filter	Blur3
	5 × 5 low-pass mean filter	Blur5
	7 × 7 low-pass mean filter	Blur7
Illumination intensity change	linear plan subtraction 0 to 100	Dark100
	linear plan subtraction 0 to 150	Dark150
	linear plan subtraction 0 to 200	Dark200
Lighting source position change	azimuth 10°, elevation 80°	LightA
	azimuth 30°, elevation 60°	LightB
	azimuth 50°, elevation 40°	LightC
Speckling	Variance 0.03	Speckle3
	Variance 0.05	Speckle5
	Variance 0.07	Speckle7

Light noise is feature-based spatially-coherent noise. The Dark noise has been produced by subtracting from the intensity image an oblique plan that linearly varied from 0 to 100, 0 to 150 and 0 to 200 (from left to right). Examples of synthetic images are shown in Figure 2. The considered noises concern typical disturbing effects that might occur in real surveys. Light noise well simulates the shadow pattern change caused by the different sun position or the presence of isolated clouds. In contrast, Dark noise represents homogeneously-varying illumination intensity. These are known issues in operative surveys [7,11,12,17,31]. Blur noise simulates image defocusing that can occur for many reasons: e.g., incorrect manual or autofocusing, low illumination, presence of haze or precipitation, condensation of the optical objective or camera vibration. To our knowledge, the influence of image defocusing has never been studied, but, according to our experience, it is a quite frequent disturbing effect. Finally, Speckle is a pseudo-random multiplicative noise that is typical of coherent waves. Probably, this phenomenon is less recurring in visual-based surveys. However, it can occur when DIC is applied to satellite SAR images [32]. We did not use combinations of different noises because the aim of our study was to evaluate the influence of specific disturbances on the DIC calculation.

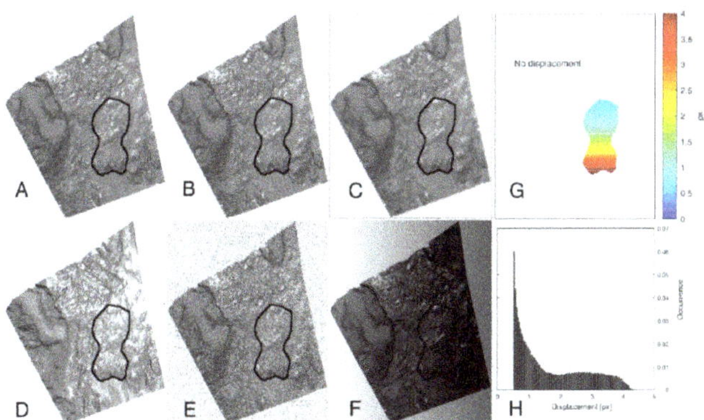

Figure 2. Examples of synthetic images with different types of noise. (**A**) Reference image; (**B**) image with displacement and without noise (motIm); (**C–F**) motIm plus Blur5, LightB, Speckle5 and Dark150 noise respectively. The region of interest (ROI) with simulated displacement is delimited in black. (**G**) Map of simulated displacement and (**H**) displacement distribution in the ROI.

In total, we analysed 13 versions of motIm with different types and levels of noise (Figure S2). The DIC was calculated on a regular grid of 8000 nodes and the ROI contained 628 nodes.

3.3. Real Images

We conducted a similar analysis using real images acquired in natural environments (Table 2). We considered two examples of image data adopted in DIC applications. We used ground-based photographs of an Alpine glacier and satellite panchromatic images of a desertic area. These sites concern diverse environments, but both characterised by high ranges of deformation, with a maximum displacement of ~10 px.

The ground-based dataset refers to oblique 18 Mpx photographs of the Montitaz Lobe of the Planpincieux Glacier (PG), in the Mont Blanc area [31]. The reference image was acquired in optimal conditions in 27 September 2019 at h12.00. To assess the different methods' performances, we calculated the DIC using three distinct search images that have been acquired in 28 September 2019 at h12.00, h08.00 and h19.00 (Figure 3). Therefore, respectively with (1) same illumination conditions, PGopt; (2) different light source position, PGlight and (3) different illumination intensity, PGblur. In PGblur, the scarce illumination also introduced a slight blurring.

Between the acquisitions of the reference and the other three images, some snow melting occurred in the upper part of the images, while a small ice break-off detached from the glacier terminus. We were able to evaluate the robustness of different DIC methods also against these kinds of disturbance. We adjusted the displacement in order to normalise the different time gaps between the three couples of photographs. The DIC processing was conducted using 64×64 template size.

Table 2. Specifics of the real images couples.

Image	Sensor	Image Dimension [px]	Template Size [px]	Dates of Acquisition	Acronym
Planpincieux Glacier	Canon EOS 700D	5184×3456	64×64	27 September 2019 h12 28 September 2019 h12	PPopt
				27 September 2019 h12 28 September 2019 h08	PPlight
				27 September 2019 h12 27 September 2019 h19	PPblur
Bodélé Depression	Sentinel 2B	2000×2000	32×32 64×64	8 October 2019 2 October 2020	BOD

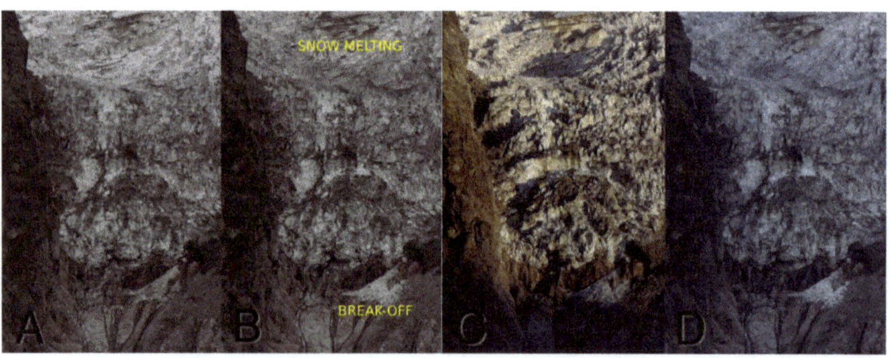

Figure 3. Images of the Planpincieux Glacier: (**A**) reference image; (**B**) image acquired in similar illumination conditions; (**C**) image acquired with different lighting position; (**D**) image acquired with diffuse illumination and slight blurring. The pairs AB, AC and AD are indicated as PGopt, PGlight and PGblur respectively.

The satellite dataset is composed of a pair of Sentinel 2B images (image codes: T33QZU_20191008T090849 and T33QZU_20201002T090719), which targeted the Bodélé Depression (BOD) in Chad [33]. We considered a portion of 400 km^2 centred approximately in 17.02° N–18.02° E. Such an area presents isolated sand dunes and clusters of small dunes and a homogeneous background (Figure 4). The environmental conditions of the first image showed strong wind and partial cloudiness. Therefore, the image appeared slightly blurred and the illumination was not uniform. We applied the DIC using templates of size 32 × 32 and 64 × 64 px.

To the real images, we applied four exemplar DIC methods (1) FFT-OR, (2) DOT-OR, (3) ZNCC-IN and (4) PC-GR.

3.4. Criteria for the Comparison

To evaluate the performances of the DIC methods, we used specific criteria for the comparison. Since the true displacement is known in the synthetic case, we defined a series of quantitative metrics. By contrast, in both the PG and BOD cases, ground-truth measurements were not available. Therefore, we conducted a qualitative analysis based on the visual identification of outliers, spatial resolution and displacement pattern.

The metrics adopted to examine the results obtained with the synthetic images are the following. Outside the ROI, where the theoretical displacement is zero, we calculated mean (μ) and standard deviation (σ) of the displacement as an estimate of accuracy and precision. In motIm, (μ) and (σ) are ideally zero. However, since the displacement is evaluated on templates that can be over the ROI limits, non-zero values could verify in the neighbourhood of the ROI. This behaviour is expected to be more evident with larger templates. Within the ROI, we calculated the mean absolute difference (MAD) and the linear correlation coefficient (CORR) between the obtained displacements and the mean theoretical displacement included in the template. We considered the mean value because the displacement is not uniform within the templates and the DIC measures a sort of averaged displacement [18]. For this reason, the DIC might misestimate the real displacement. We investigated this aspect considering the bias between the 1st, 2nd and 3rd quartiles of the distributions of the measured and theoretical displacements.

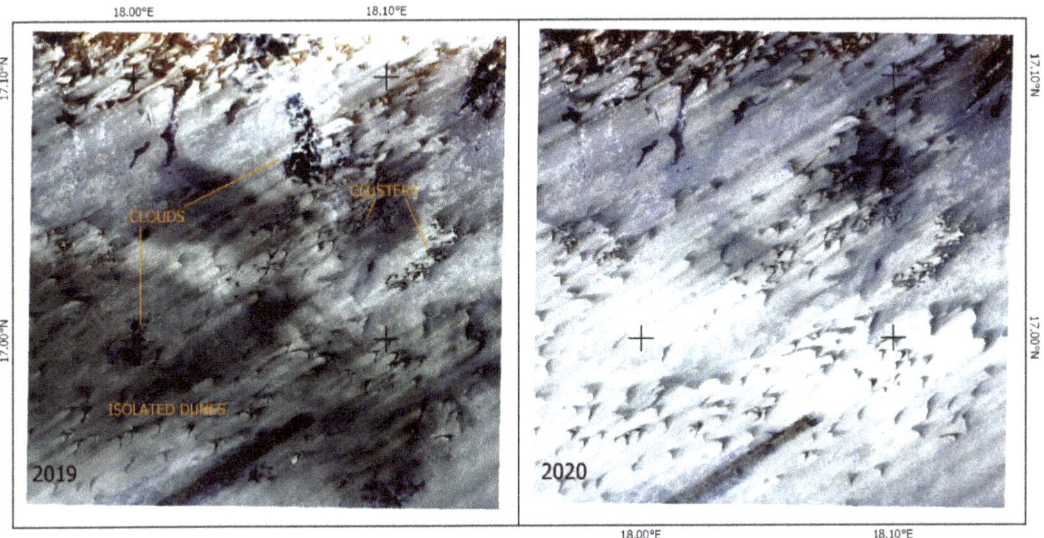

Figure 4. Sentinel 2B images of the Bodélé Depression. They were acquired in 8 October 2019 and 2 October 2020, respectively. In 2019, one can note the inhomogeneous illumination due to thin clouds and some apparent blurring caused by the strong wind.

We noticed that sometimes, even though the overall quality of the results was good, the presence of a few outliers in the residuals strongly affected in a negative way the MAD and CORR metrics. Therefore, we decided to analyse the results also excluding the outliers (MAD* and CORR*). Usually, these outliers concern areas near the ROI boundaries where the investigated templates lie over the ROI edges. When this happens, no displacement is detected near the ROI boundaries. We refer to the ability to correctly identify the displacement in the ROI limits' vicinity as "spatial resolution". Simultaneously, we reported the percentage of outliers (%OUT) in the results. We defined an outlier a value that is more than three scaled median absolute deviation (sMAD) away from the median, where sMAD is defined as $\sim$1.5 median $(x_i - \mathrm{median}(x))$. According to the outlier definition, this metrics has to be evaluated carefully. A small number of outliers could entail a very scattered distribution of the residuals and therefore results of low quality. However, a large percentage of outliers is certainly an indicator of poor performances.

Since we are presenting a certain amount of data, the reader might find hard to examine all the results. Therefore, besides the quantitative parameters, we defined a simple qualitative metrics which assigns a negative/neutral/positive score depending on the visual analysis of the results. We used the following notation: a minus $(-)$ is given when the results contain so much noise that the displacement pattern is not yet recognisable or the pattern is still recognisable, but the values are strongly over/underestimated or the spatial resolution is very poor. A zero (o) is assigned when the displacement pattern is maintained and the values are approximately correct, but many outliers are present. A plus (+) is assigned when the results show good spatial resolution, the displacement values and pattern are correct and the number of outliers is limited.

Finally, we examined the computational time required to operate the various DIC correlations. Even though the computational time is not related to the quality of the results, it is an important parameter that can be relevant in applications that require fast processing or that have to elaborate very large amount of data.

4. Results

In this chapter, we present the results of 15 DIC combinations of correlation functions and image representations. Every combination has been applied to 13 images characterised by various noise types and magnitudes. The analysis has been conducted using four distinct template sizes. In total, we compared 780 results. Furthermore, we examined the results of four model DIC approaches (i.e., ZNCC-IN, DOT-OR, FFT-OR and PC-GR) applied to four couples of images of various real environments. The results of NCC, ZNCC, SSD and ZSSD in combination with OR image provided poor results and they will be not shown. Since DOT is valid for complex numbers, it has been applied exclusively to OR images.

4.1. Methods' Comparison Using Synthetic Images

In the following, we shall present the results obtained with the considered DIC methods applied to the synthetic images. In Figures 5 and 6, we report the displacement maps, the distributions of the displacement of the stable areas and the scatterplots between the measured and theoretical data obtained with every DIC combination for (i) Speckle3 noise calculated with 32 $\times$ 32 template size and (ii) Blur5 calculated with 16 $\times$ 16 template, respectively. In these figures, we also report the values of CORR, MAD, μ and σ. In Figure S3, we report analogue figures concerning every combination of the correlation function, image representation, noise and template size (i.e., 780 outcomes).

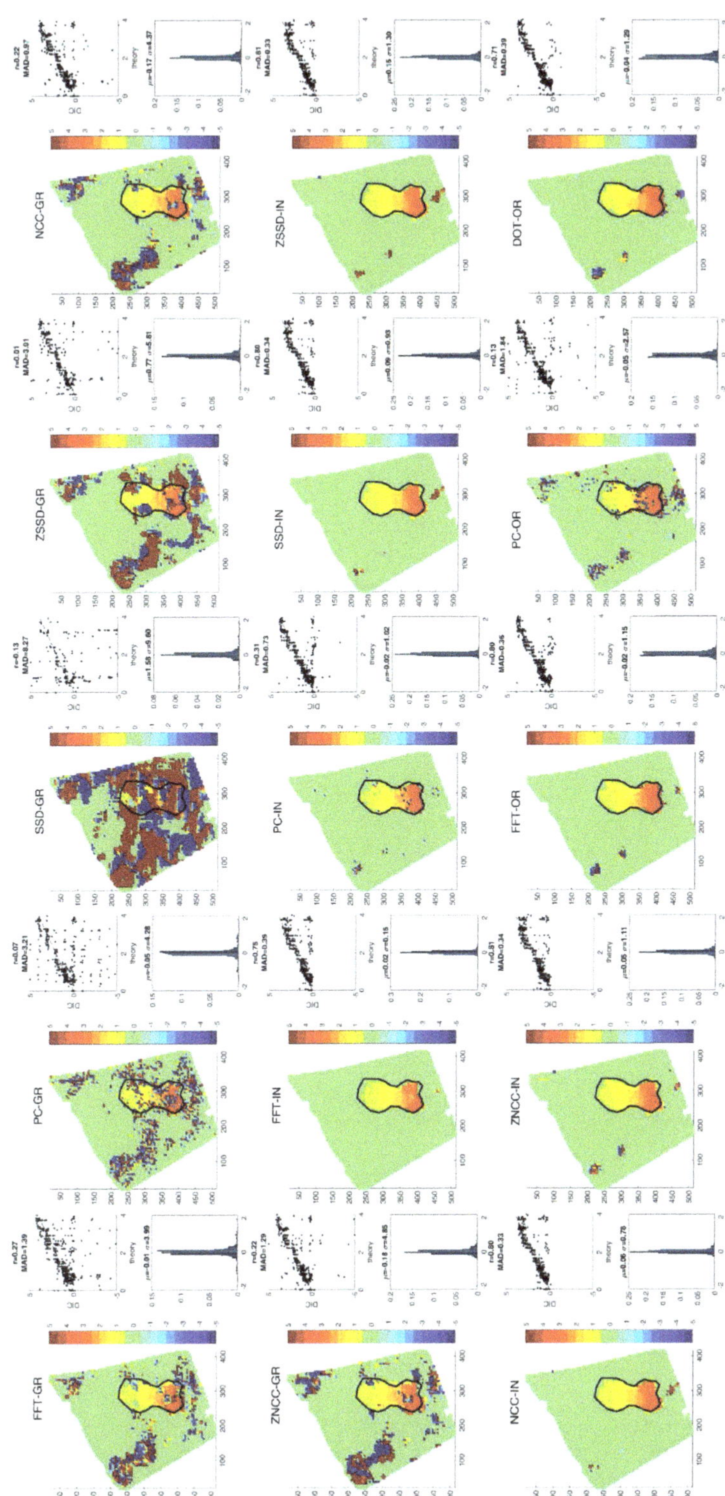

Figure 5. Displacement maps, scatterplots of theoretical vs. measured data within the ROI and displacement distributions outside the ROI for every DIC approach using 32 × 32 template size and with Speckle3 noise. The ROI is delimited in black.

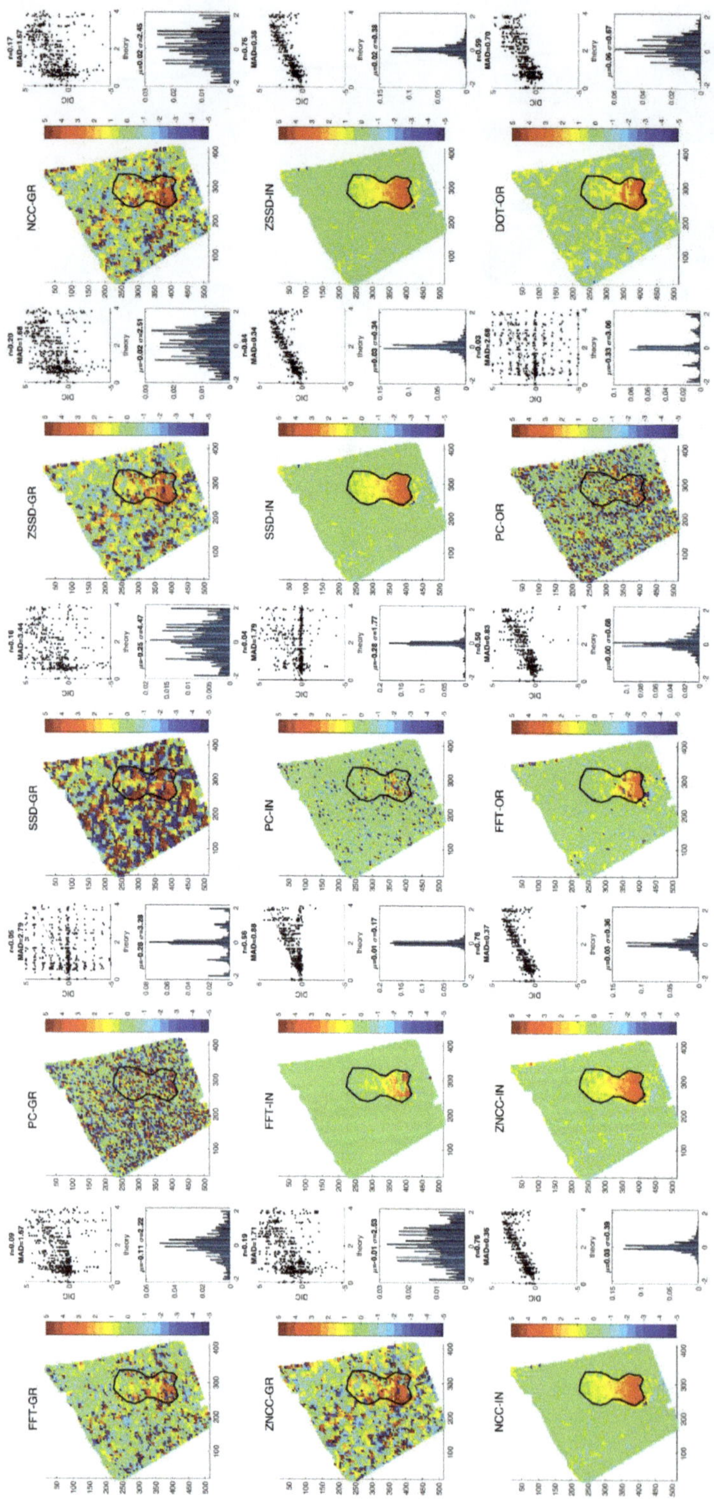

Figure 6. Displacement maps, scatterplots of theoretical *vs.* measured data within the ROI and displacement distributions outside the ROI for every DIC approach using 16 × 16 template size and with Blur5 noise. The ROI is delimited in black.

Besides, we prepared a series of figures containing the metrics present in Section 3.4. An example is presented in Figures 7 and 8, which show the outcomes of LightB noise calculated with 16 × 16 and 64 × 64 template size, respectively. We realised one similar figure for every combination of noise type and magnitude and template size (i.e., 42 figures). All the figures can be found in Figure S4.

In Table 3, we report the qualitative metrics based on the visual analysis of the displacement maps. The rightmost column contains the sum of the scores obtained with each template size.

Image intensity. DIC applied to IN image provides the best results in the presence of Blur and Speckle noise. On the other hand, Light noise can provoke large pixel differences that can influence the correlation calculus. Dark noise introduces a bias that affects non-zero-centred correlation functions.

Image gradient. The use of GR images provides poor results in the presence of Blur and Speckle noise. On the contrary, GR images appear less affected by Light noise and are almost unaffected by Dark noise. GR images return acceptable outcomes only using NCC and ZNCC correlations. With large templates, GR images provide a lower spatial resolution compared to the other image representations. The outcomes obtained with GR images are concentrated on integer displacement values, especially when combined with PC.

Image orientation. OR images yield very high performances in combination with DOT technique. Also, FFT and PC return better results when applied to OR images than IN and GR cases. OR is unaffected by Dark noise and it is resistant against Light noise.

FFT and PC. The performances of frequency-based methods with templates of size 8 × 8 are poor because the displacement is relatively large with respect to the template size. Therefore, signal decorrelates. This is more evident for IN images. However, FFT correlation performs generally better compared to PC for every template size and in combination with every image representation. A remarkable observation is that the frequency-based methods underestimate the displacement. The underestimation is more evident for IN images with Blur noise. The degree of underestimation seems independent from the template size (Figure S4).

NCC and ZNCC show similar results. However, since NCC is calculated using values that are not centred in zero, it is very affected by changes of illumination, because the difference of the mean value of the image can dominate. Therefore, NCC performs badly with Dark noise. Both NCC and ZNCC provide accurate values of displacement. However, in the presence of Blur noise, the spatial resolution decreases and they tend to spread the displacement when calculated with large templates (Figure S3).

Table 3. Performance score based on the qualitative analysis of the results. The rightmost column reports the total score obtained for every template size separately (i.e., 8 × 8/16 × 16/32 × 32/64 × 64).

	motIm	Blur3	Blur5	Blur7	Dark 100	Dark 150	Dark 200	LightA	LightB	LightC	Speckle 3	Speckle 5	Speckle 7	Score
FFT-GR	o+++	-+++	−oo	—	o+++	-+++	o+++	o+++	-+++	−oo	−o+	−oo	—	−9/1/5/6
PC-GR	o++o	-++o	—	—	o++o	o++o	o++o	o++o	-o+o	−oo	−oo	—o	—	−8/−1/3/−3
SSD-GR	+++o	o++o	—o	—	+++o	+++o	+++o	++o+	−oo	—	—	—	—	−2/−1/−1/−4
ZSSD-GR	++++	++++	−+o	—	++++	++++	++++	++++	ooo+	—o	−oo	—o	—	0/0/3/5
NCC-GR	++++	++++	−oo	—	++++	++++	++++	++++	o+++	−oo	-oo+	−o+	—o	0/2/5/8
ZNCC-GR	++++	++++	o-oo	—	++++	++++	++++	++++	oo++	−oo	-oo+	−oo	—o	1/1/5/7
FFT-IN	o+++	o++o	-ooo	-o-	o+++	o+++	o+++	o+++	oooo	—	-+++	-+++	-o++	−6/7/7/6
PC-IN	o++o	-ooo	—	—	o++o	o++o	o++o	o++o	-++o	—o	-o+o	-o+o	−o+	−8/2/5/1
SSD-IN	+++o	+++o	++oo	oooo	—	—	—	o++o	—	—	o++o	oo+o	oo+o	−2/0/1/5
ZSSD-IN	++++	+++o	+++o	-ooo	++++	++++	+++o	++++	oooo	—	o+++	-o++	-o++	3/7/9/6
NCC-IN	++++	+++o	+++o	-+oo	++++	—o	—	++++	oo+o	—	o+++	oo++	-o++	0/4/6/4
ZNCC-IN	++++	+++o	+++o	-ooo	++++	++++	+++o	++++	oo+o	—	o+++	o+++	oo++	5/8/10/6
FFT-OR	o+++	o+++	-ooo	—	o+++	o+++	o+++	o+++	o+++	—o	-o++	-o++	−++	−6/4/8/9
PC-OR	o+++	-o++	—	—	o+++	o+++	o+++	o+++	-++o	−oo	−++	−o+	-o+	−8/0/6/7
DOT-OR	++++	++++	+o+o	-o-	++++	++++	++++	++++	++++	oooo	o+++	oo++	-o++	6/7/11/9

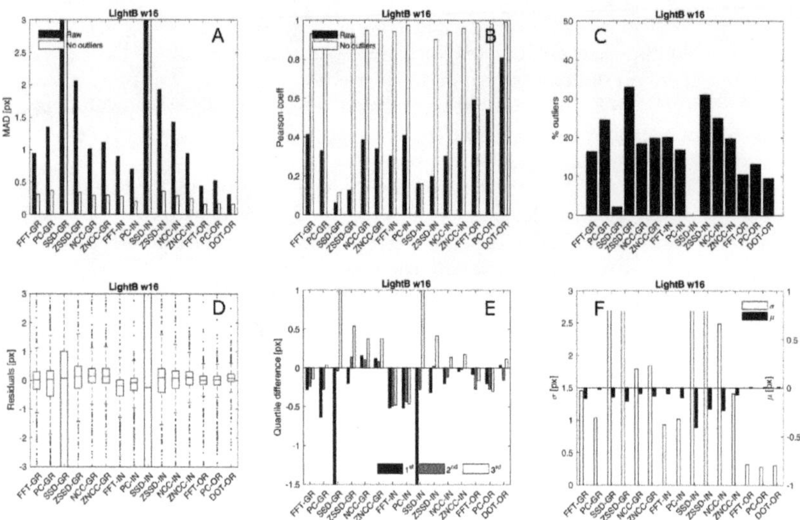

Figure 7. Outcomes of DIC processing with LightB noise using 16×16 template size. (**A**) mean absolute error with and without considering outliers, black and white bars respectively. (**B**) Pearson coefficient with and without considering outliers, black and white bars, respectively. (**C**) Percentage of outliers. (**D**) Boxplot of the residuals. (**E**) Difference between the 1st, 2nd and 3rd quartiles, black, grey and white bars, respectively. (**F**) Standard deviation and mean of the displacement in the stable area (i.e., outside the ROI), white and black bars respectively.

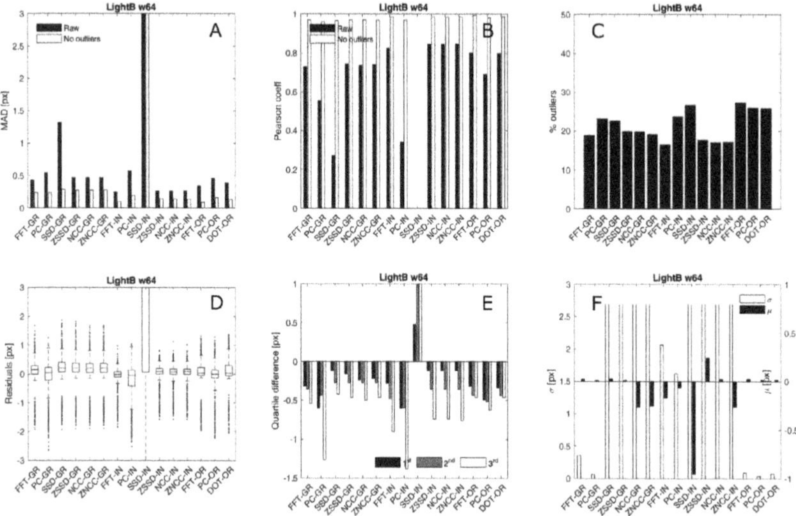

Figure 8. Outcomes of DIC processing with LightB noise using 64×64 template size. (**A**) mean absolute error with and without considering outliers, black and white bars respectively. (**B**) Pearson coefficient with and without considering outliers, black and white bars, respectively. (**C**) Percentage of outliers. (**D**) Boxplot of the residuals. (**E**) Difference between the 1st, 2nd and 3rd quartiles, black, grey and white bars, respectively. (**F**) Standard deviation and mean of the displacement in the stable area (i.e., outside the ROI), white and black bars respectively.

SSD and ZSSD. SSD is the weakest method among those considered. It performs acceptably only for IN images and slight Blur noise. However, the subtraction of the template mean value allows at reducing the impact of the quadratic term. Accordingly, ZSSD shows results similar to NCC and ZNCC. Nevertheless, it is less performing when applied to GR images. ZSSD is less robust against Speckle noise, in particular when calculated with small templates.

DOT works with OR images and it is more performing than FFT-OR and PC-OR. DOT is unaffected by Dark noise and very robust against Light noise. It performs similarly to NCC in the presence of Speckle and limited Blur noise. Poor results are obtained in the presence of strong blurring. DOT results are quite smooth and the fraction of outliers is generally lower compared to other methods.

In Figure S5, we report examples of IN images of motIm with and without noise. Besides, we show also the power spectra of each image. In this analysis, we considered only the bounding box of the ROI. We report the same analysis for GR and OR images in Figure S5. The power spectra of motIm IN and GR are similar, while in that of OR, the signal appears less defined. Blur noise smooths the edges and makes the image features less clear. This is particularly evident in GR image. The spectra of every blurred image representation have lowered high frequencies. Dark noise does not influence GR and OR images and their power spectra, while it introduced a bias in IN, as expected. Light noise tends to strengthen the features in GR image, while OR image seems to maintain the main pattern. Speckle noise introduces random scattered noise, which makes OR and GR images more disturbed. With Speckle noise, the higher frequencies strongly increase their power in IN and GR images. In general, higher frequencies of OR spectra are less influenced by any kind of noise.

4.2. Methods' Comparison Using Real Images

The maps of the vertical displacement component of the PG images are shown in Figure 9. From left to right, the columns report the results obtained with DOT-OR, ZNCC-IN, PC-GR and FFT-OR techniques. The results of PGopt, PGlight and PGblur images are reported respectively in the upper, middle and lower row. As one can note, the performances with PGopt images are approximately the same for every DIC method. PC-GR map is less smooth and several scattered outliers are present. In the upper part of the image, some noise appeared in the maps obtained with every DIC method, in correspondence of fresh snow melting (Figure 3). However, DOT-OR and FFT-OR appear less affected by this phenomenon. Similarly, DOT-OR and FFT-OR perform better with PGlight images. In this case, DOT-OR is slightly more performing compared to FFT-OR. ZNCC-IN and PC-GR are both influenced by Light noise, but the effects are different. In general, PC-GR shows scattered outliers, while ZNCC-IN decorrelates in the form of stains. DOT-OR, ZNCC-IN and FFT-OR methods performed equally well with PGblur images. They show lower displacement values compared to those measured with PGopt images. However, ZNCC-IN underestimates at a slightly less extent compared to the other methods, especially in the horizontal component (Figure S6). The underestimation confirms the observations obtained with the synthetic images, even though a slight slowing down of the glacier motion cannot be excluded. PC-GR results with PGblur images are quite poor. All the DIC methods decorrelate in the area corresponding to the break-off.

In Figure 10, we report the scatterplot of the outcomes obtained with PGopt images, which can be considered as reference, versus the outcomes of PGlight and PGblur. It is evident that DOT-OR and FFT-OR behave almost equally with all the images, while ZNCC-IN shows some random noise with PGlight images. PC-GR performs satisfactorily with PGlight, while most of the results with PGblur appear random. PC-GR outcomes present a regular grid-like distribution.

The results of BOD images partially follow those of PG. Concerning the outcomes obtained with the template of size 32×32 (Figure S6), DOT-OR and FFT-OR show similar results. The contours of the largest dune clusters are well defined and the presence of outliers very limited. Both methods detect single dunes, but with a narrow spatial resolution. On the contrary, ZNCC-IN identifies the displacement of small isolated dunes, but it presents scattered outliers. With PC-GR, the areas corresponding to the dunes appear as clusters of random noise and only the displacement of the largest groups of dunes is correctly measured. The presence of clouds strongly impacts over ZNCC-IN, while DOT-OR and FFT-OR appear almost unaffected. The inhomogeneous illumination does not alter the results of FFT-OR, PC-FR and DOT-OR significantly, while ZNCC-IN fails in correspondence of the edges of the shadows cast by the clouds.

The outcomes obtained with 64 × 64 template size are slightly different (Figure 11). DOT-OR and FFT-OR show very smooth maps. The largest groups of dunes are correctly identified with good spatial resolution. However, the smallest dunes are not entirely detected, especially by DOT-OR. Any trace of the cloud presence does not appear in these cases. In the ZNCC-IN maps, the noise is slightly lower compared to that of 32 × 32 template. However, outliers remain in correspondence of clouds and shadows. The displacement of the isolated dunes is identified, but it is largely spread around in the form of stains. PC-GR performs better compared to the case with 32 × 32 template. All the dunes are identified and the spatial resolution is fair, but the displacement appears slightly underestimated. The presence of outliers is significantly reduced and are much less than in ZNCC-IN results. Like ZNCC-IN, the displacement appears slightly spread around the dunes, but at a smaller degree.

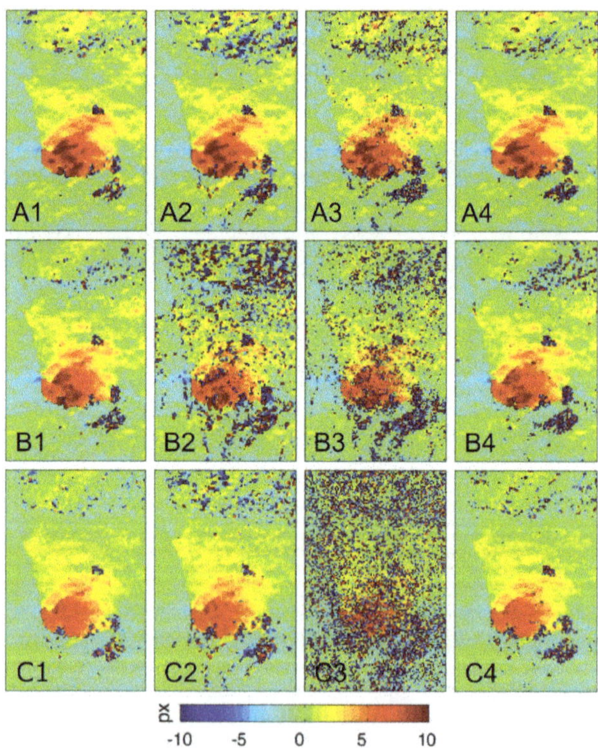

Figure 9. Displacement values obtained with the images of the Planpincieux Glacier. Rows A–C refer to PGopt, PGlight and PGblur images pairs respectively. Results *1 (**A1**–**C1**) are obtained with DOT-OR method, while *2 (**A2**–**C2**), *3 (**A3**–**C3**) and *4 (**A4**–**C4**) are obtained with ZNCC-IN, PC-GR and FFT-OR respectively.

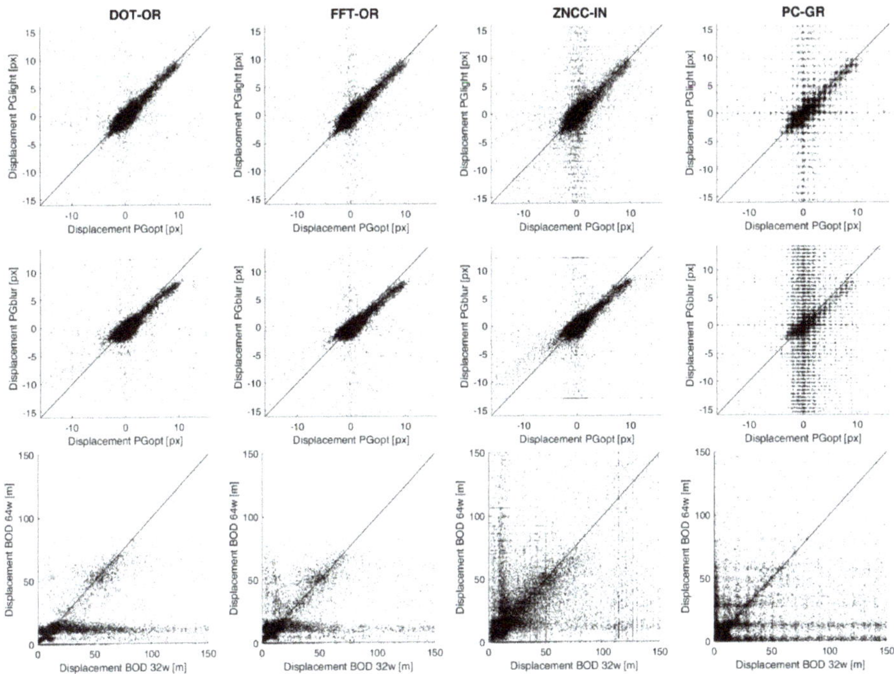

Figure 10. First row: scatterplot between the outcomes of PGopt vs. those of PGlight. Second row: same as the first row, but PGopt is shown against PGblur. Third row: outcomes of BOD images obtained with the template size of 32 × 32 vs. template of 64 × 64. From left to right, the results have been obtained with DOT-OR, FFT-OR, ZNCC-IN and PC-GR method, respectively.

The scatterplots in Figure 10 show that the adoption of 64 × 64 template with DOT-OR, FFT-OR and PC-GR reduces notably the occurrence of displacements greater than 70–80 m, which are very likely errors. On the contrary, the displacement distributions of ZNCC-IN remain almost the same and show a strong presence of outliers using both 32 × 32 and 64 × 64 templates. Like in PG images, PC-GR outcomes present a regular grid-like shape.

4.3. Computational Time

According to our analysis, the computational complexity varies largely depending on the domain where the correlation is calculated. We examined the aspect of the computational time using random templates of size 8 × 8, 16 × 16, 32 × 32 and 64 × 64 px. For the space-based correlation methods, we used search bands of half the size of the template. The results are shown in Figure 12. We used an eight-core CPU AMD Ryzen 7 1800X. SSD is the fastest method among the space-based ones, while NCC, ZSSD, DOT and ZNCC are approximately 2, 3, 4 and 6 times slower, respectively. We observed that the computational time of NCC, ZSSD and DOT tend to uniform for large sizes. On the other hand, frequency-based functions require a computational time three to five orders of magnitude lower, depending on the template size.

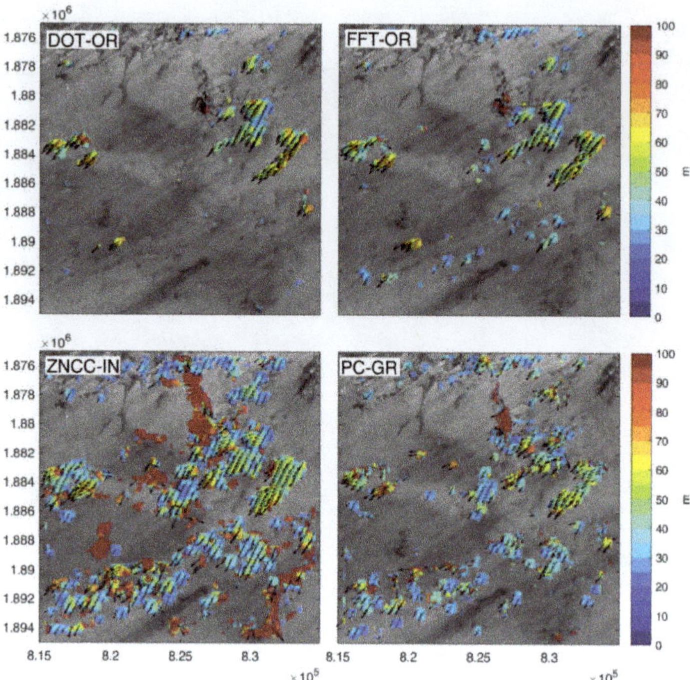

Figure 11. Displacement magnitude and direction obtained with images of the Bodélé Depression using 64×64 template size. Only the areas with displacement >20 m year^{-1} are shown. DOT-OR and FFT-OR fail to detect isolated dunes. By contrast, ZNCC-IN spreads the displacement around and it suffers the presence of thin clouds and slight blurring. PC-GR performs well, it shows the best spatial resolution, even though it slightly underestimates the displacement.

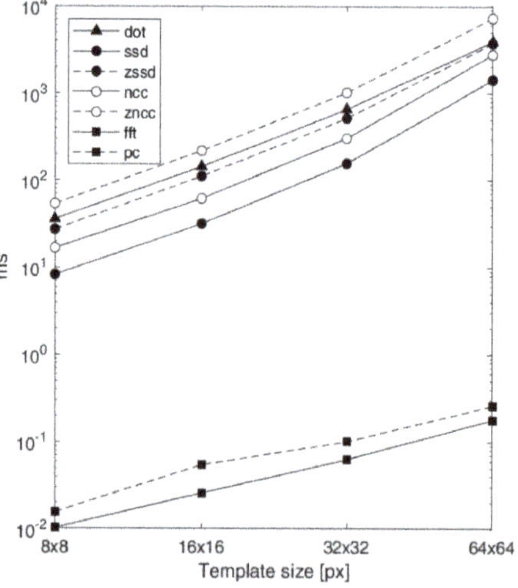

Figure 12. Computational time (in milliseconds) necessary to process one template of various sizes using different correlation functions.

It is worth noting that the computational time of space-based methods depends mostly by the size of the search band. Probably, most of the circumstances do not require as large search bands as those we used in this experiment.

5. Discussion

The results of our experiment using synthetic images demonstrated that IN images are quite resistant to Blur and Speckle noise. Nevertheless, frequency-based methods still fail in the presence of blurring, probably because it introduces strong low-frequency components that dominate over the high-frequency features. On the other hand, IN images are more sensitive to Light noise, because it can produce coherent areas of large pixel difference. GR images do not suffer changes in illumination and they provide acceptable results in the presence of Light noise. However, they are more sensitive to Blur and Speckle noise, in accordance to the observation of Sjödahl [25]. This could appear in contrast with the findings of Martin and Crowley [20], which observed more stable correlation indexes using GR images. However, their study was only focused on index stability, rather than on the results' effectiveness. We noticed a peculiar behaviour of PC-GR method. The obtained displacements were concentrated on integer values (the dimension of one pixel of Sentinel-2 images is 10 m). This occurs because the autocorrelation function of GR images has a narrow prominent peak which makes the signal well-defined [25]. However, such a sharp peak probably concentrates all the information into integer shift values. As a consequence, GR images are less suitable for detecting sub-pixel displacements. OR images behave similarly to GR, but they return much better results. Since OR images are complex, a single element of the matrix is composed of two components. Therefore, OR images contain more information compared to IN and GR. Consequently, the original pattern information is less corrupted by noise and the signal-to-noise ratio maintains higher.

Concerning the correlation methods, space-based correlations non-centred in zero (i.e., SSD and NCC) fail in the presence of a change in illumination. Results strongly improve by considering the zero-mean values and adopting normalised function. As a matter of fact, ZNCC outperforms the other space-based methods. In general, SSD and ZSSD are more sensitive to large pixel variations, because the quadratic form enhances the largest differences. Among the space-based correlations, SSD appears the least effective method, particularly when adopted with GR images. This result differs from the finding of Martin & Crowley (1995) [20]. A possible reason, as also suggested by Pust [22], is that they analysed the performances of SSD without any source of noise. NCC and ZNCC return similar results, but NCC strongly suffers changes of average intensity using IN images. Space-based methods fail when applied to OR images. Probably, this happens because OR images are complex and the space-based functions are well defined only for real numbers. In general, we observed that space-based methods (excluding SSD in general and NCC for Dark noise) are more robust compared to frequency-based ones, confirming the observations of Pust [22]. On the other hand, Bickel et al. [19] obtained slightly better results with FFT compared to ZNCC using large template size.

Formally, DOT correlation is calculated in the space domain, but we describe its behaviour individually for two reasons: first, it differs from classical space-based functions as it operates with complex numbers; second, we want to highlight the high performances of this original method that we developed. In principle, DOT may be applied to whatever image representation composed of two separate components: e.g., the first derivatives of the image intensity (like in this study) or two different bands of multispectral images. This opens new possibilities of investigation for defining original DIC approaches. In general, DOT-OR appears quite robust against every source of considered noise. It is unaffected by dark noise and it provides the best outcomes in the presence of Light noise. We observed acceptable results with Speckle and weak Blur noise. The higher performances of DOT probably depend on two elements: (i) DOT applies to OR images, which are less corrupted by noise and (ii) DOT correlation operates in Lagrangian specification. Therefore, it is less sensitive to decorrelation respect to frequency-based methods [30].

Compared to space-based methods, normalisation worsens the performances of frequency-based correlations. Indeed, FFT outperforms PC in every situation we examined. Probably, this happens because in PC, the contribution of the amplitude signal is ignored and all the phase components have the same weight in the correlation calculus. Therefore, the presence of noise that alters the frequency spectrum can easily limit the performance of PC. Similarly, Heid and Kääb [18] observed better outcomes from FFT, especially in the presence of homogeneous texture. Probably, as suggested by Lewis [29], since in PC all the frequencies are weighted equally, it fails when a predominant frequency is lacking, which is the case of homogeneous areas. As a consequence, PC is more prone to artefacts. We observed a general underestimation of frequency-based methods. Heid and Kääb [18] noticed the same behaviour in areas with the presence of displacement gradients. They suggested that this occurs because the features that move slower change less their texture (i.e., they decorrelate to a lesser extent). Therefore, the slower features' contribution weights more in the correlation calculus. This hypothesis seems confirmed by the fact that the underestimation is less evident in OR images, whose intrinsic normalisation limits this effect. Frequency-based correlations suffer limitation concerning the template size. In our experiment, both FFT and PC failed when applied with 8×8 templates. Theoretically, the dimension must be at least twice larger as the maximum displacement for correctly detecting it. In practice, Bickel et al. [19] suggested as a rule of thumb to use a template size 6 to 17 times larger than the maximum displacement. We observed acceptable results with templates four times larger than the maximum displacement.

It is worth noting that normalised functions (i.e., PC, NCC, ZNCC and DOT) have the advantage of outputting an image-independent similarity index. This permits to compare the results between different attempts of the DIC calculus [18] and makes the similarity matrix much easier to understand. For example, this may allow to define heuristic criteria of data quality and help in the outlier detection [7].

The outcomes obtained with PG images agree with those observed with synthetic images. DOT-OR and FFT-OR return similar high-quality results. ZNCC-IN performs well with PGopt and PGblur, but it suffers the Light noise in the PGlight images. This confirms the lower robustness of the IN images against Light noise, as also observed by Dematteis et al. [17], who analysed the effects of lighting change using FFT-IN approach. Besides, ZNCC-IN fails in the areas where snow melting occurs, as already noticed by Heid & Kääb (2012) [18]. As expected, PC-GR shows acceptable results with PGlight images, as it is less affected by Light noise, while it totally fails with PGblur.

On the other hand, the outcomes of BOD dataset have some interesting difference. DOT-OR and FFT-OR perform well with 32×32 template. They both show smooth results with few outliers. However, they cannot detect small isolated dunes adopting 64×64 template. Probably, small features disappear in OR images. ZNCC-IN correctly detects single dunes, but it spreads the displacement. Such behaviour is more evident with a larger template. We observed a similar effect with blurred synthetic images. Finally, PC-GR performs differently depending on the template size. With 32×32 template it fails, while it returns fair results with 64×64 template. This behaviour was unexpected, as PC-GR has never shown robust outcomes. However, it confirmed that the adoption of larger templates ensures a lower impact of the noise [18,30].

5.1. Influence of the Template Size

In general, larger template sizes provide better results in the presence of strong noise. This behaviour was expected, because the signal-to-noise ratio increases. As a matter of fact, using synthetic images, most of the considered methods perform better with 32×32 template size and a minority returns better outcomes with 64×64 template. However, larger sizes tend to produce underestimated values and coarser spatial resolution. It is possible to appreciate such differences from the comparison of Figures 7 and 8. Similar considerations also hold for the BOD case study (Figure 11 and S6). There, the reduced spatial resolution is particularly evident with template sizes of 32×32 and 64×64.

GR images and OR images provide the worst and the best spatial resolution, respectively. Frequency-based methods do not detect displacement near the ROI boundaries using wide templates. When applied to blurred images, SSD, ZSSD, ZNCC and NCC spread the displacement out of the ROI and underestimate the results.

As mentioned before, FFT and PC perform very badly when applied to templates of size 8×8. However, FFT-IN and FFT-OR provide good results with larger sizes. DOT-OR, ZNCC-IN and ZSSD-IN returned similar high-quality outcomes for every template size. In general, the performances of PC correlation and GR images appear more dependent on the template size.

5.2. Possible Uses of Specific DIC Methods

Applying exemplar DIC methods to couples of real images that present some forms of noise, we observed the same effects that we obtained using the synthetic dataset where we introduced known and controlled types of noise. This demonstrates that the analysis conducted on the synthetic images was able to successfully describe the influence of typical disturbances that can occur in operative surveys in the geoscience field. Our study shows that various kinds and levels of noise affect differently the outcomes of the most commonly adopted DIC methods. However, we noticed that specific methods generally perform better.

The results of our analysis evidence that ZNCC-IN is quite robust against many kinds of noise, even though it suffers a change of the shadow pattern (i.e., Light noise). ZNCC-IN is sensitive to weak features and it is suitable to use when dealing with homogeneous images (e.g., river flow, dune migration, ice caps), although it might spread the displacement around the objects. However, scattered changes of image intensity, like those caused by the presence of snow or isolated shadows, might hamper the ZNCC-IN calculus (Figures 9 and 11). The use of GR images could be a possible alternative to overcome light noise. But ZNCC-GR largely suffers blurring and speckling and the use of GR images could hamper the detection of sub-pixel displacement.

Instead, DOT-OR is generally more resistant against every noise source and more performing in images with high feature contrast. Since it considers only the gradient orientation, it is insensitive to changes of illumination and colour. Therefore, its use might be suggested in glaciological applications, where the presence of surface debris is possible and snow melting and unfavourable illumination may occur frequently. On the other hand, it is less performing with homogeneous images that do not show a well-defined pattern.

On the contrary, in general, frequency-based methods revealed worse performances. However, using large templates, FFT-OR is comparable with ZNCC-IN and DOT-OR techniques and its computational time is several orders of magnitude lower. Therefore, FFT-OR is a good option when fast processing is required, e.g., during an emergency or when dealing with large datasets. FFT-IN could be an alternative when dealing with Speckle noise, even though speckling is expected to occur less frequently. Besides, it must be considered that frequency-based methods tend to underestimate the displacement and provide a coarser spatial resolution of the displacement maps.

The remaining DIC methods we examined revealed various disadvantages compared to the abovementioned approaches. PC performed much worse than FFT with every image representation and it is more time-consuming. Instead, NCC, SSD and ZSSD require less computational costs compared to ZNCC, but they generally perform worse. If fast processing is really necessary, FFT-OR is a much more valid alternative.

6. Conclusions

Our study analyses the effects of four types of noise (i.e., Blur, Dark, Light and Speckle) and various template sizes on the performances of different DIC algorithms applied to synthetic and real images. We simulated the noise on a shaded relief into which we artificially introduced displacement. We considered six popular correlation functions (i.e., NCC, ZNCC, SSD, ZSSD, FFT and PC) plus an original correlation function (referred as

DOT) based on the scalar product of complex numbers. We applied every DIC method to three image representations (IN, GR and OR). In total, we analysed 800 combinations of correlation function and image representation (780 using synthetic images and 20 using real images). Concerning the image representations, GR images returned the poorest outcomes, although they appeared less sensitive by Light noise. In general, we observed that space-based methods produced better results compared to frequency-based methods. Among the former ones, ZNCC-IN and DOT-OR are the most performing methods. Nevertheless, ZNCC-IN suffers Light noise and DOT-OR is less sensitive in homogeneous areas. The original DOT-OR method proved to be quite resistant against every type of considered noise and we registered low-quality results only with strong blurring. An advantage of DOT is that it employs two distinct components. This permits to exploit more information of the main signal and thus to be less affected by noise.

Frequency-based methods fail using small templates, but FFT-IN and FFT-OR provide good results with larger templates. However, they are in general less robust than ZNCC-IN and DOT-OR. Among the frequency-based methods, FFT-OR returned the best results, while PC underperformed with just weak noise. Nevertheless, we registered satisfactorily outcomes with PC-GR applied to Sentinel-2B images. We observed that frequency-based methods tend to underestimate the actual displacement.

We compared the results of ZNCC-IN, DOT-OR, FFT-OR and PC-GR applied to terrestrial images of the Planpincieux Glacier (Italy) and Sentinel 2B images of the Bodélé Depression (Chad). In general, the results obtained with the real photos confirmed those observed with the synthetic images.

The study shows that space-based methods should be adopted in applications where strong noise is expected. However, the change in the shadow pattern might negatively influence them. DOT-OR seems the more robust method, as it provides smooth displacement maps. Nevertheless, it may fail in homogeneous areas. FFT-OR and FFT-IN are more sensitive to noise, but they require much less computational time and they should be preferred in applications that need fast processing.

Therefore, the choice of the most suitable DIC method should be led by the expected environmental conditions and noise types. However, the results of our analysis suggest that such a choice should be made among one of these methods: ZNCC-IN, DOT-OR and FFT-OR.

Supplementary Materials: The following are available online at https://www.mdpi.com/2072-4292/13/2/327/s1, The interested reader can find supplementary material attached to this paper. Figure S1: The details of the DTM adopted to produce the synthetic images; Figure S2: the synthetic images used in the comparison analysis; Figure S3: the displacement map, scatterplot of theoretical vs. measured displacement and the distributions of the displacement of the stable areas for every combination of method, noise and template size; Figure S4: The metrics explained in Section 3.4. Every figure reports the metrics of all the considered methods for a specific noise and template size; Figure S5: the IN, GR and OR images of the ROI and the corresponding power spectra and Figure S6: the results obtained with the PG images (both horizontal and vertical displacement components) and with the BOD images (displacement fields with 32×32 and 64×64 template size).

Author Contributions: N.D. conceived the study, implemented the methodology and wrote the manuscript. D.G. contributed to writing the manuscript and supervised the research. All authors have read and agreed to the published version of the manuscript.

Funding: This research received no external funding.

Institutional Review Board Statement: Not applicable.

Informed Consent Statement: Not applicable.

Data Availability Statement: Data is contained within the article or Supplementary Materials.

Conflicts of Interest: The authors declare no conflict of interest.

References

1. Oliveira, F.P.M.; Tavares, J.M.R.S. Medical image registration: A review. *Comput. Methods Biomech. Biomed. Engin.* **2014**, *17*, 73–93. [CrossRef]
2. Westerweel, J. Fundamentals of digital particle image velocimetry. *Meas. Sci. Technol.* **1997**, *8*, 1379–1392. [CrossRef]
3. Willert, C.E.; Gharib, M. Digital particle image velocimetry. *Exp. Fluids* **1991**, *10*, 181–193. [CrossRef]
4. Chu, T.C.; Ranson, W.F.; Sutton, M.A. Applications of digital-image-correlation techniques to experimental mechanics. *Exp. Mech.* **1985**, *25*, 232–244. [CrossRef]
5. Evans, A.N. Glacier surface motion computation from digital image séquences. *IEEE Trans. Geosci. Remote Sens.* **2000**, *38*, 1064–1072. [CrossRef]
6. Leprince, S.; Berthier, E.; Ayoub, F.; Delacourt, C.; Avouac, J.P. Monitoring earth surface dynamics with optical imagery. *EOS* **2008**, *89*, 1–2. [CrossRef]
7. Giordan, D.; Allasia, P.; Dematteis, N.; Dell'Anese, F.; Vagliasindi, M.; Motta, E. A low-cost optical remote sensing application for glacier deformation monitoring in an alpine environment. *Sensors* **2016**, *17*, 1750. [CrossRef]
8. Sun, X.; Shiono, K.; Chandler, J.H.; Rameshwaran, P.; Sellin, R.H.J.; Fujita, I. Discharge estimation in small irregular river using LSPIV. *Proc. Inst. Civ. Eng. Water Manag.* **2010**, *163*, 247–254. [CrossRef]
9. Fujita, I.; Aya, S. Refinement of LSPIV technique for monitoring river surface flows. In Proceedings of the Joint Conference on Water Resource Engineering and Water Resources Planning and Management 2000: Building Partnerships, 2004, Minneapolis, MN, USA, 30 July–2 August 2000; Volume 104.
10. Gabrieli, F.; Corain, L.; Vettore, L. A low-cost landslide displacement activity assessment from time-lapse photogrammetry and rainfall data: Application to the Tessina landslide site. *Geomorphology* **2016**, *269*, 56–74. [CrossRef]
11. Travelletti, J.; Delacourt, C.; Allemand, P.; Malet, J.P.; Schmittbuhl, J.; Toussaint, R.; Bastard, M. Correlation of multi-temporal ground-based optical images for landslide monitoring: Application, potential and limitations. *ISPRS J. Photogramm. Remote Sens.* **2012**, *70*, 39–55. [CrossRef]
12. Ahn, Y.; Box, J.E. Glacier velocities from time-lapse photos: Technique development and first results from the Extreme Ice Survey (EIS) in Greenland. *J. Glaciol.* **2010**, *56*, 723–734. [CrossRef]
13. Dematteis, N.; Giordan, D.; Zucca, F.; Luzi, G.; Allasia, P. 4D surface kinematics monitoring through terrestrial radar interferometry and image cross-correlation coupling. *ISPRS J. Photogramm. Remote Sens.* **2018**, *142*, 38–50. [CrossRef]
14. Westerweel, J.; Scarano, F. Universal outlier detection for PIV data. *Exp. Fluids* **2005**, *39*, 1096–1100. [CrossRef]
15. Brinkerhoff, D.; O'Neel, S. Velocity variations at Columbia Glacier captured by particle filtering of oblique time-lapse images. *arXiv* **2017**, arXiv:1711.05366.
16. Hadhri, H.; Vernier, F.; Atto, A.M.; Trouvé, E. Time-lapse optical flow regularization for geophysical complex phenomena monitoring. *ISPRS J. Photogramm. Remote Sens.* **2019**, *150*, 135–156. [CrossRef]
17. Dematteis, N.; Giordan, D.; Allasia, P. Image Classification for Automated Image Cross-Correlation Applications in the Geosciences. *Appl. Sci.* **2019**, *9*, 2357. [CrossRef]
18. Heid, T.; Kääb, A. Evaluation of existing image matching methods for deriving glacier surface displacements globally from optical satellite imagery. *Remote Sens. Environ.* **2012**, *118*, 339–355. [CrossRef]
19. Bickel, V.T.; Manconi, A.; Amann, F. Quantitative assessment of digital image correlation methods to detect and monitor surface displacements of large slope instabilities. *Remote Sens.* **2018**, *10*, 865. [CrossRef]
20. Martin, J.; Crowley, J.L. Comparison of Correlation Techniques. In *Proceedings of the Intelligent Autonomous Systems*; IOS Press: Karlsruhe, Germany, 1995; ISBN 90-5199-213-0.
21. Merzkirch, W.; Gui, L. A comparative study of the MQD method and several correlation-based PIV evaluation algorithms. *Exp. Fluids* **2000**, *28*, 36–44. [CrossRef]
22. Pust, O. PIV: Direct cross-correlation compared with FFT-based cross-correlation. In Proceedings of the 10th International Symposium on Applications of Laser Techniques to Fluid Mechanics, Lisbon, Portugal, 10–13 July 2000; Volume 27, pp. 1–12.
23. Leprince, S.; Barbot, S.; Ayoub, F.; Avouac, J.P. Automatic and precise orthorectification, coregistration, and subpixel correlation of satellite images, application to ground deformation measurements. *IEEE Trans. Geosci. Remote Sens.* **2007**, *45*, 1529–1558. [CrossRef]
24. Guizar-Sicairos, M.; Thurman, S.T.; Fienup, J.R. Efficient subpixel image registration algorithms. *Opt. Lett.* **2008**, *33*, 156. [CrossRef] [PubMed]
25. Sjödahl, M. Gradient correlation functions in digital image correlation. *Appl. Sci.* **2019**, *9*, 2127. [CrossRef]
26. Fitch, A.J.; Kadyrov, A.; Christmas, W.J.; Kittler, J. Orientation Correlation. In Proceedings of the British Machine Vision Conference, Cardiff, UK, 2–5 September 2002; pp. 133–142.
27. Gui, L.C.; Merzkirch, W. Method of tracking ensembles of particle images. *Exp. Fluids* **1996**, *21*, 465–468. [CrossRef]
28. Kuglin, C.D.; Hines, D.C. The phase correlation image alignment method. *IEEE Int. Conf. Cybern. Soc.* **1975**, *6*, 163–165.
29. Lewis, J.P. Fast normalized cross-correlation. *Vis. Interface* **1995**, *10*, 120–123.
30. Thielicke, W.; Stamhuis, E.J. PIVlab–Towards User-friendly, Affordable and Accurate Digital Particle Image Velocimetry in MATLAB. *J. Open Res. Softw.* **2014**, *2*, e30. [CrossRef]
31. Giordan, D.; Dematteis, N.; Allasia, P.; Motta, E. Classification and kinematics of the Planpincieux Glacier break-offs using photographic time-lapse analysis. *J. Glaciol.* **2020**, *66*, 188–202. [CrossRef]

32. Singh, P.; Shree, R. Analysis and effects of speckle noise in SAR images. In Proceedings of the 2016 International Conference on Advances in Computing, Communication and Automation (Fall), ICACCA 2016, Bareilly, India, 30 September–1 October 2016.
33. Baird, T.; Bristow, C.S.; Vermeesch, P. Measuring sand dune migration rates with COSI-Corr and landsat: Opportunities and challenges. *Remote Sens.* **2019**, *11*, 2423. [CrossRef]

MDPI
St. Alban-Anlage 66
4052 Basel
Switzerland
Tel. +41 61 683 77 34
Fax +41 61 302 89 18
www.mdpi.com

Remote Sensing Editorial Office
E-mail: remotesensing@mdpi.com
www.mdpi.com/journal/remotesensing